Phosphoinositides

About the Cover

The cover art shows a side-by-side view of the topologies of BPI–PLC (left picture) and LPI–PLC (right picture) with bound *myo*-inositol (ball-and-stick, labeled Ins). β-Strands are symbolized by arrows and labeled with Roman numerals (I–VIII), α-helices with letters (A–G).

ACS SYMPOSIUM SERIES **718**

Phosphoinositides

Chemistry, Biochemistry, and Biomedical Applications

Karol S. Bruzik, EDITOR
University of Illinois at Chicago

American Chemical Society, Washington, DC

Library of Congress Cataloging-in-Publication Data

Phosphoinositides : chemistry, biochemistry, and biomedical applications /
Karol S. Bruzik, editor.

    p.   cm.—(ACS symposium series , ISSN 0097–6156 ; 718)

    "Developed from a symposium sponsored by the Division of Carbohydrate
Chemistry at the 214th National Meeting of the American Chemical Society,
Las Vegas, Nevada, September 7–11, 1997."

    Includes bibliographical references and index.

    ISBN 0–8412–3628–3 (alk. paper)

    1. Phosphoinositides—Congresses.

    I. Bruzik, K. S. II. Series.

QP752.P52P49   1998
572'.57—dc21
                                 98–38729
                                      CIP

The paper used in this publication meets the minimum requirements of American National Standard for
Information Sciences—Permanence of Paper for Printed Library Materials, ANSI Z39.48–1984.

Advisory Board

ACS Symposium Series

Foreword

THE ACS SYMPOSIUM SERIES was first published in 1974 to provide a mechanism for publishing symposia quickly in book form. The purpose of the series is to publish timely, comprehensive books developed from ACS-sponsored symposia based on current scientific research. Occasionally, books are developed from symposia sponsored by other organizations when the topic is of keen interest to the chemistry audience.

Before agreeing to publish a book, the proposed table of contents is reviewed for appropriate and comprehensive coverage and for interest to the audience. Some papers may be excluded in order to better focus the book; others may be added to provide comprehensiveness. When appropriate, overview or introductory chapters are added. Drafts of chapters are peer-reviewed prior to final acceptance or rejection, and manuscripts are prepared in camera-ready format.

As a rule, only original research papers and original review papers are included in the volumes. Verbatim reproductions of previously published papers are not accepted.

ACS BOOKS DEPARTMENT

Contents

Inositol Phosphate Metabolism and Inositol Phospholipid Binding Proteins

Mechanism and Structure of Phosphatidylinositol-Specific Phospholipases C

Synthesis and Biological Properties of Analogs of Phosphoinositides

Indexes

Preface

Inositol phospholipids are minor components of biological membranes. Yet as precursors of intracellular second messengers—generated in response to neurotransmitter, growth factor, or hormonal signals—they are very important from the point of view of homeostasis of higher organisms. The receptors and effectors participating in inositol signal transduction also constitute potential targets of pharmacological intervention in states of disease. The progress of research in this area heavily depends on collaboration among cellular biologists, biochemists, and chemists. Recognizing the need for closer interactions among those groups, Ching-Shih Chen and I have organized the ACS Symposium on "Advances in Phosphoinositides" held in Las Vegas. This volume, which is based on the content of the Symposium, provides an up-to-date summary of several areas in this field.

During the seven years that elapsed from the previous ACS Symposium on the related topic, and the edition of the publication on *Inositol Phosphates: Synthesis, Biochemistry, and Therapeutic Potential,* a large qualitative change occurred in both the global view of the role of inositol phosphate signaling in cells and understanding the molecular mechanisms of inositol-related signal transduction pathways.

First, we have come to realize an overwhelming complexity of inositol signaling pathways. New pathways of signal transduction emerged, most notably those originating from further phosphorylation of phosphatidylinositol 4,5-bisphosphate. Although the details of the pathways involving phosphorylation of inositol phospholipids at their 3-position still remain unclear, this area promises to be as important as the older pathways involving phospholipase C-mediated generation of inositol 1,4,5-trisphosphate. Highly phosphorylated inositol phospholipids, such as 4,5-bisphosphate and 3,4,5-trisphosphate, appear to have second messenger functions of their own because of their specific interactions with multiple proteins, resulting in cell transformation, cytoskeletal rearrangement, membrane association of signaling proteins, and vesicular trafficking. Inositol 1,4,5-trisphosphate does not remain a sole second messenger derived from inositol phospholipids anymore; 3,4,5,6-tetrakisphosphate second messenger appears to regulate many cellular functions as well. Although the degree of inositol phosphorylation is limited by the number of the available hydroxyl groups, the ever-increasing complexity of inositol metabolism has recently entered a new phase with identification of inositol pyrophosphates. On the other hand, certain earlier proposed pathways of signal transduction, such as insulin-mediators derived from glycosylphosphatidylinositols, have slowly faded away.

Second, a spectacular progress has been achieved in unraveling molecular details of signal transduction, starting with structures of the key effector, phosphatidylinositol-specific phospholipases C. The available crystallographic data for several types of PI-PLC, including both mammalian and bacterial enzymes, and results of kinetic studies of site-directed mutants of these enzymes employing both natural substrates and their analogs provided highly detailed pictures of the mechanisms of these enzymes. The recent studies address the effect of interactions of these enzymes with lipid bilayers as the mechanisms regulating activity of these enzymes.

Third, a new emerging area is the cross-talk and integration of several pathways originating from the action of phospholipases A, C, and D, as well as activation of various forms of phospholipases C by 3-phosphorylated PI.

Finally, although in the past the progress in studies of inositol metabolism has been hampered by the limited availability of enzyme substrates, receptor agonists, and their analogs, a considerable progress has been achieved within the past decade in the field of chemical synthesis of phosphoinositides, to the extent that most if not all, naturally occurring molecules have been synthesized. This fact is reflected in that most current synthetic efforts are directed toward synthesis of inositol-based molecular probes for testing specific interactions of phosphoinositides with intracellular proteins, rather than naturally occurring phosphoinositides.

In summary, the state of knowledge in the field of inositol signaling pathways has reached the level where, on one hand, a global picture of inositol metabolism integrated into whole cellular metabolism is starting to emerge, while, on the other hand, atomic details of individual steps of metabolism are becoming clear. It is hoped that in the end, knowing both, one will be able to intervene into these pathways to obtain a narrow physiological effect. It is also my hope that by providing the summary of the current state of the art, this volume will enhance understanding of signaling events at both cellular and molecular levels and will foster further progress through close collaborative interactions between research groups.

I gratefully acknowledge the effort of the international group of authors who contributed their chapters to this volume. I also thank Ching-Shih Chen of the University of Kentucky for his contribution to the Symposium organization and to many participants who have encouraged us to go ahead with the Symposium. Finally, I extend many thanks to all those colleagues who reviewed the manuscripts of the contributed papers.

KAROL S. BRUZIK
Department of Medicinal Chemistry and Phatmacognosy
University of Illinois at Chicago
Chicago, IL 60612

Inositol Phosphate Metabolism and Inositol Phospholipid Binding Proteins

Chapter 1

The Structural and Functional Versatility of Inositol Phosphates

James J. Caffrey, Stephen T. Safrany, Xiaonian Yang, Masako Yoshida, and
Stephen B. Shears

Inositide Signaling Group, National Institute of Environmental Health Sciences,
NIH, Research Triangle Park, P.O. Box 12233, NC 27709

This review presents a summary of our current knowledge of inositol
phosphate turnover in animal cells. In the Figure 1, the various
metabolic reactions are sub-divided into several distinct groups; each of
these is discussed in turn in this review, in an effort to illustrate the
functional versatility of these polyphosphates.

The $Ins(1,4,5)P_3$ / $Ins(1,3,4,5)P_4$ cycle

The discovery of inositol phosphate-mediated cellular Ca^{2+} mobilization was a pivotal
development in the field of signal transduction (*1*), but this breakthrough also owes
much to the persistence of some other workers who had the foresight to develop and
promote the idea of receptor-activated inositide signaling. It was the Hokins who first
discovered that muscarinic agonists accelerate the turnover of PtdIns in tissue slices
from brain and pancreas (*2*). It was unfortunate that their dedication to characterizing
this effect over the following 20 years did not lead them to ascribe its biological
function. Indeed, there was little general interest in this so-called "phosphoinositide
effect", until Michell (*3*) proposed that it might drive receptor-dependent Ca^{2+} influx
into the cell. Later, Michell and his colleagues showed that an accelerated
$PtdIns(4,5)P_2$ turnover immediately followed receptor-occupation by Ca^{2+}-mobilizing
hormones (*4,5*). The modern era of inositide research was now born: Berridge (*6*)
recognized that the $Ins(1,4,5)P_3$ formed by hydrolysis of $PtdIns(4,5)P_2$ was a
candidate intracellular signal, and so it was discovered that $Ins(1,4,5)P_3$ released Ca^{2+}
from endoplasmic reticulum (ER) (*1*). This intracellular release of Ca^{2+} was found to
be co-ordinated with an increase in the rate of Ca^{2+} influx into the cell (*7*) - which
finally confirmed the essence of the original idea (*3*) that inositol lipid turnover regulated
Ca^{2+} entry.

$Ins(1,4,5)P_3$-induced Ca^{2+} release mediates an abundance of cellular responses
as diverse as fertilization, cell growth and differentiation, neuronal signaling, secretion,
and phototransduction. The distinct signaling requirements of individual cells are
served by cell- and agonist-specific spatiotemporal patterns of Ca^{2+} signaling. These
arise not just from the binding of $Ins(1,4,5)P_3$ to its receptor, but also by modulation of
this ligand/protein interaction by both cytosolic and ER-luminal Ca^{2+}, and there is also
input from the process of Ca^{2+} entry, and $Ins(1,4,5)P_3$ metabolism (*8*). There is also
regulatory diversity between the three major forms of the $Ins(1,4,5)P_3$ receptors that are
currently recognized (see reference 9 for a review).

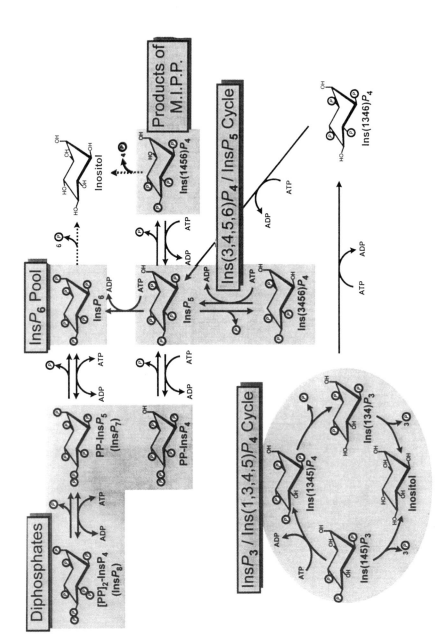

Figure 1. Overview of the cellular metabolism of inositol polyphosphates.

In unstimulated mammalian cells, the levels of Ins$(1,4,5)P_3$ are frequently about 0.1 - 0.2 µM, increasing up to 10-fold upon receptor activation (10). The cellular concentrations of Ins$(1,4,5)P_3$ are kept within this relatively narrow range through a complex and carefully-coordinated balance between its rates of synthesis (the signaling "on-switch") and metabolism (the signaling "off-switch"). The latter is analogous to the phosphodiesterase-mediated inactivation of cAMP. This comparison originally led us to expect that the metabolism of inositol phosphates would be equally simple. Indeed, by 1985 a series of enzyme reactions were identified which sequentially dephosphorylated Ins$(1,4,5)P_3$ to Ins$(1,4)P_2$, InsP and Ins (11-13). The InsP monophosphatase was soon appreciated (e.g. ref.14) to be the same enzyme that, 20 years earlier, had been shown to dephosphorylate the Ins$3P$ that is formed by cyclization of glucose 6-phosphate (15). This monophosphatase is therefore particularly important for the synthesis of Ins for those tissues - such as brain - that have limited access to extracellular Ins.

Any satisfaction that might have initially been gained from seemingly characterizing the pathway of mammalian inositol phosphate metabolism was short-lived. By 1984 it was already known that receptor-mediated PtdIns$(4,5)P_2$ hydrolysis yielded far more Ins$(1,3,4)P_3$ than Ins$(1,4,5)P_3$ (16). Then, it was found that an Ins$(1,4,5)P_3$ 3-kinase competed with the 5-phosphatase in a bifurcating pathway of Ins$(1,4,5)P_3$ metabolism that led to the synthesis of Ins$(1,3,4,5)P_4$ and Ins$(1,3,4)P_3$ (17-19). Although primarily dephosphorylated to Ins, Ins$(1,3,4)P_3$ was also itself found to be phosphorylated (20,21) to a series of higher polyphosphates that, even today, remain incompletely characterized.

The Ins$(1,4,5)P_3$ 3-kinase activity not only helps regulate levels of Ins$(1,4,5)P_3$, but, in addition, this enzyme control the synthesis of a product, Ins$(1,3,4,5)P_4$, that is itself believed to be physiologically active. Due in large part to the efforts of Irvine and colleagues, there is now a strong case that Ins$(1,3,4,5)P_4$ contributes to the control of receptor-regulated Ca^{2+} entry into cells (22). Yet, despite a large body of literature showing effects of Ins$(1,3,4,5)P_4$ upon Ca^{2+} fluxes in many experimental systems, a sound molecular explanation for its mechanism of action *in vivo* still eludes us. A candidate Ins$(1,3,4,5)P_4$ receptor (GAP1[IP4BP]) has been isolated which is also a GTPase-activating protein (GAP) of the Ras family (23). Ins$(1,3,4,5)P_4$ by itself did not affect Ras-GAP activity, but Ins$(1,3,4,5)P_4$ was able to act relatively specifically to restore the activity that, *in vitro*, was inhibited by phospholipids (23).

GAP1[IP4BP] is now considered to be a member of the GAP1 family that is represented by at least one other Ins$(1,3,4,5)P_4$-binding protein, GAP1[m] (24). The Ins$(1,3,4,5)P_4$-binding domain of GAP1[IP4BP] and GAP1[m] have been mapped to their pleckstrin homology (PH) domains (24-26) which is also conserved in the otherwise unrelated Bruton's tyrosine kinase, another highly-specific Ins$(1,3,4,5)P_4$-binding protein (27). It is believed that PH domains provide inositides with the opportunity to regulate protein-protein and protein-phospholipid interactions within multimeric signal transduction complexes (28). It is therefore possible that there is some general contribution of Ins$(1,3,4,5)P_4$ to signal transduction, as a consequence of it being a ligand for the GAP1 family of PH domains. The question as to whether the interaction of Ins$(1,3,4,5)P_4$ with GAP1[IP4BP] has specific relevance to the control of cellular Ca^{2+} entry remains to be directly demonstrated (25). Incidentally, mutations in the gene for Bruton's tyrosine kinase cause a decrease in the number of mature B-lymphocytes, leading to X-linked agammaglobulinemia; some of these mutations are expressed as impaired binding of Ins$(1,3,4,5)P_4$ to the PH domain (27,29). The possibility that a defect in the function of a specific inositol phosphates lies at the center of a disease state is of considerable interest.

Molecular techniques were originally used to demonstrate the existence of two isoforms of 3-kinase (type A (30) and type B (31)). A further kinase (type C) may occur in platelets (32). The 3-kinases have a high affinity for $Ins(1,4,5)P_3$ (0.2 to 1.5 μM, see (33) for refs). These enzymes were originally reported to be largely soluble enzymes following subcellular fractionation of various cell types (see (33) for refs). However, the cDNA for kinase B was recently transfected into cells and its subcellular distribution was analyzed immunocytochemically (34). Substantial amounts of enzyme were found to be attached to the outer (cytosolic) face of the ER (34). This new development probably owes much to the analysis of a full-length protein; the hydrophobic region of the N-terminus, which is a candidate membrane-anchoring region, could have been removed by proteolysis during the earlier subcellular fractionation experiments - thereby liberating the protein from the membranes (34).

The amino acid sequences of kinases A and B are relatively conserved within the C-terminal catalytic domains; their N-terminal regulatory domains are more diverse (35), which provides the cell with a number of mechanisms for regulating the conversion of $Ins(1,4,5)P_3$ to $Ins(1,3,4,5)P_4$. The best characterized process is the stimulation of 3-kinase activity by Ca^{2+}/calmodulin (35). Recently, the 3-kinase type A was also found to be phosphorylated on Thr311 by calmodulin dependent protein kinase II (36). This phosphorylation greatly amplifies the activation of enzyme activity by Ca^{2+}/calmodulin. Thus, the synthesis and metabolism of the inositol phosphates ($Ins(1,4,5)P_3$ and $Ins(1,3,4,5)P_4$) that regulate Ca^{2+} mobilization, is regulated by Ca^{2+} itself. The 3-kinase has also been shown to be inhibited by protein kinase C and slightly activated by protein kinase A (37). Note that the role of the 3-kinase, and hence the significance of its regulation, is not exclusively restricted to control over Ca^{2+} mobilization. The 3-kinase initiates a metabolic pathway that yields $Ins(1,3,4)P_3$, which in turn sets in motion a chain of signaling events which ultimately leads to down-regulation of Ca^{2+}-activated Cl^- channels (see below).

Both $Ins(1,4,5)P_3$ and $Ins(1,3,4,5)P_4$ are degraded by hydrolysis of their 5-phosphate groups, forming $Ins(1,4)P_2$ and $Ins(1,3,4)P_3$ respectively. All 5-phosphatases contain the PXWCDRXL catalytic signature motifs (38-41). These enzymes have been subdivided into four distinct groups (42), on the basis of their size and substrate specificity. The 43 kDa (type I) 5-phosphatase hydrolyzes both $Ins(1,4,5)P_3$ (K_m approx 5-10 μM) and $Ins(1,3,4,5)P_4$ (K_m approx 1 μM) (38,41). The 75 kDa (type 2) $Ins(1,4,5)P_3$ 5-phosphatase also attacks $Ins(1,4,5)P_3$ (K_m approx 20 μM) and $Ins(1,3,4,5)P_4$ (K_m approx 10 μM), but in addition both $PtdIns(3,4,5)P_3$ and $PtdIns(4,5)P_2$ are substrates (40,43). Type III 5-phosphatases are grouped together by virtue of their only dephosphorylating $PtdIns(3,4,5)P_3$ (43), and type IV 5-phosphatases hydrolyze $Ins(1,3,4,5)P_4$ and $PtdIns(3,4,5)P_3$ (42). We will concentrate here on the enzymes that dephosphorylate $Ins(1,4,5)P_3$ and/or $Ins(1,3,4,5)P_4$.

There is little evidence that the type I and II $Ins(1,4,5)P_3$/$Ins(1,3,4,5)P_4$ 5-phosphatase activities are regulated by cross-talk from other cellular signals. This is an unusual situation for enzymes that de-activate intracellular signals, and in particular contrast to the complexity of the processes that regulate the $Ins(1,4,5)P_3$ receptor and the $Ins(1,4,5)P_3$ 3-kinase (see above). Early reports that the type I 5-phosphatase from platelets was directly phosphorylated and regulated by protein kinase C (44) have now been discounted (45). It turns out that the original preparations of 5-phosphatase were contaminated with pleckstrin, which was in reality the target for protein kinase C (46). However, in vitro at least, 5-phosphatase activity is increased from 3- to 7-fold by the presence of phosphorylated pleckstrin (46), although the extent to which this protein/protein interaction might regulate 5-phosphatase during PKC activation in vivo still needs to be evaluated.

Type I and II 5-phosphatases are directed to the plasma membrane by isoprenylation of the carboxy terminus (40,47). In polarized epithelial cells, 5-

phosphatase activity may subsequently be redistributed, by vesicle trafficking, to specific domains of the plasma membrane (48). It is of interest that over-expression of membrane-bound 5-phosphatase activity can prevent the proper expression of Ca^{2+} oscillations (47). Ca^{2+} signaling was not similarly affected when the over expressed 5-phosphatase was directed to the cytosol (by mutating its isoprenylation motif) (47). Thus, an appropriate, localized metabolism of $Ins(1,4,5)P_3$ seems to be an important aspect of the spatiotemporal characteristics of Ca^{2+} signaling.

The under expression of type 1 5-phosphatase has also yielded some interesting results. These data were obtained using an antisense strategy in rat kidney cells (49). The antisense-transfected cells had 2-4 fold elevated levels of both $Ins(1,4,5)P_3$ and $Ins(1,3,4,5)P_4$, and there was a corresponding 2-fold increase in cytosolic $[Ca^{2+}]$. These cells demonstrated a transformed phenotype that formed tumors in mice (49).

The type II family of 5-phosphatases includes the synaptojanins, which have been implicated in regulating synaptic vesicle transport (50). The prevailing view at present is that $PtdIns(3,4,5)P_3$ and $PtdIns(4,5)P_2$ are the more physiologically important substrates of these proteins (51,52). Another interesting type II 5-phosphatase is deficient in Lowe's oculocerebrorenal (OCRL) syndrome, a developmental disorder which, as its name suggests, affects the lens, kidney and central nervous system (53,54). Both $PtdIns(3,4,5)P_3$ and $PtdIns(4,5)P_2$ are metabolized by the normal OCRL protein more efficiently than are $Ins(1,4,5)P_3$ and $Ins(1,3,4,5)P_4$, the affinities for which are relatively low (K_m values of 70 μM and 30 μM, respectively (42,54)). In other words, any defects in ORCL activity are more likely to impact the metabolism of the $PtdIns(3,4,5)P_3$ and/or $PtdIns(4,5)P_2$ lipids than the soluble inositol phosphates.

The 5-phosphatase activities have also been suggested to be perturbed by some environmental agents. For example, $Ins(1,4,5)P_3$ and $Ins(1,3,4,5)P_4$ metabolism by the 5-phosphatases were proposed to be particularly susceptible to toxic insult by Al^{3+} (55). However, we could found no evidence that this was the case (56). The HIV gp120 coat protein has been proposed to target 5-phosphatase activity (57,58), but when we investigated this issue we saw no such effects of gp120, despite our preparations of this protein being biologically competent in several other respects (59).

An intriguing new group of 5-phosphatases (type IV) have been found to hydrolyze $Ins(1,3,4,5)P_4$ and $PtdIns(3,4,5)P_3$, but not $Ins(1,4,5)P_3$ nor $PtdIns(4,5)P_2$ (42,60). This 5-phosphatase has an SH2 domain which promotes its association with phosphotyrosines within multimeric signaling complexes, thus it is also known by the acronyms "SHIP" (for SH2-containing inositol phosphatase (61)) and "SIP" (for signaling inositol polyphosphate 5-phosphatase (42)). Most current interest in this 5-phosphatase is directed at its likely impact on regulating $PtdIns(3,4,5)P_3$ metabolism. However, it actually metabolizes $Ins(1,3,4,5)P_4$ with greater catalytic efficiency than the type I and II 5-phosphatases, even though the K_m of the type IV enzyme for $Ins(1,3,4,5)P_4$ is 16 μM (42). Tyrosine phosphorylation of the type IV 5-phosphatase down-regulates its activity (62), so it is possible that this may influence the signaling capacity of $Ins(1,3,4,5)P_4$.

$Ins(1,3,4,5)P_4$ is dephosphorylated to $Ins(1,3,4)P_3$. Until recently $Ins(1,3,4)P_3$ had been considered as merely being a downstream metabolite of no particular functional significance. However, we now know that $Ins(1,3,4)P_3$ is a potent inhibitor of the $Ins(3,4,5,6)P_4$ 1-kinase (63,64). By inhibiting this kinase, and altering the poise of the 1-kinase/1-phosphatase cycle, $Ins(1,3,4)P_3$ provides a link between activation of PLC and increases in levels of $Ins(3,4,5,6)P_4$ (64), an inhibitor of Ca^{2+}-regulated Cl^- secretion (65).

$Ins(1,3,4)P_3$ sits at a branch-point of inositol phosphate metabolism; one of its fates is to be dephosphorylated to replenish the pool of Ins (see below). In addition, $Ins(1,3,4)P_3$ serves as a precursor for the higher inositol polyphosphates. It is phosphorylated first to Ins(1,3,4,6)P_4 and then to Ins(1,3,4,5,6)P_5 (reviewed in ref (33)). *In vitro* at least, the activity of the Ins(1,3,4)P_3 6-kinase is constrained by product inhibition (66) and by its near complete inhibition by physiologically relevant levels of Ins(3,4,5,6)P_4 (67,68). It is therefore remarkable that, upon receptor-activation, *so much* Ins(1,3,4,6)P_4 is formed so quickly. It therefore seems likely that there is something important about the short-term regulation of 6-kinase activity *in vivo* that has not been revealed in experiments with cell-free systems. There are no clues as to what this regulatory process might be in the predicted amino acid sequence of a cDNA clone of the 6-kinase (69). However, the cDNA that was obtained was probably missing 500-600 bp at the 5'-end (69), so there may still be additional regulatory properties to be discovered. There was also some evidence from Northern blots of an additional, larger mRNA transcript, so there may be isoforms of the 6-kinase (69).

There may be longer-term control over the *levels* of the Ins(1,3,4)P_3 6-kinase, perhaps through cAMP-activated synthesis of the protein (70). It may also be worth studying if control over the levels of 6-kinase could also be mediated by regulation of its rate of proteolysis. In this respect, it may be relevant that residues 237-273 of the 6-kinase are similar to a calpain-sensitive domain of PKC (69). The Ins(1,3,4)P_3 6-kinase has another remarkable characteristic; about 20% of the Ins(1,3,4)P_3 that this enzyme phosphorylates is converted to Ins(1,3,4,5)P_4 (66,69,71). The synthesis of both Ins(1,3,4,5)P_4 and Ins(1,3,4,6)P_4 from Ins(1,3,4)P_3 has been hypothesized to reflect the formation of a 5,6-cyclic intermediate during the reaction (69). The conversion of a single substrate to more than one product is unusual for kinases in general, and unique for an inositol phosphate kinase in particular.

As mentioned above, Ins(1,3,4)P_3 is also dephosphorylated; two classes of phosphatases are responsible, a 4-phosphatase that forms Ins(1,3)P_2 (72,73) and a 1-phosphatase that yields Ins(3,4)P_2 (74). Both Ins(1,3)P_2 and Ins(3,4)P_2 have attracted attention by virtue of their being metabolized by the same enzymes that dephosphorylate the 3-phosphorylated inositol lipids. Thus, Ins(1,3)P_2 and PtdIns3P are competing substrates for a 3-phosphatase (75) while Ins(3,4)P_2 and PtdIns(3,4)P_2 compete for a 4-phosphatase (76,77). The physiological value of this competition continues to be explored.

With regards to the Ins(1,3,4)P_3 1-phosphatase, this is the same enzyme that is also responsible for catalyzing Ins(1,4)P_2 1-phosphatase activity (74,78). Although the primary role of this enzyme is considered to be to recycle inositol phosphates to free inositol, it may have additional, more complex functions, since some of the cell's 1-phosphatase is found in the nucleus (79). One possible explanation for this location is that the enzyme is a more precise "off-switch" for nuclear Ins(1,4)P_2, which has been reported to activate a DNA polymerase (80). Consistent with this idea, transient over-expression of the 1-phosphatase did inhibit DNA synthesis (79). Clearly, further work is needed in order to pursue these interesting proposals. It is unclear how the 1-phosphatase is retained in the nucleus, since it does not have a known nuclear localization signal (79).

The Ins(1,4)P_2 / Ins(1,3,4)P_3 1-phosphatase, and the inositol monophosphatase, are both inhibited by lithium (74,81), an observation that provides a theoretical basis behind the clinical use of lithium as a treatment for some manic disorders that might be due to excessive neuronal PLC activity (82). By inhibiting inositol phosphate phosphatases, lithium can decrease the size of the free Ins metabolic pool. Thus, in patients undergoing lithium therapy, it can be envisaged that the over-stimulated neurons can be down-regulated by draining them of the Ins that the lipid signaling system requires. This proposed action of lithium is suggested not to target cells where PLC activity is normal, but instead is believed to be selective for the overactive cells, since lithium inhibits inositol phosphatases uncompetitively. That is, the degree of inhibition increases with substrate concentration. Thus, lithium is believed to be a more effective pharmacological agent in cells containing the higher levels of InsP substrate that result from overactive PLC activity (82). However, it should be emphasized that this theory is still a matter of some dispute. Lithium has many other actions that might contribute to its pharmacological properties; it can affect adenylate cyclase activity for example (83). Moreover, the basic three-dimensional core structure that is shared by the 1-phosphatase and the monophosphatase is quite widespread; there may therefore be a surprisingly large family of these lithium-sensitive proteins (84). Indeed, it has been proposed that this group of proteins might provide additional targets, not only for lithium's therapeutic abilities, but possibly also its toxic effects (84).

The Ins(3,4,5,6)P_4 / Ins(1,3,4,5,6)P_5 cycle

The first demonstration of Ins(1,3,4,5,6)P_5 being a constituent of an animal cell came from a study with avian erythrocytes (85). This report also contains the first identification of Ins(3,4,5,6)P_4 (or its enantiomer, Ins(1,4,5,6)P_4). Interest in these higher polyphosphates initially arose from their ability to reduce hemoglobin's affinity for oxygen in erythrocytes of birds, air-breathing fish and possibly turtles (86). Developmental changes in blood oxygen affinity, such as those that occur during hatching, have been correlated with appropriate alterations in erythrocyte Ins(1,3,4,5,6)P_5 levels (86). It is, perhaps, surprising that there is no evidence for any shorter-term metabolic regulation of the Ins(1,3,4,5,6)P_5 pool in mature erythrocytes. However, one interesting adaptation has been found in a strain of high-altitude chicken (*Gallus gallus*). These birds have adapted to the low oxygen environment of the Peruvian Andes with the help of a modification to their hemoglobin, the oxygen affinity of which is less sensitive to binding of inositol polyphosphates (87).

Ins(1,3,4,5,6)P_5 is not present in mammalian erythrocytes, where 2,3-bisphosphoglycerate instead regulates hemoglobin's oxygen affinity (86). However, in 1985 Ins(1,3,4,5,6)P_5 began to be recognized to be a constituent of most other mammalian cells (88). Subsequently, Stephens and colleagues devised a stereospecific, enzymatically-based method for distinguishing Ins(3,4,5,6)P_4 from Ins(1,4,5,6)P_4 (89). This assay enabled them to demonstrate that Ins(3,4,5,6)P_4 was also present in mammalian cells. Moreover, they showed that its levels increased during receptor-regulated activation of PLC (90). These observations were subsequently confirmed by several other laboratories using a variety of cell types (70,91-94). Levels of Ins(3,4,5,6)P_4 in unstimulated cells are approximately 1 µM, rising to 10-15 µM when PLC is activated (65,95).

Whenever receptor-activation elevates the cellular levels of Ins(1,4,5)P_3, the concentrations of many downstream metabolites inevitably increase in parallel, due primarily to mass-action effects. This metabolic "domino effect" was at first generally felt to be the explanation for receptor-dependent changes in the concentration of Ins(3,4,5,6)P_4 (90,91,96,97). That is, Ins(3,4,5,6)P_4 and Ins(1,4,5)P_3 were considered to be contained within the same metabolic pool. Since Ins(3,4,5,6)P_4 is phosphorylated by a 1-kinase (91,98,99), Ins(3,4,5,6)P_4 was further considered to be a precursor for *de novo* Ins(1,3,4,5,6)P_5 synthesis (90,91,96,97). It was a study by Menniti et al., (93) that led to a reinterpretation of our understanding of both the nature and the significance of Ins(3,4,5,6)P_4 metabolism. An important feature of this study (93) was the evidence that Ins(3,4,5,6)P_4 and Ins(1,3,4,5,6)P_5 belong to a metabolic pool that is distinct from that of Ins(1,4,5)P_3 and its more closely-related derivatives. This conclusion has been independently verified (94). Moreover, Ins(3,4,5,6)P_4 and Ins(1,3,4,5,6)P_5 were suggested to be interconverted by a metabolically autonomous 1-kinase/1-phosphatase substrate cycle (93). Now, Ins(1,3,4,5,6)P_5 is suggested to be the precursor for *de novo* Ins(3,4,5,6)P_4 synthesis (93). PLC-mediated elevations in

$Ins(3,4,5,6)P_4$ levels were then envisaged as a manifestation of receptor-regulation of the 1-kinase/1-phosphatase cycle. So $Ins(3,4,5,6)P_4$ was elevated to the status of a candidate intracellular signal (93).

It took a few more years before we began to understand the molecular mechanisms by which the 1-kinase/1-phosphatase cycle was regulated. We now know that $Ins(1,3,4)P_3$ is a potent inhibitor of the $Ins(3,4,5,6)P_4$ 1-kinase ($63,64$). $Ins(1,3,4)P_3$ is a downstream metabolite of $Ins(1,4,5)P_3$; indeed, the cellular levels of $Ins(1,3,4)P_3$ closely mirror both the extent and the duration of PLC activation. By inhibiting the 1-kinase, and altering the poise of the 1-kinase/1phosphatase cycle, $Ins(1,3,4)P_3$ provides a link between activation of PLC and increases in levels of $Ins(3,4,5,6)P_4$ (64). Indeed, if $Ins(1,3,4)P_3$ dephosphorylation inside cells is inhibited by treatment with lithium during receptor activation, the increased levels of $Ins(1,3,4)P_3$ promote a larger elevation of $Ins(3,4,5,6)P_4$ levels (M. Yoshida and S. B. Shears, unpublished data).

The search for the physiological significance of receptor-dependent increases in $Ins(3,4,5,6)P_4$ involved the combined efforts of several laboratories, beginning with the interest of Barrett and Traynor-Kaplan in the regulation of Ca^{2+}-dependent Cl^- secretion from intestinal epithelia (see (100)). This laboratory has for some time been studying how positive and negative control over Cl^- secretion is co-ordinated during PLC activation. As well as being relevant to salt and fluid secretion, cellular Cl^- flux across the plasma membrane may have several other important physiological functions, including osmoregulation, pH balance, and smooth muscle contraction (101-104).

The mechanism by which PLC activates Cl^- secretion was already established. Receptor-dependent $Ins(1,4,5)P_3$-mediated mobilization of cellular Ca^{2+} stores (8) activates calmodulin-dependent protein kinase II (CaMK II). This kinase increases the ionic conductance through the Cl^- channels (101-$103,105,106$). However, negative signals have also been shown to contribute to the overall regulation of Cl^- secretion. Perhaps the first experimental indication of this comes from a 1990 study (107) of the HT-29 colonic epithelial cell line. These workers obtained whole-cell current recordings using the perforated patch technique. They observed that there was a rapid fall in the size of the Cl^- current following particularly large agonist-mediated rises in intracellular $[Ca^{2+}]$, but not when similar changes in $[Ca^{2+}]$ were induced by ionophores. They postulated that activation of PLC promoted a homeostatic protective mechanism that prevented the potentially deleterious consequences of large Cl^- conductance changes upon ionic composition, and hence cell volume.

Barrett and Traynor-Kaplan (108) considerably strengthened the case for there being a PLC-linked inhibitory regulator of Ca^{2+}-dependent Cl^- secretion, from experiments with T_{84} cells, which, like HT-29, is a colonic epithelial cell-line. In this study (108), Cl^- secretion across epithelial layers was measured in Ussing chambers. It emerged that carbachol elicited a biphasic effect. This agonist initially stimulated Cl^- secretion. However, carbachol strongly inhibited the ability of a subsequent elevation in cytosolic $[Ca^{2+}]$ to also activate Cl^- secretion, irrespective of whether this later mobilization of Ca^{2+} was brought about by either histamine or thapsigargin ($65,108$). Other workers have also shown that even 20 minutes after carbachol is washed off T_{84} cells, the Cl^- secretory response to a second dose of this agonist is blunted by 50% (109). It was proposed that one or more of the $InsP_4$ isomers in cells might inhibit Ca^{2+}-dependent Cl^- secretion (108).

Studies we performed in collaboration with Barrett and Traynor-Kaplan (65) provided the first evidence that, among the various $InsP_4$ isomers, $Ins(3,4,5,6)P_4$ was the most likely regulator of Cl^- secretion. Together, we showed that there was a close correlation between the development of the carbachol-dependent inhibition of Cl^- secretion and increases in cellular levels of $Ins(3,4,5,6)P_4$, and not the other $InsP_4$ isomers (65). More strikingly, when carbachol's effects were reversed with atropine, the inhibition of Cl⁻ secretion persisted for more than 15 min; at this time point, $Ins(3,4,5,6)P_4$ was the only inositol phosphate whose concentration remained elevated above basal levels (65).

This correlation between levels of $Ins(3,4,5,6)P_4$ and inhibition of Ca^{2+}-dependent Cl⁻ secretion was pursued further using a cell permeant analogue of $Ins(3,4,5,6)P_4$ which was synthesized by Tsien and Schultz (65). This compound has the hydroxyl groups hidden by butyrates, and the phosphates are masked by acetoxymethyl esters. This enables the analogue to diffuse across the plasma membrane and into the cell, whereupon intracellular esterases liberate free $Ins(3,4,5,6)P_4$. Treatment of T_{84} cells with this analogue did not affect cellular Ca^{2+} pools, nor did it alter agonist-mediated Ca^{2+}-fluxes. However, cell-permeant $Ins(3,4,5,6)P_4$ inhibited both histamine-activated and thapsigargin-dependent Cl⁻ secretion (65). Cell-permeant $Ins(1,4,5,6)P_4$ was an important control that had no effect, indicating that Cl⁻ secretion was not being perturbed by non-specific effects of the "charge-masking" groups. This work provided the first evidence for a physiological function for $Ins(3,4,5,6)P_4$.

The next question to be addressed was the nature of the ion channel that was targeted by $Ins(3,4,5,6)P_4$. Net Cl⁻ secretion across an epithelial cell layer requires more than the action of Cl⁻ channels. There is also a cooperative activation of Ca^{2+}-activated K⁺ efflux across the basolateral cell membrane, in order to facilitate a co-ordinated recirculation of K⁺ ions that enables co-transporter mediated Na^+-K^+-$2Cl^-$ uptake (101,102). Thus, $Ins(3,4,5,6)P_4$ could have acted by inhibiting either Cl⁻ or K⁺ channels. In collaboration with Nelson's laboratory (110) we measured whole-cell Cl⁻ current in T84 cells. Autophosphorylated and constitutively active CaMK II was perfused into the cell through the patch pipette. This brought about a 20-fold increase in the Cl⁻ current (110). The CaMK II-activated current was completely blocked by 10 μM $Ins(3,4,5,6)P_4$, even though the polyphosphate had no effect upon CaM KII activity itself. Thus, the polyphosphate was shown to inhibit Cl⁻ channels. This effect was mediated downstream of CaMK II, the activity of which was not affected by $Ins(3,4,5,6)P_4$. This is a very specific effect, in T84 cells and elsewhere: $Ins(1,3,4)P_3$, $Ins(1,3,4,6)P_4$, $Ins(1,3,4,5)P_4$, $Ins(1,4,5,6)P_4$ and $Ins(1,3,4,5,6)P_5$ all have no effect upon Cl⁻ current even when added at concentrations 15 times greater than the IC_{50} value for $Ins(3,4,5,6)P_4$ (110-112).

$Ins(3,4,5,6)P_4$ also blocked a recombinant Ca^{2+}-activated Cl⁻ channel (Cl_{Ca}) from bovine trachea (111). In the latter studies, recombinant Cl_{Ca} was expressed in Xenopus oocytes, from which membrane fragments were extracted for incorporation into lipid bilayers. The Cl⁻ conductance through the bilayer was inhibited by $Ins(3,4,5,6)P_4$ (111). This demonstration of a direct effect of $Ins(3,4,5,6)P_4$ upon a recombinant ion channel was an important development. However, two aspects of these data indicate that the bilayer experiments only partially reconstitute the mechanism of action of $Ins(3,4,5,6)P_4$ in intact cells. First, Ismailov et al., (111) found that nanomolar concentrations of $Ins(3,4,5,6)P_4$ blocked Cl_{Ca}. In contrast, 2-10 μM $Ins(3,4,5,6)P_4$ is required inside cells to inhibit Cl⁻ current (65,110,112). Second, in the bilayer experiments a single site dose/response relationship for $Ins(3,4,5,6)P_4$ was obtained, without any evidence for co-operativity (111). $Ins(3,4,5,6)P_4$ behaves differently inside cells, where a particularly steep slope in the dose/response curve suggests a highly co-operative mechanism of action (110,112). These quantitative differences in the results obtained in the two experimental paradigms suggest that important auxiliary factors also participate in the action of $Ins(3,4,5,6)P_4$ in vivo. One of the goals of our laboratory is to pursue the nature of these putative regulators. It is also exciting to consider if there are any other CaM KII-activated physiological processes that might be attenuated by $Ins(3,4,5,6)P_4$.

Another objective is to elucidate the structural determinants of $Ins(3,4,5,6)P_4$ action. These are not only physiologically-significant goals, but they also have pathological and therapeutic implications. This information could aid the rational design of $Ins(3,4,5,6)P_4$ antagonists, which are candidate drugs for up-regulating Cl^- secretion in the genetic disease of cystic fibrosis (113), where cAMP-regulated Cl^- channels are defective (114,115). The resultant thickened and poorly cleared airway secretions leads to progressive lung disease and early death.

$InsP_6$

Research into $InsP_6$ and its metabolism can be traced back 90 years to a report that this polyphosphate was dephosphorylated by calf liver (116). However, most of the ensuing, early studies in animal systems focused upon *extracellular* metabolism of dietary $InsP_6$ by the digestive tract. As a result, we know that intestinal epithelial cells possess a battery of relatively non-specific cell-surface phosphatases that can digest $InsP_6$ to release inositol for absorption (117-119).

With the exception of the digestive studies mentioned above, it was the case until the mid 1980s that work with plants and micro-organisms had generated most of our information on $InsP_6$, where it is considered to be a major store of both inositol and phosphate. Its importance can be judged from the fact that 50-80% of plant seed total phosphorous may be in the form of $InsP_6$-cation complexes, deposited as globular inclusions in membrane-bound storage bodies (for reviews, see (120,121)). At least one of the phosphates of this $InsP_6$ store appears able to support ATP synthesis; Biswas *et al.*, (122) proposed that, in plants at least, the 2-phosphate of $InsP_6$ could be transferred directly to ADP through the reversibility of the $Ins(1,3,4,5,6)P_5$ 2-kinase; in support of this idea, this enzyme in soybean seeds was determined to have an equilibrium constant of 14 (123). Dense $InsP_6$-cation deposits have also been found in the animal kingdom; they may comprise half the weight of the mature dispersal larvae of most species of dicyemid mesozoans. The ensuing negative buoyancy may keep these minute parasites close to the sea bottom where they can encounter their invertebrate hosts (124).

This well-known tendency of $InsP_6$ to form insoluble complexes with divalent cations *in vivo* presented us with a problem, when reports of apparently *soluble* $InsP_6$ being present in mammalian cells first emerged (88,125,126): where in the cell could $InsP_6$ be prevented from precipitating? The idea that $InsP_6$ might be sequestered into a vesicular compartment (127) has not been confirmed experimentally (128). Instead, there is evidence that much of the cell's $InsP_6$ is "wallpapered" to cellular membranes (129). The polyphosphate is proposed to be held in place by the formation of an electrostatically bonded $InsP_6$ - Al^{3+} - phospholipid sandwich (129). Consistent with this idea, the binding of $InsP_6$ to cellular membranes is disrupted by chelation of divalent and trivalent cations by EDTA (128,129). As a consequence, even though total cellular levels of $InsP_6$ might range from 5-100 μM (130-132), it is not clear how much of this will be free in the cytosol. This lack of a good understanding of the free cellular $InsP_6$ concentration creates great uncertainty concerning what is the relevance *in vivo* of some effects of $InsP_6$ obtained *in vitro*, particularly those that require μM levels.

Two observations initially generated the misleading impression that the turnover of $InsP_6$ *in vivo* was rather sluggish: First, in experiments with isolated cells, there is a slow rate of incorporation of exogenous [^{3}H]inositol into cellular $InsP_6$, and second, the levels of this compound do not generally respond dramatically to short-term receptor activation (e.g. (126)). The perception of slow $InsP_6$ turnover inside cells led to the extrapolation from the plant to the animal kingdom of the idea that it might be a storage depot for inositol and phosphate (96,127). In one sense, this is true, although not quite in the way this was originally envisaged; $InsP_6$ comprises a relatively large precursor pool for the diphosphorylated inositol polyphosphates, namely, $PP\text{-}InsP_5$ and $[PP]_2\text{-}InsP_4$ (133). Nevertheless, the metabolism of $InsP_6$ is much more dynamic than was at first appreciated. Now we know that up to 50% of the $InsP_6$ pool cycles through the diphosphates every hour (133,134).

It has been considered that $InsP_6$ may function independently of its interconversion with the diphosphorylated polyphosphates. $InsP_6$ is, for example, a powerful anti-oxidant in vitro (135). Hawkins et al., (136) have extended this hypothesis, by demonstrating that the remarkable affinity of $InsP_6$ for iron totally inhibited this metal's ability to catalyze the formation of hydroxyl radicals. These authors therefore suggested that a physiological function of $InsP_6$ was to transport iron within the cell in a form that protects against the potentially lethal consequences of free radical formation (136). Although these chemical properties of $InsP_6$ are indisputable, there is no direct evidence that the polyphosphate either transports iron or acts as an antioxidant in vivo. Incidentally, an early proposal that $InsP_6$ might be transported out of cells to act as an extracellular signal (125), has more recently fallen out of favor (137).

The very nature of $InsP_6$ presents a problem for interpreting its effects in vitro. As this compound is highly phosphorylated, it can influence protein function by non-specific charge effects (138). One way to exclude this possibility is to show that an effect of $InsP_6$ is neither imitated by inositol hexasulphate ($InsS_6$), nor reversed by adding excess divalent cations (138). Unfortunately, these negative controls are only useful when they yield just such a result. In contrast, experiments where $InsS_6$ mimics an effect of $InsP_6$ are equivocal, such as was the case with their both activating L-type Ca^{2+} channels and insulin secretion in the H1T T15 insulinoma (139,140). We cannot tell, just from this direct comparison, whether $InsS_6$ was imitating a physiological or non-physiological effect of $InsP_6$. In such circumstances, it is especially important to determine if effects of $InsP_6$ are preserved in a physiologically-relevant ionic milieu. This particular action of $InsP_6$ upon Ca^{2+} channels has been attributed to inhibition of protein phosphatase activity. In the same study, $InsP_6$, with IC_{50} values of 4-13 μM, inhibited three different species of serine/threonine protein phosphatases in vitro (140). It is difficult to envisage this effect could have a specific signaling consequence. It is a broad spectrum effect upon the activities of several different protein phosphatases, and not even specific for $InsP_6$, since $Ins(1,3,4,5,6)P_5$ acted with approximately the same potency.

Another approach that has been taken in order to determine if $InsP_6$ is physiologically active has been to isolate and characterize proteins which bind this polyphosphate with higher affinity than other inositides. Several such proteins have been discovered, and typically their affinities for $InsP_6$ lie in the nanomolar range. However, there are few accompanying demonstrations that ligand binding alters the function of the target protein. For example, there is no indication as to the possible physiological consequences of $InsP_6$ binding tightly to either vinculin, a component of platelet cytoskeleton (141) or to myelin proteolipid protein, which participates in myelin deposition (142) or to coatomer, which regulates vesicle traffic between the ER and the Golgi (143,144).

Studies of some other $InsP_6$-binding proteins may be considered more informative. For example $InsP_6$ binds tightly to the clathrin assembly proteins, AP-2 (145-147) and AP-3 (148,149). These particular studies are significant in that they also show what the physiological consequences of ligand binding might be. Thus, $InsP_6$ has been found to inhibit the ability of these adaptor proteins to promote clathrin assembly, which is an early event in the process of endocytosis. This has led to the suggestion that $InsP_6$ might be a partial antagonist (or "clamp") of endocytic vesicle traffic. This hypothesis seems unlikely to develop further unless evidence can be found to indicate that such a "clamp" might be regulated. Cellular levels of $InsP_6$ do not respond dramatically to short-term receptor activation (see above). However, they do rise and fall at certain points in the cell-cycle and during cellular differentiation (131,150-152), when there are also dramatic changes in vesicle trafficking processes (153). It is possible that extracellular agonists might also act upon $InsP_6$ in a more subtle manner, by perhaps altering its distribution between cellular membranes (see above) and the cytosol. Another possibility is that covalent modification of AP-2 and AP-3 may alter their affinity for $InsP_6$ and thereby relax the trafficking clamp. These ideas remain to be tested. Demonstrations that microinjection of $InsP_6$ into cells can inhibit vesicle traffic (154) is consistent with the participation in this process of an inositide-binding site. However, such experiments do not prove that the physiologically relevant ligand is $InsP_6$ itself. In fact, the inositol lipid, $PtdIns(3,4,5)P_3$, has recently been found to be a more potent ligand than $InsP_6$ for AP-2 (155) and AP-3 (156). As pointed out in these studies, the levels of $PtdIns(3,4,5)P_3$ also increase acutely upon receptor activation. Thus, general attention has begun to shift towards this lipid as being the more important, short-term regulatory ligand for these particular proteins.

Although $InsP_6$ is actively phosphorylated (see above) it seems to be dephosphorylated relatively slowly; the only mammalian enzyme so far shown to hydrolyze $InsP_6$, i.e. MIPP (157), has a very low V_{max} for this substrate. The major pathway of $InsP_6$ dephosphorylation in mammals is not well-characterized, but several $InsP_5$ isomers are generated (157), and $Ins(1,2,3,4)P_4$ / $Ins(1,2,3,6)P_4$, $Ins(1,2,3)P_3$ and $Ins(1,2)P_2$ / $Ins(2,3)P_2$ also appear to be formed during the degradation to free Ins (93,150,158). There is some evidence that the activity of this catabolic pathway may change both during the cell cycle, or upon cell differentiation (131,150-152), but the cause/effect relationship between these events is not known. Some workers have reported that $InsP_6$, or one of its metabolites, has an anti-neoplastic action (159). This effect of $InsP_6$ has been patented as being of potential therapeutic benefit; its biological basis is unknown.

Products of M.I.P.P.

The Multiple Inositol Polyphosphate Phosphatase (MIPP) provides the mammalian cell with the only known means of dissipating the cellular pools of $Ins(1,3,4,5,6)P_5$ and $InsP_6$ (157,160). In liver at least, much of the cell's complement of MIPP is restricted to the interior of the endoplasmic reticulum (ER) (161). To date, we have been unable to demonstrate the uptake of any inositol phosphate into isolated microsomes. It is therefore not clear how this enzyme encounters its substrates, which appear not to be compartmentalized inside a vesicular pool (see (128)).

If small amounts of hepatic MIPP are present outside the ER, this would have been missed in our earlier experiments with subcellular fractions (161). In fact, MIPP is known to be present in plasma membranes of mature, mammalian erythrocytes (160,162,163). This location would presumably provide the enzyme with ample opportunity to encounter its substrates. While mature erythrocytes do not contain inositol phosphates, $Ins(1,3,4,5,6)P_5$ and $InsP_6$ seem likely to accumulate during erythropoiesis; erythropoietic cells express receptor-regulated phospholipase C activity (164), and so would synthesize $Ins(1,3,4)P_3$, the precursor for the higher inositol polyphosphates. There may be a good reason for MIPP activity outliving the ability of these cells to produce the enzyme's substrates. If $Ins(1,3,4,5,6)P_5$ and $InsP_6$ were to be retained by the mature erythrocyte, they would then bind tightly to hemoglobin and impair the normal regulation of its affinity for oxygen by 2,3-diphosphoglycerate (165). The persistence of MIPP activity as the developing erythrocyte attains maturity may be part of the process by which the removal of any residual $Ins(1,3,4,5,6)P_5$ and

InsP_6 can be assured. There is one other confirmed sighting of a MIPP-like enzyme being present in plasma membranes, in the slime-mould *Dictyostelium* (*166*). Interestingly, this form of MIPP can convert Ins(1,3,4,5,6)P_5 to Ins(1,4,5)P_3 *in vivo* (*167*) (mammalian MIPP also has this ability (*167*), but there is as yet no evidence that this occurs in intact cells).

MIPP-like Ins(1,3,4,5,6)P_5/Ins(1,3,4,5)P_4 phosphatase activity was briefly reported to be present on the surface of NIH3T3 cells (*168*). Although these data have never been fully published, it has been reiterated that such an enzyme may be a physiologically relevant inactivator of putative extracellular pools of Ins(1,3,4,5,6)P_5, and possibly also InsP_6 (*137*). However, there is an alternative interpretation of these workers' conclusions. The determination by Carpenter *et al.*, (*168*) that "intact" 3T3 cells could metabolize Ins(1,3,4,5,6)P$_5$ and Ins(1,3,4,5)P$_4$ could have been due to the unavoidable presence in the culture of a small proportion of "damaged" cells. Indeed, how else can we explain why these same cells also dephosphorylated Ins(1,4,5)P$_3$ (albeit more slowly) even though Ins(1,4,5)P$_3$ 5-phosphatase is entirely intracellular (*48,169*)? Carpenter et al., (*168*) went on to state that the rate of Ins(1,3,4,5,6)P$_5$ metabolism was "unchanged" when the cells were permeabilized with digitonin. Digitonin treatment is often used to specifically permeabilize the plasma membrane (e.g. reference *128*). This detergent is relatively inefficient at permeabilizing the membranes of intracellular organelles such as ER, where considerable MIPP activity resides (*161*). With this hindsight, it can now be appreciated that ER-based Ins(1,3,4,5)P$_4$ / Ins(1,3,4,5,6)P$_5$ phosphatase activity would not increase when the plasma membrane was preferentially permeabilized - without our having to invoke an extracellular location for this enzyme activity.

Despite not knowing how MIPP accesses its substrates inside mammalian cells, there are several reasons for believing it can. Ins(1,4,5,6)P_4 is one of the signature products of MIPP activity, produced by 3-phosphatase attack on Ins(1,3,4,5,6)P_5 (*157,160*). Ins(1,4,5,6)P_4 is not only present in cells, at a concentration of about 1 μM (*65*), but its levels change under certain conditions. For example, increased concentrations of Ins(1,4,5,6)P_4 have been found in *src*-transformed fibroblasts (*170*). Also, following stimulation of WRK1 cells with vasopressin, levels of Ins(1,4,5,6)P_4 rose about 2-fold (*94*). These phenomena suggest that ongoing MIPP activity is not only occurring inside cells, but in fact is carefully regulated. This idea is even more dramatically illustrated by the response of T84 colonic epithelial cells to infection with *Salmonella*: within 30 min of the addition of these enteric bacteria, cellular levels of Ins(1,4,5,6)P_4 increase up to 14-fold (*171*).

The idea that Ins(1,4,5,6)P_4 is a physiologically-relevant metabolite is also supported by the observation that cells contain a 3-kinase that re-phosphorylates it back to Ins(1,3,4,5,6)P_5 (*63,172*). Moreover, when antimycin A was used to deplete cellular ATP so as to inhibit Ins(1,4,5,6)P_4 3-kinase activity, levels of Ins(1,4,5,6)P_4 increased 4-fold within 1 hr (*172*). The latter result suggests that there is normally a significant ongoing metabolic cycling between Ins(1,3,4,5,6)P_5 and Ins(1,4,5,6)P_4.

Ins(1,4,5,6)P_4 may even have some important cellular functions. For example, it has high affinity for the pleckstrin homology (PH) domain of a protein, p130, that has considerable similarity to PLC but does not hydrolyze inositol lipids (*173*). PH domains may in general provide inositides with the opportunity to regulate protein-protein and protein-phospholipid interactions within multimeric signal transduction complexes (*28*). There may also be circumstances under which Ins(1,4,5,6)P_4 can antagonize the inositol lipid 3-kinase signaling pathway (*171*), possibly because Ins(1,4,5,6)P_4 is a rather good analogue of the Ins(1,3,4,5)P_4 headgroup (*167*) of PtdIns(3,4,5)P_3. These considerations present us with additional motivation to understand the cellular regulation of MIPP activity.

We recently described the molecular cloning and expression of a catalytically active form of rat hepatic MIPP (*174*). It is interesting that the ER of avian chondrocytes contains a MIPP homologue, the expression of which is up-regulated when these cells become hypertrophic (*175,176*). Hypertrophic chondrocytes are part of a specialized developmental structure, namely, the advancing ossification front that divides newly synthesized bone from the remaining cartilage (*177*). It is an intriguing possibility that up-regulation of a MIPP-like protein, and a corresponding increase in the cell's capacity to hydrolyze Ins(1,3,4,5,6)P_5 and InsP_6, are functionally important during a key period of bone development. The similarity between MIPP and the chick chondrocyte protein extends to the tetrapeptide carboxy terminus of both proteins comprising a signal that retains them in the lumen of the ER (*174,175*).

A further notable region of MIPP is the 18 amino acid sequence that aligns with 61% and 55% identity to a corresponding region of isoforms of phytase A (InsP_6 phosphatase) from, respectively, *Aspergillus* and *Myceliophthora* (*174*). These aligned sequences include the histidine acid phosphatase RHGXRXP catalytic motif (*178*) which is a feature not yet found in any other mammalian inositol phosphate metabolizing enzyme. An interesting evolutionary question concerns the nature of the selective pressures that led to the specific conservation in MIPP of only this 18 amino-acid portion of phytase A. Outside of this region the MIPP sequence does not show any significant similarity with the fungal enzymes.

Now that we have obtained a cDNA clone of MIPP that encodes a catalytically active protein, we are an important step closer to analyzing the metabolic and physiological consequences of manipulating the expression of MIPP activity inside cells. Such studies should improve our insight into the actions of both the substrates and the products of this enzyme.

Diphosphorylated Inositides

Throughout the 1980's and early 1990's we were trying to keep abreast of the consequences of a rapidly expanding family of inositol polyphosphates. At that time, we could at least console ourselves that InsP_6 represented an upper limit to the structural diversity of these compounds. That illusion faded in 1993 with the discovery of the diphosphorylated inositol polyphosphates (*133,179*). This subgroup is itself now growing in its complexity. Several members of this family have been identified. One of them is a diphosphoinositol tetrakisphosphate (PP-InsP_4) that is synthesized by the kinase-mediated addition of a diphosphate group to Ins(1,3,4,5,6)P_5 (*133*). PP-InsP_4 is rapidly dephosphorylated back to Ins(1,3,4,5,6)P_5 *in vivo* (*133*). Little else is known about this polyphosphate.

We have a little more information on the diphosphoinositol pentakisphosphate that is present in mammalian cells (PP-InsP_5, a.k.a. "InsP_7"). A PP-InsP_5 isomer with the diphosphate group attached to the 5-carbon has, so far, been the only one found to occur in a variety of mammalian cell types (*134*). It has not yet been determined what is the structure of the mammalian bis-diphosphoinositol tetrakisphosphate ([PP]$_2$-InsP_4, a.k.a. "InsP_8"), although it is possible it is the 5,6-diphosphate (*134,180,181*). PP-InsP_5 and [PP]$_2$-InsP_4 are synthesized by sequential phosphorylation of InsP_6 (*133,179,182,183*). The InsP_6 kinase has a K_m value for ATP of 1.4 mM, which is sufficiently close to physiological levels so that changes in cellular [ATP] could impact on the concentration of PP-InsP_5 (*182*).

Mammalian cells also contain a 5-PP-InsP_5 5-β-phosphatase and a [PP]$_2$-InsP_4 β-phosphatase (tentatively a 6-β-phosphatase, contingent on the characterization of the structure of [PP]$_2$-InsP_4). These phosphatases degrade [PP]$_2$-InsP_4 back to InsP_6 (*133,183*). The diphosphorylated inositides appear to have arisen at an early evolutionary stage: 5-PP-InsP_5 has also been identified in *Entomoeba histolytica* (*184*) and *Phreatamoeba balamuthi* (*185*). Although 5-PP-InsP_5 is also present in the slime mould, *Dictyostelium* (*134*), that particular organism contains much higher levels of 6-PP-InsP_5 (*134,180,181*). The significance of this isomeric variability is unknown, but it serves as a warning that it as too early to generalize about either the structures or the functions of these compounds, particularly when crossing phylogenetic barriers.

An approximate estimate of the total levels of diphosphoinositol polyphosphates in mammalian cells can be obtained by labeling them *in situ* with [^{3}H]inositol, since PP-InsP_5 and [PP]$_2$-InsP_4 seem to be in isotopic equilibrium with InsP_6 (*133,186*). The levels of PP-[^{3}H]InsP_5 and [PP]$_2$-[^{3}H]InsP_4 are typically 0.5 to 5% of the level of [^{3}H]InsP_6 (*133,134,179,183,186*). Estimates of the total cellular concentration of InsP_6 range from 5 to 100 µM (*130-132*). This puts the levels of PP-InsP_5 and [PP]$_2$-InsP_4 in the low micromolar range, at most.

The relatively low cellular levels of the diphosphorylated inosites make it difficult to study their metabolic turnover *in vivo*. Nevertheless, it is now clear that there is considerable ongoing metabolic flux through PP-InsP_5 and [PP]$_2$-InsP_4. The latter information was initially obtained by exploiting the ability of F$^-$ to inhibit the phosphatase(s) that attack PP-InsP_5 and [PP]$_2$-InsP_4 (*133,183*). By blocking the dephosphorylation of PP-InsP_5 and [PP]$_2$-InsP_4 in intact cells and exposing their ongoing rate of synthesis, F$^-$ caused the levels of these compounds to increase many-fold within minutes. This application of F$^-$ as a metabolic trap has shown that 30-50% of the cell's InsP_6 pool cycles through these diphosphorylated inosites every hour (*133,134,183*). The discovery of this effect (*133*) swept away another dogma of inositide metabolism that was prevalent in the 1980s: the idea that InsP_6 was metabolically lethargic (*96*). Now, it is generally appreciated how dynamic is the turnover of InsP_6. Experiments with the purified PP-InsP_5 and [PP]$_2$-InsP_4 phosphatases have shown that the K_i for F$^-$ is approximately 10 µM (S.T. Safrany and S. B. Shears, unpublished data). Serum levels of F$^-$ may reach or even exceed this level during the administration of fluorinated anesthetics (*187*), or when using monofluorophosphate therapy for osteoporosis (*188,189*). In these circumstances, it is possible that the normal physiological functions of the diphosphates may be perturbed.

It is generally considered that the hydrolysis of the diphosphates is accompanied by a substantial free energy change (*179-182*). The high-energy potential of the β-phosphate of PP-InsP_5 is evident from the demonstration that it can be transferred to ADP to form ATP by the InsP_6 kinase acting "in reverse" (*182*). Nevertheless, *in vivo* ADP might not be the most functionally significant acceptor of the β-phosphate of PP-InsP_5 (*182*). Indeed, some preliminary evidence indicates that PP-InsP_5 may act as a "molecular switch" by transferring a phosphate group to proteins (*190*).

The importance of the diphosphorylated inosites to signal transduction now seems clear, following the demonstration that levels of [PP]$_2$-InsP_4 are specifically regulated by both cAMP and cGMP (*191*). It is of particular interest that these cyclic nucleotides act in an unusual manner, in that they regulate [PP]$_2$-InsP_4 levels independently of the activation of both A-kinase and G-kinase (*191*). From a cell-signaling perspective, another interesting feature of the turnover of diphosphoinositol polyphosphates is its sensitivity to perturbation of cellular Ca^{2+} pools by thapsigargin. The latter is a skin-irritating and tumour promoting sesquiterpene lactone produced by the umbelliferous plant, *Thapsia garganica L.* (see *192* for refs). Thapsigargin acts on mammalian cells by specifically inhibiting the Ca^{2+}-ATPase of endoplasmic reticulum (*192*), which depletes intraluminal Ca^{2+} stores, elevates cytosolic [Ca^{2+}], and thereby promotes Ca^{2+} entry into the cell (see *193*). Thapsigargin also reduces steady-state levels of the diphosphoinositol polyphosphates by up to 50%, by apparently inhibiting their rates of synthesis (*186*). A similar response was obtained when cells were treated with a structurally unrelated inhibitor of Ca^{2+}-ATPase, cyclopiazonic acid (N. Ali and S. B. Shears, unpublished data). Therefore, it seems likely that the effect common to both thapsigargin and cyclopiazonic acid, namely, changes in the status of cellular Ca^{2+} pools, is responsible for the inhibition of synthesis of the diphosphoinositol polyphosphates. There is currently considerable interest in the nature of cytosolic effectors that can sense changes in the Ca^{2+} concentration of the endoplasmic reticulum (*193*). However, mobilization of Ca^{2+} through the activation of extracellular receptors has, at best, only a marginal effect on levels of the diphosphoinositol polyphosphates (S. T. Safrany and S. B. Shears, unpublished data, see also refs *133* and *186*). It is possible that steady-state levels of diphosphoinositol polyphosphates are influenced by subtle differences in the pattern of Ca^{2+} mobilization induced by agonists as compared to Ca^{2+}-ATPase inhibitors.

References Cited

1. Streb, H.; Irvine, R. F.; Berridge, M. J; Schulz, I. *Nature* **1983**, *306*, 67-68.
2. Hokin, L. E; Hokin, M. R. *Biochim. Biophys. Acta* **1955**, *18*, 102-110.
3. Michell, R. H. *Biochim. Biophys. Acta* **1975**, *415*, 81-147.
4. Kirk, C. J.; Creba, J. A.; Downes, C. P; Michell, R. H. *Biochem. Soc. Trans.* **1981**, *9* , 377-379.
5. Creba, J.; Downes, C. P.; Hawkins, P. T.; Brewster, G.; Michell, R. H; Kirk, C. J. *Biochem. J* . **1983**, *212*, 733-747.
6. Berridge, M. J. *Biochem. J* **1983**, *212*, 849-858.
7. Putney, J. W.; Jr. *Cell Calcium* **1986**, *7*, 1-12.
8. Berridge, M. J. *Nature* **1993**, *361*, 315-325.
9. Joseph, S. K. *Cell. Signal.* **1996**, **8**, 1-7.
10. Palmer, S.; Hughes, K. T.; Lee, D. Y; Wakelam, M. J. O. *Cell. Signal.* **1988**, *1*, 147-156.
11. Seyfred, M. A.; Farrell, L. E; Wells, W. W. *J. Biol. Chem.* **1984**, *259*, 13204-13208.
12. Storey, D. J.; Shears, S. B.; Kirk, C. J; Michell, R. H. *Nature* **1984**, *312*, 374-376.
13. Joseph, S. K; Williams, R. J. *FEBS Lett.* **1985**, *180*, 150-154.
14. Ackermann, K. E.; Gish, B. G.; Honchar, M. P; Sherman, W. R. *Biochem. J.* **1987**, *242* , 517-524.
15. Eisenberg, F. *J. Biol. Chem.* **1967**, *242*, 1375-1382.
16. Irvine, R. F.; Letcher, A. J.; Lander, D. J; Downes, C. P. *Biochem. J* . **1984**, *223*, 237-243.
17. Downes, C. P.; Hawkins, P. T; Irvine, R. F. *Biochem. J* . **1986**, *238*, 501-506.
18. Batty, I. H.; Nahorski, S. R; Irvine, R. F. *Biochem. J.* **1985**, *232*, 211-215.
19. Irvine, R. F.; Letcher, A. J.; Heslop, J. P; Berridge, M. J. *Nature* **1986**, *320*, 631-634.
20. Shears, S. B.; Parry, J. B.; Tang, E. K. Y.; Irvine, R. F.; Michell, R. H; Kirk, C.J. *Biochem. J.* **1987**, *246*, 139-147.
21. Balla, T.; Guillemette, G.; Baukal, A. J; Catt, K. *J. Biol. Chem.* **1987**, *262*, 9952-9955.
22. Irvine, R. F. in *Advances in Second Messenger and Phosphoprotein Research* (Putney, J. W., Jr., ed.), New York, Raven Press, **1992**, pp. 161-185.
23. Cullen, P. J.; Hsuan, J. J.; Truong, O.; Letcher, A. J.; Jackson, T. R.; Dawson, A. P; Irvine, R. F. *Nature* **1995**, *376*, 527-530.
24. Fukuda, M; Mikoshiba, K. *J.Biol.Chem.* **1996**, *271*, 18838-18842.
25. Cullen, P. J.; Loomis-Husselbee, J.; Dawson, A. P; Irvine, R. F. *Biochem. Soc. Trans.* **1997**, **25**, 991-996.
26. Kojima, T.; Fukuda, M.; Watanabe, Y.; Hamazato, F; Mokoshiba, K. *Biochem. Biophys. Res. Commun.* **1997**, *236*, 333-339.

27. Fukuda, M.; Kojima, T.; Kabayama, H; Mikoshiba, K. *J. Biol. Chem.* **1996**, *271*, 30303-30306.
28. Lemmon, M. A.; Falasca, M.; Ferguson, K. M; Schlessinger, J. *Trends Cell Biol.* **1997**, *7*, 237-242.
29. Fukuda, M; Mikoshiba, K. *Bioessays* **1997**, *19*, 593-603.
30. Takazawa, K.; Vandekerckhove, J.; Dumont, J.E; Erneux, C. *Biochem. J.* **1990**, *272*, 107-112.
31. Takazawa, K.; Perret, J.; Dumont, J. E; Erneux, C. *Biochem. J.* **1991**, *278*, 883-886.
32. Communi, D.; Vanweyenberg, V; Erneux, C. *Biochem. J.* **1994**, *298*, 669-673.
33. Shears, S. B. in *Advances in Second Messenger and Phosphoprotein Research* (Putney, J.W., Jr., ed.), New York , Raven Press, **1992**, pp. 63-92.
34. Soriano, S.; Thomas, S.; High, S.; Griffiths, G.; D'Santos, C.; Cullen, P; Banting, G. *Biochem. J.* **1997**, *324*, 579-589.
35. Communi, D.; Vanweyenberg, V; Erneux, C. *Cell. Signal.* **1995**, *7*, 643-650.
36. Communi, D.; Vanweyenberg, V; Erneux, C. *EMBO J.* **1997**, *16*, 1943-1952.
37. Sim, S. S.; Kim, J. W; Rhee, S. G. *J. Biol. Chem.* **1990**, *265*, 10367-10372.
38. Verjans, B.; De Smedt, F.; Lecocq, R.; Vanweyenberg, V.; Moreau, C; Erneux, C. *Biochem. J.* **1994**, *300*, 85-90.
39. Communi, D.; Lecocq, R; Erneux, C. *J. Biol. Chem.* **1996**, *271*, 11676-11683.
40. Jefferson, A. B; Majerus, P. W. *J. Biol. Chem.* **1995**, *270*, 9370-9377.
41. Laxminarayan, K.; Chan, B. K.; Tetaz, T.; Bird, P. I; Mitchell, C. A. *J. Biol. Chem.* **1994**, *269*, 17305-17310.
42. Jefferson, A. B.; Auethavekiat, V.; Pot, D. A.; Williams, L. T; Majerus, P. W. *J. Biol. Chem.* **1997**, *272*, 5983-5988.
43. Jackson, S. P.; Schoenwaelder, S. M.; Matzaris, M.; Brown, S; Mitchell, C. A. *EMBO J.* **1995**, *14*, 4490-4500.
44. Connolly, T. M.; Lawing, W. J. Jr; Majerus, P. W. *Cell* **1986**, *46*, 951-958.
45. Erneux, C.; De Smedt, F.; Moreau, C.; Rider, M; Communi, D. *Eur. J. Biochem.* **1995**, *234*, 598-602.
46. Auethavekiat, V.; Abrams, C. S; Majerus, P. W. *J. Biol. Chem.* **1997**, *272*, 1786-1790.
47. De Smedt, F.; Missiaen, L.; Parys, J. B.; Vanweyenberg, V.; de Smedt, H; Erneux, C. *J. Biol. Chem.* **1997**, *272*, 17367-17375.
48. Shears, S. B.; Evans, W. H.; Kirk, C. J; Michell, R. H. *Biochem. J.* **1988**, *256*, 363-369.
49. Speed, C. J.; Little, P. J.; Hayman, J. A; Mitchell, C. A. *EMBO J.* **1996**, *15*, 4852-4861.
50. McPherson, P. S.; Garcia, E. P.; Slepnev, V. I.; David, C.; Zhang, X.; Grabs, D.; Sossin, W. S.; Bauerfeind, R.; Nemoto, Y; De Camilli, P. *Nature* **1996**, *379*, 353-357.
51. Chung, J. K.; Sekiya, F.; Kang, H. S.; Lee, C.; Han, J.-S.; Kim, S. R.; Bae, Y. S.; Morris, A. J; Rhee, S. G. *J. Biol. Chem.* **1997**, *272*, 15980-15985.
52. Woscholski, R.; Finan, P. M.; Radley, E.; Totty, N. F.; Sterling, A. E.; Hsuan, J. J.; Waterfield, M. D; Parker, P. J. *J. Biol. Chem.* **1997**, *272*, 9625-9628.
53. Attree, O.; Olivos, I. M.; Okabe, I.; Bailey, L. C.; Nelson, D. L.; Lewis, R. A.; McInnes, R. R; Nussbaum, R. L. *Nature* **1992**, *358*, 239-242.
54. Zhang, X.; Jefferson, A. B.; Auethavekiat, V; Majerus, P. W. *Proc. Nat. Acad. Sci.USA* **1995**, *92*, 4853-4856.
55. Birchall, J. D; Chappell, J. S. *Lancet* **1988**, *ii*, 1008-1010.
56. Shears, S. B.; Dawson, A. P.; Loomis-Husselbee, J. W; Cullen, P. J. *Biochem.J.* **1990**, *270*, 837.

57. Nye, K. E.; Knox, K. A; Pinching, A. J. *Aids* **1991**, *5*, 413-417.
58. Nye, K.E.; Riley, G.A; Pinching, A.J. *Clin. Exp. Immunol.* **1992**, *89*, 89-93.
59. Sumner, M; Shears, S. B. *FEBS Lett.* **1997**, *413*, 75-80.
60. Lioubin, M. N.; Algate, P. A.; Tsai, S.; Carlberg, K.; Aebersold, R; Rohrschneider, L. R. *Genes Dev.* **1996**, **10**, 1084-1095.
61. Damen, J. E.; Liu, L.; Rosten, P.; Humphries, R. K.; Jefferson, A. B.; Majerus, P. W; Krystal, G. *Proc. Nat. Acad. Sci. USA* **1996**, *93*, 1689-1693.
62. Osborne, M. A.; Zenner, G.; Lubinus, M.; Zhang, X.; Songyang, Z.; Cantley, L. C.; Majerus, P.; Burn, P; Kochan, J. P. *J. Biol. Chem.* **1997**, *271*, 29271-29278.
63. Craxton, A.; Erneux, C; Shears, S. B. *J. Biol. Chem.* **1994**, *269*, 4337-4342.
64. Tan, Z.; Bruzik, K. S; Shears, S. B. *J. Biol. Chem.* **1997**, *272*, 2285-2290.
65. Vajanaphanich, M.; Schultz, C.; Rudolf, M. T.; Wasserman, M.; Enyedi, P.; Craxton, A.; Shears, S. B.; Tsien, R. Y.; Barrett, K. E; Traynor-Kaplan, A. E. *Nature* **1995**, *371*, 711-714.
66. Abdullah, M.; Hughes, P. J.; Craxton, A.; Gigg, R.; Desai, T.; Marecek, J. F.; Prestwich, G. D; Shears, S. B. *J. Biol. Chem.* **1992**, *267*, 22340-22345.
67. Hughes, P. J.; Hughes, A. R.; Putney, J. W.; Jr; Shears, S. B. *J. Biol. Chem.* **1989**, *264*, 19871-19878.
68. Hildebrant, J.-P; Shuttleworth, T. J. *Biochem. J.* **1992**, *282*, 703-710.
69. Wilson, M. P; Majerus, P. W. *J. Biol. Chem.* **1996**, *271*, 11904-11910.
70. Balla, T.; Baukal, A. J.; Hunyady, L; Catt, K. J. *J. Biol. Chem.* **1989**, *264*, 13605-13611.
71. Shears, S. B. *J. Biol. Chem.* **1989**, *264*, 19879-19886.
72. Bansal, V. S.; Inhorn, R. C; Majerus, P. W. *J. Biol. Chem.* **1987**, *262*, 9444-9447.
73. Shears, S. B.; Kirk, C. J; Michell, R. H. *Biochem. J.* **1987**, *248*, 977-980.
74. Inhorn, R. C; Majerus, P. W. *J. Biol. Chem.* **1988**, *262*, 15946-15952.
75. Caldwell, K. K.; Lips, D. L.; Bansal, V. S; Majerus, P. W. *J. Biol. Chem.* **1991**, *266* , 18378-18386.
76. Norris, F. A.; Atkins, R. C; Majerus, P. W. *J. Biol. Chem.* **1997**, *272*, 10987-10989.
77. Norris, F. A.; Auethavekiat, V; Majerus, P. W. *J.Biol.Chem.* **1995**, *270*, 16128-16133.
78. Inhorn, R. C; Majerus, P. W. *J. Biol. Chem.* **1988**, *263*, 14559-14565.
79. York, J. D.; Saffitz, J. E; Majerus, P. W. *J. Biol. Chem.* **1994**, *269*, 19992-19999.
80. Sylvia, V.; Curtin, G.; Norman, J.; Stec, J; Busbee, D. *Cell* **1988**, *54*, 651-658.
81. Hallcher, L. M; Sherman, W. R. *J. Biol. Chem.* **1960**, *255*, 10896-10901.
82. Nahorski, S. R.; Ragan, C. I; Challiss, R. A. J. *Trends Pharmacol. Sci.* **1991**, *12*, 297-301.
83. Avissar, S.; Schreiber, G.; Danon, A; Belmaker, R. H. *Nature* **1988**, *331*, 440-442.
84. York, J. D.; Ponder, J. W; Majerus, P. W. *Proc. Nat. Acad. Sci. USA* **1995**, *92*, 5149-5153.
85. Johnson, L. F; Tate, M. E. *Can. J. Biochem.* **1969**, *47*, 63-73.
86. Isaacks, R. E.; Harkness, D. R.; Adler, J. L; Goldman, P. H. *Arch. Biochem. Biophys.* **1973**, *173*, 114-120.
87. Mejía, O.; León-Velarde, F; Monge-C.C. *Comp. Biochem. Physiol.* **1995**, *109B*, 437-441.
88. Heslop, J. P.; Irvine, R. F.; Tashjian, A. H; Berridge, M. J. *J. Exp. Biol.* **1985**, *119*, 395-401.

89. Stephens, L. R.; Hawkins, P. T.; Carter, N.; Chahwala, S. B.; Morris, A. J.; Whetton, A. D; Downes, P. C. *Biochem. J.* **1988**, *249*, 271-282.
90. Stephens, L. R.; Hawkins, P. T.; Barker, C. J; Downes, C. P. *Biochem.J.* **1988**, *253* , 721-733.
91. Balla, T.; Hunyady, L.; Baukal, A. J; Catt, K. J. *J. Biol. Chem.* **1989**, *264*, 9386-9390.
92. Li, G.; Pralong, W.-F.; Pittet, D.; Mayr, G. W.; Schlegel, W; Woolheim, C. B. *J. Biol. Chem.* **1992**, *267*, 4349-4356.
93. Menniti, F. S.; Oliver, K. G.; Nogimori, K.; Obie, J. F.; Shears, S. B; Putney, J. W.; Jr. *J. Biol. Chem.* **1990**, *265*, 11167-11176.
94. Barker, C. J.; Wong, N. S.; Maccallum, S. M.; Hunt, P. A.; Michell, R. H; Kirk, C. J. *Biochem. J.* **1992**, *286*, 469-474.
95. Pittet, D.; Lew, D. P.; Mayr, G. W.; Monod, A; Schlegel, W. *J. Biol. Chem.* **1989**, *264*, 7251-7261.
96. Berridge, M. J; Irvine, R. F. *Nature* **1989**, *341*, 197-205.
97. Stephens, L. R; Downes, C. P. *Biochem. J.* **1990**, *265*, 435-452.
98. Stephens, L. R.; Hawkins, P. T.; Morris, A. J; Downes, P. C. *Biochem. J.* **1988**, *249* , 283-292.
99. McConnell, F. M.; Stephens, L. R; Shears, S. B. *Biochem. J.* **1991**, *280*, 323-329.
100. Barrett, K. E. *Am. J. Physiol.* **1993**, *265*, C859-C868.
101. Nauntofte, B. *Am. J. Physiol.* **1992**, *263*, G823-G837.
102. Petersen, O. H. *J. Physiol. (Lond.)* **1992**, *448*, 1-51.
103. Large, W. A; Wang, Q. *Am. J. Physiol.* **1996**, *271*, C435-C454.
104. Barrett, K. E. *Am. J. Physiol.* **1997**, *272*, C1069-C1076.
105. Chan, H. C.; Goldstein, J; Nelson, D. J. *Am. J. Physiol.* **1992**, *262*, C1273-C1283.
106. Jentsch, T. J; Günther, W. *Bioessays* **1996**, *19*, 117-126.
107. Morris, A. P.; Kirk, K. L; Frizzell, R. A. *Cell Regulation* **1990**, *1*, 951-963.
108. Kachintorn, U.; Vajanaphanich, M.; Barrett, K. E; Traynor-Kaplan, A. E. *Am. J. Physiol.* **1993**, *264*, c671-c676.
109. McEwan, G. T. A.; Hirst, B. H; Simmons, N. L. *Biochim. Biophys. Acta* **1994**, *1220*, 241-247.
110. Xie, W.; Kaetzel, M. A.; Bruzik, K. S.; Dedman, J. R.; Shears, S.B; Nelson, D. J. *J. Biol. Chem.* **1996**, *271*, 14092-14097.
111. Ismailov, I. I.; Fuller, C. M.; Berdiev, B. K.; Shlyonsky, V. G.; Benos, D. J; Barrett, K. E. *Proc. Nat. Acad. Sci. USA* **1996**, *93*, 10505-10509.
112. Ho, M. W. Y.; Shears, S. B.; Bruzik, K. S.; Duszyk, M; French, A. S. *Am. J. Physiol.* **1997**, *272*, 1160-1168.
113. Roemer, S.; Stadler, C.; Rudolf, M. T.; Wolfson-E; Traynor-Kaplan, A.E. *Gastroenterology* **1997**, *112*, A401.
114. Welsh, M. J. *FASEB J.* **1997**, *4*, 2718-2725.
115. Knowles, M. R.; Clarke, L. L; Boucher, R. C. *Chest* **1992**, *101*, 60S-63S.
116. McCollum, E. V; Hart, E. B. *J. Biol. Chem.* **1908**, *4*, 497-500.
117. Moore, R. J; Veum, T. L. *Br. J. Nutr.* **1983**, *49*, 145-152.
118. Rao, R. K; Ramakrishnan, C. V. *Enzyme* **1985**, *33*, 205-215.
119. Rubiera, C.; Velasco, G.; Michell, R. H.; Lazo, P. S; Shears, S. B. *Biochem. J.* **1988**, *255*, 131-137.
120. Gibson, D. M; Ullah, A. B. J. in *Inositol metabolism in plants* (Morre, D. J.; Boss, W. F; Loewus, F. A.; eds.), **1990**, pp. 77-92, Wiley-Liss, New York.
121. Raboy, V. in *Inositol metabolism in plants* (Morre, D. J.; Boss, W. F; Loewus, A. L.; eds), **1990**, pp. 55-76, Wiley-Liss, New York.
122. Biswas, S.; Maity, I. B.; Chakrabarti, S; Biswas, B. B. *Arch. Biochem. Biophys.* **1978**, *185*, 557-566.

123. Phillippy, B. Q.; Ullah, A. H. J; Ehrlich, K. C. *J. Biol. Chem.* **1994**, *269*, 28393-28399.
124. Lapan, E. A. *Exp. Cell Res.* **1975**, *94*, 277-282.
125. Vallejo, M.; Jackson, T.; Lightman, S; Hanley, M. R. *Nature* **1987**, *330*, 656-658.
126. Michell, R. H.; King, C. E.; Piper, C. J.; Stephens, L. R.; Bunce, C. M.; Guy, G. R; Brown, G. *J. Gen. Physiol.* **1988**, *43*, 345-355.
127. Irvine, R. F.; Moor, R. M.; Pollock, W. K.; Smith, P. M; Wreggett, K. A. *Philos. Trans. R. Soc. Lond. [Biol]* **1988**, *320*, 281-298.
128. Stuart, J. A.; Anderson, K. L.; French, P. J.; Kirk, C. J; Michell, R. H. *Biochem. J.* **1994**, *303*, 517-525.
129. Poyner, D. R.; Cooke, F.; Hanley, M. R.; Reynolds, D. J. M; Hawkins, P. T. *J. Biol. Chem.* **1993**, *268*, 1032-1038.
130. Szwergold, B. S.; Graham, R. A; Brown, T. R. *Biochem. Biophys. Res. Commun.* **1987**, *149*, 874-881.
131. French, P. J.; Bunce, C. M.; Stephens, L. R.; Lord, J. M.; McConnell, F. M.; Brown, G.; Creba, J. A; Michell, R. H. *Philos. Trans. R. Soc. Lond. [Biol]* **1991**, *245*, 193-201.
132. Bunce, C. M.; French, P. J.; Allen, P.; Mountford, J. C.; Moor, B.; Greaves, M. F.; Michell, R. H; Brown, G. *Biochem. J.* **1993**, *289*, 667-673.
133. Menniti, F. S.; Miller, R. N.; Putney, J. W.; Jr; Shears, S. B. *J. Biol. Chem.* **1993**, *268*, 3850-3856.
134. Albert, C.; Safrany, S. T.; Bembenek, M. E.; Reddy, K. M.; Reddy, K. K.; Falck, J. R.; Bröker, M.; Shears, S. B; Mayr, G. W. *Biochem.J.* **1997**, *327*, 553-560.
135. Graf, E; Empson, K. L. *J. Biol. Chem.* **1987**, *262*, 11647-11650.
136. Hawkins, P. T.; Poyner, D. R.; Jackson, T. R.; Letcher, A. J.; Lander, D. A; Irvine, R. F. *Biochem. J.* **1993**, *294*, 929-934.
137. Sasakawa, N.; Sharif, M; Hanley, M. R. *Biochem. Pharmacol.* **1995**, *50*, 137-146.
138. Palmer, R. H.; Lodewijk, V. D.; Woscholski, R.; Le Good, J. A.; Gigg, R; Parker, P. J. *J. Biol. Chem.* **1995**, *270*, 22412-22416.
139. Efanov, A. M.; Zaitsev, S. V; Berggren, P.-O. *Proc. Nat. Acad. Sci. USA* **1997**, *94*, 4435-4439.
140. Larsson, O.; Barker, C. J.; Sjöholm, A.; Carlqvist, H.; Michell, R. H.; Bertorello, A.; Nilsson, T.; Honkanen, R.E .; Mayr, G. W.; Zwiler, J; Berggren, P.-O. *Science* **1997**, *278*, 471-474.
141. O'Rouke, F.; Matthews, E; Feinstein, M. B. *Biochem. J.* **1996**, *315*, 1027-1034.
142. Yamaguchi, Y.; Ikenaka, K.; Niinobe, M.; Yamada, H; Mikoshiba, K. *J. Biol. Chem.* **1996**, *271*, 27838-27846.
143. Fleischer, B.; Xie, J.; Mayrleitner, M.; Shears, S. B.; Palmer, D. J; Fleischer, S. *J. Biol. Chem.* **1994**, *269*, 17826-17832.
144. Ali, N.; Duden, R.; Bembenek, M. E; Shears, S. B. *Biochem.J.* **1995**, *310*, 279-284.
145. Beck, K. A; Keen, J. H. *J. Biol. Chem.* **1991**, *266*, 4442-4447.
146. Timerman, A. P.; Mayrleitner, M. M.; Lukas, T. J.; Chadwick, C. C.; Saito, A.; Watterson, D. M.; Schindler, H; Fleischer, S. *Proc. Nat. Acad. Sci. USA* **1992**, *89*, 8976-8980.
147. Voglmaier, S. M.; Keen, J. H.; Murphy, J.-E.; Ferris, C. D.; Prestwich, G. D.; Snyder, S. H; Theibert, A. B. *Biochem. Biophys. Res. Commun.* **1992**, *187*, 158-163.

148. Norris, F. A.; Ungewickell, E; Majerus, P. W. *J. Biol. Chem.* **1995**, *270*, 214-218.
149. Ye, W.; Ali, N.; Bembenek, M. E.; Shears, S. B; Lafer, E. M. *J. Biol. Chem.* **1995**, *270*, 1564-1568.
150. Barker, C. J.; French, P. J.; Moore, A. J.; Nilsson, T.; Berggren, P.-O.; Bunce, C. M.; Kirk, C. J; Michell, R. H. *Biochem. J.* **1995**, *306*, 557-564.
151. Guse, A. H.; Greiner, E.; Emmrich, F; Brand, K. *J. Biol. Chem.* **1993**, *268*, 7129-7133.
152. Balla, T.; Sim, S. S.; Baukal, A. J.; Rhee, S. G; Catt, K. J. *Molecular Biology of the Cell* **1994**, *5*, 17-28.
153. Warren, G; Wickner, W. *Cell* **1996**, *84*, 395-400.
154. Llinás, R.; Sugimori, M.; Lang, E. J.; Morita, M.; Fukuda, M.; Niinobe, M; Mikoshiba, K. *Proc. Nat. Acad. Sci. USA* **1994**, *91*, 12990-12993.
155. Gaidarov, I.; Chen, Q.; Falck, J. R.; Reddy, K. K; Keen, J. H. *J. Biol. Chem.* **1996**, *271*, 20922-20929.
156. Hao, W.; Tan, Z.; Prasad, K.; Reddy, K. K.; Chen, J.; Prestwich, G. D.; Falck, J. R.; Shears, S. B; Lafer, E. M. *J. Biol. Chem.* **1997**, *272*, 6393-6398.
157. Nogimori, K.; Hughes, P. J.; Glennon, M. C.; Hodgson, M. E.; Putney, J. W.; Jr; Shears, S. B. *J. Biol. Chem.* **1991**, *266*, 16499-16506.
158. McConnell, F. M.; Shears, S. B.; Lane, P. J. L.; Scheibel, M. S; Clark, E. A. *Biochem.J.* **1992**, *284*, 447-455.
159. Vucenik, I; Shamsuddin, A. M. *J. Nutr.* **1994**, *124*, 861-868.
160. Craxton, A.; Ali, N; Shears, S. B. *Biochem. J.* **1995**, *305*, 491-498.
161. Ali, N.; Craxton, A; Shears, S. B. *J. Biol. Chem.* **1993**, *268*, 6161-6167.
162. Doughney, C.; McPherson, M. A; Dormer, R. L. *Biochem. J.* **1988**, *251*, 927-929.
163. Estrada-Garcia, T.; Craxton, A.; Kirk, C. J; Michell, R. H. *Proc. R. Soc. Lond.* **1991**, *244*, 63-68.
164. Ren, H.-Y.; Komatsu, N.; Shimizu, R.; Okada, K; Miura, Y. *J. Biol. Chem.* **1994**, *269*, 19633-19638.
165. Arnone, A; Perutz, M. F. *Nature* **1974**, *249*, 34-36.
166. Van Dijken, P.; Bergsma, J. C. T; van Haastert, P. J. M. *Eur. J. Biochem.* **1997**, *244*, 113-119.
167. Van Dijken, P.; de Haas, J.-R.; Craxton, A.; Erneux, C.; Shears, S. B; van Haastert, P. J. M. *J. Biol. Chem.* **1995**, *270*, 29724-29731.
168. Carpenter, D.; Hanley, M. R.; Hawkins, P. T.; Jackson, T. R.; Stephens, L. R; Vallejo, M. *Biochem. Soc. Trans.* **1989**, *17*, 3-5.
169. Shears, S. B.; Storey, D. J.; Morris, A. J.; Cubitt, A. B.; Parry, J. B.; Michell, R. H; Kirk, C. J. *Biochem. J.* **1987**, *242*, 393-402.
170. Mattingly, R. R.; Stephens, L. R.; Irvine, R. F; Garrison, J. C. *J. Biol. Chem.* **1991**, *266*, 15144-15153.
171. Eckmann, L.; Rudolf, M. T.; Ptasznik, A.; Schultz, C.; Jiang, T.; Wolfson, N.; Tsien, R.; Fierer, J.; Shears, S. B.; Kagnoff, M. F; Traynor-Kaplan, A. *Proc. Natl. Acad. Sci. USA* **1997**, *94*,
172. Oliver, K. G.; Putney, J. W.; Jr.; Obie, J. F; Shears, S. B. *J. Biol. Chem.* **1995**, *267*, 21528-21534.
173. Takeuchi, H.; Kanematsu, T.; Misumi, Y.; Yaakob, H.B.; Yagisawa, H.; Ikehara, Y.; Watanabe, Y.; Tan, Z.; Shears, S. B; Hirata, M. *Biochem. J.* **1995**, *318*, 561-568.
174. Craxton, A.; Caffrey, J.; Burkhart, W; Shears, S. B. *Biochem. J.* **1997**, *328*, 75-81.
175. Reynolds, S. D.; Johnston, C.; Leboy, P. S.; O'Keefe, R. J.; Puzas, J. E.; Rosier, R. N; Reynolds, P. R. *Exp. Cell Res.* **1996**, *226*, 197-207.

176. Romano, P.; Wang, J.; O'Keefe, R. J.; Puzas, J. E.; Rosier, R. N; Reynolds, P. R. *J.Cell Sci.* **1998**, *111*, 803-813.
177. Gilbert, S. F. *Developmental Biology*, Sinauer Associates Inc.; Sunderland, New York, 1994.
178. Ostanin, K.; Saeed, A; Van Etten, R. L. *J. Biol. Chem.* **1994**, *269*, 8971-8978.
179. Stephens, L. R.; Radenberg, T.; Thiel, U.; Vogel, G.; Khoo, K.-H.; Dell, A.; Jackson, T. R.; Hawkins, P. T; Mayr, G. W. *J. Biol. Chem.* **1993**, *268*, 4009-4015.
180. Laussmann, T.; Eujen, R.; Weisshuhn, C. M.; Thiel, U.; Falck, J. R; Vogel, G. *Biochem. J.* **1996**, *315*, 715-725.
181. Laussmann, T.; Reddy, K. M.; Reddy, K. K.; Falck, J. R; Vogel, G. *Biochem. J.* **1997**, *322*, 31-33.
182. Voglmaier, S. M.; Bembenek, M. E.; Kaplin, A. I.; Dormán, G.; Olszewski, J. D.; Prestwich, G. D; Snyder, S. H. *Proc. Nat. Acad. Sci. USA* **1996**, *93*, 4305-4310.
183. Shears, S. B.; Ali, N.; Craxton, A; Bembenek, M. E. *J. Biol. Chem.* **1995**, *270*, 10489-10497.
184. Martin, J. B.; Bakker-Grunwald, T; Klein, G. *Eur. J. Biochem.* **1993**, *214*, 711-718.
185. Martin, J.-B.; Bakker-Grunwald, T; Klein, G. *J. Euk. Microbiol.* **1995**, *42*, 183-191.
186. Glennon, M. C; Shears, S. B. *Biochem. J.* **1993**, *293*, 583-590.
187. Campbell, C.; Andreen, M.; Battito, M. F.; Camporesi, E. M.; Goldberg, M. E.; Grounds, R. M.; Hobbhahn, J.; Murray, J. M.; Solanki, D. R.; Heard, S.O; Coriat, P. *J. Clin. Anesth.* **1996**, *8*, 557-563.
188. Battmann, A.; Resch, H.; Libanati, C. R.; Ludy, D.; Fischer, M.; Farley, M; Baylink, D. J. *Osteoporosis Int.* **1997**, *7*, 48-51.
189. Patel, S.; Chan, J. K; Hosking, D. J. *Bone* **1996**, *19*, 651-655.
190. Voglmaier, S. M.; Kaplin, A.I.; Bembenek, M. E.; Dormán, G.; Prestwich, G.D; Snyder, S. H. *Abstracts of the 24th annual Meeting of the Society for Neuroscience* **1994**, 383.6.
191. Safrany, S. T; Shears, S. B. *EMBO J.* **1998**,
192. Jackson, T. R.; Patterson, S. I.; Thastrup, O. I; Hanley, M. R. *Biochem. J.* **1998**, *253*, 81-86.
193. Putney, J. W.; Jr. *Cell Calcium* **1991**, *11*, 611-624.

Chapter 2

Probing Phosphoinositide Polyphosphate Binding to Proteins

Glenn D. Prestwich, Anu Chaudhary, Jian Chen, Li Feng, Bharat Mehrotra, and Jirong Peng

Department of Medicinal Chemistry, The University of Utah, 30 South, 2000 East, Room 201, Salt Lake City, UT 84112–5820

Polyphosphoinositides play key roles in endo- and exocytosis, signal transduction, and cytoskeletal remodeling. To identify and characterize protein targets of these signaling molecules, we have synthesized affinity probes of soluble inositol phosphates and, more recently, all naturally-occurring lipid-modified phosphoinositides: PtdIns(3)P, PtdIns(4)P, PtdIns(5)P, PtdIns(3,4)P_2, PtdIns(4,5)P_2, PtdIns(3,5)P_2, and PtdIns(3,4,5)P_3. Fluorescent, biotinylated, resin-attached, and photoaffinity probes have been used to characterize the (Ptd)InsP_n binding sites of phospholipases, PI kinases and transfer proteins, synaptotagmin C2B domains, profilin, PH domains, assembly proteins AP-2 and AP-3, and several new PtdIns(3,4,5)P_3-target proteins.

Phosphoinositide polyphosphates (PtdInsP_ns) are key signaling molecules in cellular communication via protein kinases, in endo- and exocytosis, in vesicular trafficking of proteins *(1-4)*, and in signaling in leukocytes *(5)*. As the central molecule (Figure 1), L-α-phosphatidyl-D-*myo*-inositol 4,5-bisphosphate (PtdIns(4,5)P_2) serves three different cellular roles, each utilizing different portions of the total molecule. First, PtdIns(4,5)P_2 is a substrate for phospholipase C (PLC), releasing the calcium-mobilizing second messenger Ins(1,4,5)P_3 *(6-9)* (Figure 1).

Second, PtdIns(4,5)P_2 itself functions in the recruitment of proteins to membranes via pleckstrin homology (PH) domains *(10)*. Indeed, the PH domains of PLCs allow membrane localization via the 4,5-bisphosphate of PtdIns(4,5)P_2 *(11)*, while the catalytic domain cleaves the P-O bond of an adjacent PtdIns(4,5)P_2 molecule. Three-dimensional structures of the β-spectrin-Ins(1,4,5)P_3 complex *(12)*, dynamin *(13)*, and the PLC-δ_1 PH domain-complex *(14)* have verified the importance of the 4,5-bisphosphate interaction with hydrogen bonding and protonated basic residues. In addition, PtdIns(4,5)P_2 sequesters profilin to the inner surface of the phospholipid bilayer, facilitating the polymerization of actin *(1,15)*. Release of profilin following PtdIns(4,5)P_2 hydrolysis suppresses further cytoskeletal reorganization. PtdIns(4,5)P_2 also regulates the GTPase function of ARF and the phospholipase activity of PLD *(16,17)*.

Third, PtdIns(4,5)P$_2$ can be converted by agonist-stimulated, receptor-mediated activation of phosphoinositide 3-kinase (PI 3-K) (18) to PtdIns(3,4,5)P$_3$, the key element in a new intracellular signaling system (19,20). The PI 3-K pathway and the corresponding D-3 PtdInsP$_n$ products (21) have been linked to mechanisms of oncogene transformation, cytoskeletal rearrangements, membrane association of signaling proteins, and trafficking of proteins by coated vesicles (22,23). A structurally-diverse set of isoforms and splice variants of PI 3-K (24) catalyzes the transfer of the γ phosphate of ATP to the D-3 hydroxyl of PtdIns, PtdIns(4)P, and PtdIns(4,5)P$_2$ to form PtdIns(3)P, PtdIns(3,4)P$_2$, and PtdIns(3,4,5)P$_3$, respectively (22,25). Characteristics of the three families of PI 3-Ks were recently reviewed (26).

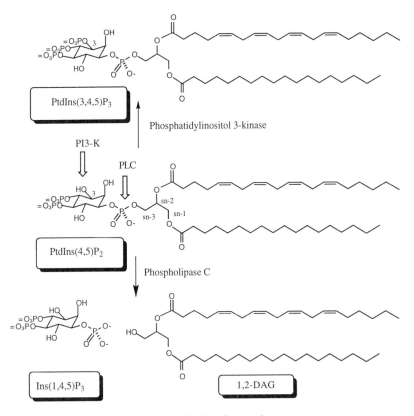

Figure 1. PtdIns(4,5)P$_2$ is central to cell signaling pathways

PtdIns(3,4,5)P$_3$ appears to be a second messenger (27,28), and several putative targets for this phosphoinositide have been reported: the serine/threonine kinase Akt/PKB (29,30), protein kinase C (31,32) a mediator of the phosphorylation of platelet p47 (pleckstrin) (33,34), interaction with PH domain such as GRP1 and Btk (35), and activation of a protein related kinase PRK1 (36). In addition, a new 46-kDa protein, centaurin (37), has been purified from rat brain using a Ins(1,3,4,5)P$_4$ affinity column (38) and labeled by a photolabile Ins(1,3,4,5)P$_4$ analog (39,40). Centaurin has optimal binding to PtdIns(3,4,5)P$_3$ relative to other soluble and lipid inositol polyphosphates, and the cloned cDNA indicates that this novel protein contains modules involved in mediating interactions between the actin cytoskeleton and membrane-associated proteins. The yeast homolog GCS1 (I. Blader, A.A. Profit, T.R. Jackson, G.D. Prestwich, A.B. Theibert, submitted for publication) also appears to be a high affinity PtdIns(3,4,5)P$_3$ binding protein, which is puzzling since these polyphosphoinositides are unknown in yeast. PtdIns(3,4,5)P$_3$ also shows high affinity binding to assembly protein AP-2, specifically α-adaptin, and inhibited clathrin coating of vesicles. High affinity binding to AP-3 *(41)*, profilin *(42)*, and synaptotagmin *(43)* have also been recently described. In some cases, PtdIns(3,4)P$_2$ appears to be the preferred activator, e.g., for protein kinases such as Akt *(31,44)* and profilin *(42)*. A new bisphosphate, PtdIns(3,5)P$_2$ was recently discovered *(45)* . but its biological role is as yet unknown.

For both PtdIns(4,5)P$_2$ and PtdIns(3,4,5)P$_3$, the question remains as to what role the diacylglycerol moiety might have in the interaction of these phosphoinositides with hydrophobic regions of PtdInsP$_n$ binding proteins. Figure 2 shows that PH domains appear to interact primarily with the head group, allowing membrane localization (left). In contrast, binding studies, cell or protein activation studies, and photoaffinity labeling suggest the importance of protein-diacylglycerol contacts (right). One key participant in the PI pathway is PI transfer protein (PI-TP) which is proposed to bind PI via its diacylglycerol chain during the installation of the 4- and 5-phosphates by their respective kinases *(46,47)*.

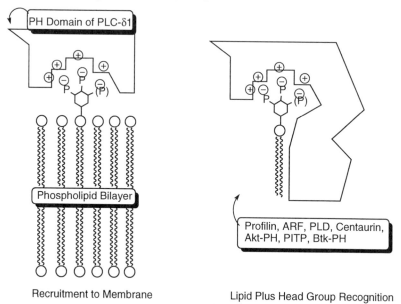

Recruitment to Membrane   Lipid Plus Head Group Recognition

Figure 2. Modes of binding to PtdInsP$_n$

A Unified Synthetic Approach to Phosphoinositide Polyphosphates

To meet the needs for InsP$_n$ and PtdInsP$_n$ analogs and affinity probes needed in cell biology, biochemistry, structural biology, and biophysics, efficient chemical synthesis of these materials was necessary. We recently described the broad application of the Ferrier rearrangement *(48)* of an inexpensive starting material, methyl α-D-glucopyranoside (Figure 3), to most of the possible head groups in the PtdInsP$_n$ probes *(49)*.

Furthermore, we have presented versatile synthetic routes to PtdIns(3,4)P$_2$, PtdIns(4,5)P$_2$, PtdIns(3,4,5)P$_3$ that allow introduction of affinity probes as well as introduction of most desired diacylglycerol moieties *(50-52)*. Earlier syntheses of PtdInsP$_n$ derivatives and these synthetic compounds have been employed to verify specific biological roles for these ligands. The *sn*-1,2-*O*-distearoyl analog of PtdIns(3,4,5)P$_3$ was prepared using (-)-quinic acid as the chiral starting material *(53,54)*, while an enzymatically-resolved 1-D-dicyclohexylidene-*myo*-inositol *(55,56)* or camphor ketal of *myo*-inositol has also been used *(57)*. Routes to both the racemic and optically-pure forms of PtdIns(3,4,5)P$_3$ used 1-D-1,2-*O*-cyclohexylidene-*myo*-inositol *(58,59)*.

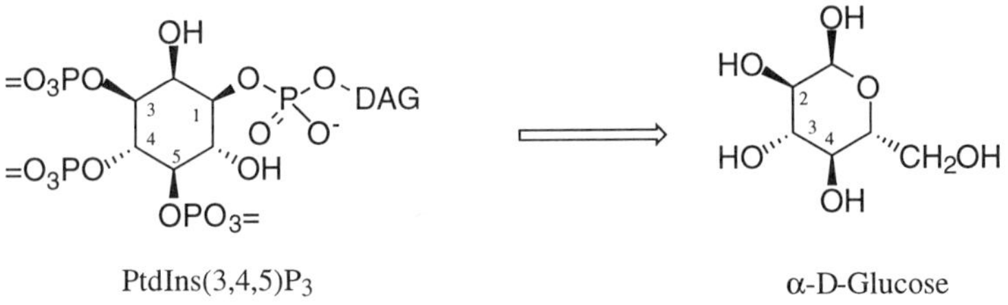

PtdIns(3,4,5)P$_3$ α-D-Glucose

Figure 3. Retrosynthetic derivation of PtdIns(3,4,5)P$_3$ from glucose

The syntheses of modified PtdInsP$_2$ and PtdInsP$_3$ analogs developed in our labs offer three important features. First, the Ferrier rearrangement of an inexpensive α-D-glucose derivative provides the enantiomerically-pure inositol framework without the need for an enzymatic or chemical resolution. Second, selective protection schemes have been devised and implemented for biologically-relevant PtdInsP$_n$ regioisomers. Third, the syntheses permit introduction of tritium-labeled photoaffinity probes, biotin, or fluorophores to an aminoacyl ester in the *sn*-1 position. Moreover, we can attach these functionalized PtdInsP$_n$s to affinity matrices, surfaces, or proteins, thus permitting exploration of a variety of potential PtdInsP$_n$ targets.

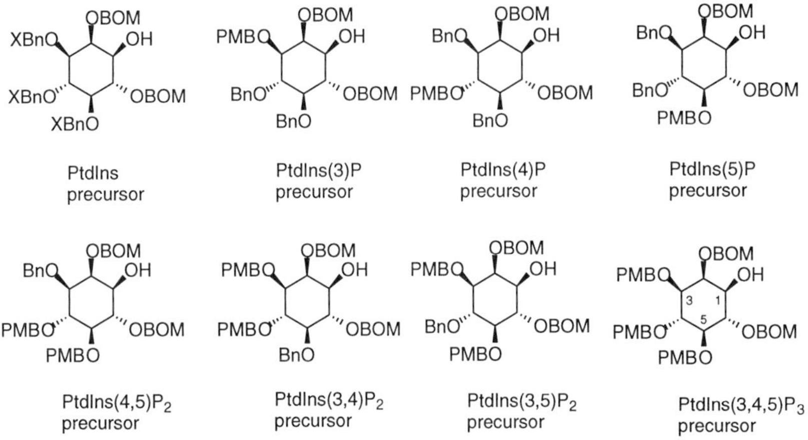

Figure 4. Inositol precursors required for PtdInsP$_n$ set. **Key:** Bn, benzyl; PMB, *p*-methoxybenzyl; BOM, benzyloxymethyl; XBn, either Bn or PMB

Studies with photoaffinity, biotinylated, and fluorescent probes based on PtdInsP, PtdInsP$_2$ and PtdInsP$_3$ probes have been initiated. For example, a fluorescent probe,

NBD-PtdIns(4,5)P$_2$ *(50)* was synthesized for studies on the interaction of a highly basic peptide from MARCKS with the acidic phospholipids PtdIns(4,5)P$_2$ and phosphatidyl-serine (PS) in bilayers. The basic peptides caused sequestration of PtdIns(4,5)P$_2$ and PS into lateral domains, thus preventing hydrolysis of PtdIns(4,5)P$_2$ by PLC *(60)*. PH domains, but not the catalytic site of PLC, can be labeled with PtdIns(4,5)P$_2$ probes *(61)*. An MBP fusion with the PH domain of c-Akt showed selective photolabeling by PtdIns(3,4)P$_2$-derived triester (A. Chaudhary, unpublished results). Labeling of PI-TP, Golgi coatomer, AP-2, AP-3, CAPS, ARF, PLD, Syt C2B domains *(62)*, PI 3-K, and other proteins are currently underway.

The PtdInsP$_n$ probes are constructed from two sets of building blocks: the inositol precursors and the diacylglyceryl modules. Figure 4 illustrates the necessary inositol precursors, and the syntheses of the PtdInsP$_2$ and PtdInsP$_3$ modules have been published *(40,50-52,63)*. The PtdIns precursor can be virtually any of the other precursors, since the absence of ring phosphorylations allows all protecting groups to be removed simultaneously in the final step. Figure 5 illustrates the range of final products available.

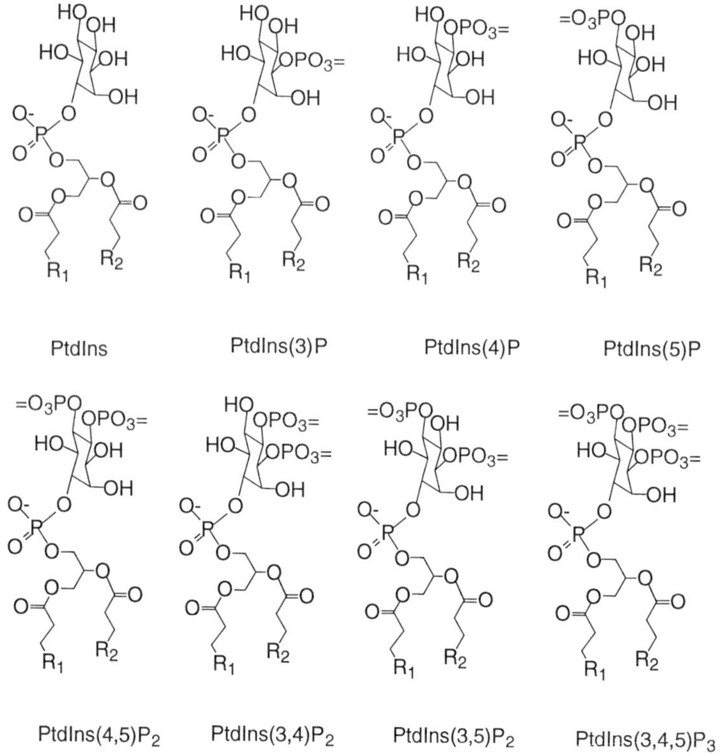

PtdIns PtdIns(3)P PtdIns(4)P PtdIns(5)P

PtdIns(4,5)P$_2$ PtdIns(3,4)P$_2$ PtdIns(3,5)P$_2$ PtdIns(3,4,5)P$_3$

$R_1 = R_2 = CH_3$, di-C$_4$ PtdInsP$_n$
$R_1 = R_2 = (CH_2)_4CH_3$, di-C$_8$ PtdInsP$_n$
$R_1 = R_2 = (CH_2)_{12}CH_3$, di-C$_{16}$ PtdInsP$_n$
$R_1 = (CH_2)_{12}CH_3$, $R_2 = (CH_2)_3NH_2$, PtdInsP$_n$ affinity probe precursor

Figure 5. Chemical structures of phosphoinositide polyphosphates (PtdInsP$_n$s)

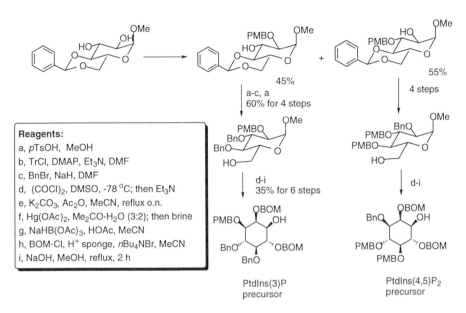

Reagents:
a, pTsOH, MeOH
b, TrCl, DMAP, Et₃N, DMF
c, BnBr, NaH, DMF
d, (COCl)₂, DMSO, -78 °C; then Et₃N
e, K₂CO₃, Ac₂O, MeCN, reflux o.n.
f, Hg(OAc)₂, Me₂CO-H₂O (3:2); then brine
g, NaHB(OAc)₃, HOAc, MeCN
h, BOM-Cl, H⁺ sponge, nBu₄NBr, MeCN
i, NaOH, MeOH, reflux, 2 h

Figure 6. Synthesis of both PtdIns(3)P and PtdIns(4,5)P₂ precursors from a common intermediate

The PtdIns(3)P precursor and PtdIns(4,5)P₂ precursors are now obtainable from two separable regioisomers of a single reaction (Figure 6; reagents and yields shown for PtdIns(3)P). This greatly facilitates preparing multiple scaffoldings by a synthetic scheme that uses both intermediates. We have developed similar routes (not shown) to prepare the precursors and PtdIns(4)P (J. Chen, L. Feng, and G. D. Prestwich, submitted for publication), PtdIns(5)P and PtdIns(3,5)P₂ (J. Peng and G. D. Prestwich, submitted for publication).

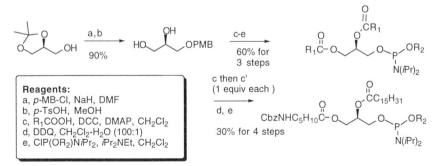

Reagents:
a, p-MB-Cl, NaH, DMF
b, p-TsOH, MeOH
c, R₁COOH, DCC, DMAP, CH₂Cl₂
d, DDQ, CH₂Cl₂-H₂O (100:1)
e, CIP(OR₂)NiPr₂, iPr₂NEt, CH₂Cl₂

Figure 7. Synthesis of preformed diacylglycerol modules

The mixed diacylglycerol, symmetric diacylglycerol, and isopropylidene-protected diacylglycerol moieties are each prepared from (+)-1,2-O-isopropylidene glycerol as shown in Figure 7, following modifications *(50)* of the method of Martin *et al.* *(64)* for sequential introduction of *sn*-1 and *sn*-2-O-acyl groups. The PMB protecting group can be oxidatively removed (DDQ, CH_2Cl_2), and no acyl migration occurs under these conditions. Reaction of the diacylglycerols with (benzyloxy) *N,N*-diisopropyl-amino)chlorophosphine gave relatively labile phosphitylating reagents that must be stored at -20 °C and used within several days. From a practical viewpoint, the recrystallized diacylglycerols can be stored for months at -20 °C and activated just prior to use.

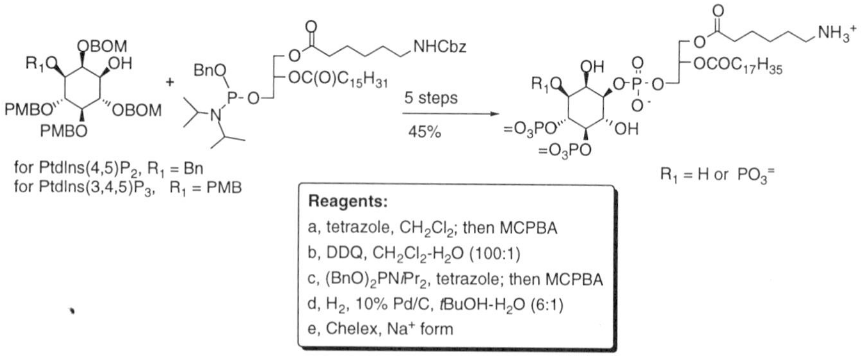

5 steps
45%
for PtdIns(4,5)P2, R1 = Bn
for PtdIns(3,4,5)P3, R1 = PMB
R1 = H or PO3=
Reagents:
a, tetrazole, CH2Cl2; then MCPBA
b, DDQ, CH2Cl2-H2O (100:1)
c, (BnO)2PNiPr2, tetrazole; then MCPBA
d, H2, 10% Pd/C, tBuOH-H2O (6:1)
e, Chelex, Na+ form

Figure 8. Synthesis of acyl-modified PtdIns(4,5)P_2 and PtdIns(3,4,5)P_3 using preformed diacylglyceryl module

In our published synthetic routes, we have used the mode shown in Figure 8. That is, we have used diacylated glyceryl phosphites for coupling with the inositol precursors, as illustrated for PtdIns(4,5)P_2 and PtdIns(3,4,5)P_3 *(50)*. In this route, the phosphite coupling occurs in anhydrous solvent using tetrazole followed by low-temperature oxidation to the phosphate. Then, removal of PMB groups with ceric ion or DDQ, followed by phosphitylation/oxidation gives a fully-protected PtdInsP_n derivative, which is purified to >98% homogeneity and fully characterized. Hydrogenolysis of all benzyl ethers, esters, and the Cbz group (if present) occurs in essentially quantitative yield in a single step. Purification of the sodium form of the PtdInsP_n is accomplished on a Chelex resin. Purification of fluorescent, biotin, or photolabile adducts is achieved as the triethylammonium species by DEAE anion exchange chromatography.

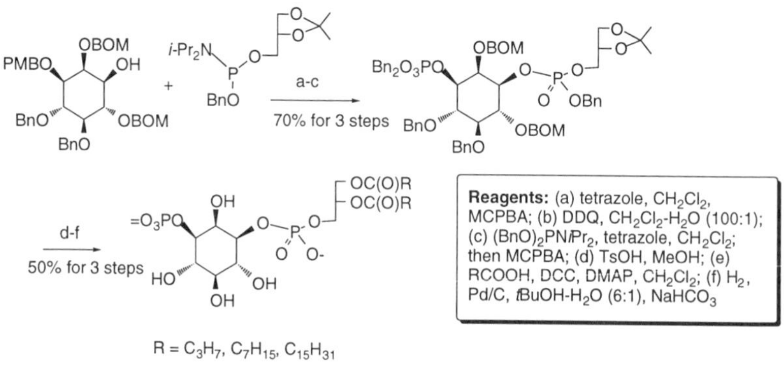

a-c
70% for 3 steps
d-f
50% for 3 steps
Reagents: (a) tetrazole, CH2Cl2, MCPBA; (b) DDQ, CH2Cl2-H2O (100:1); (c) (BnO)2PNiPr2, tetrazole, CH2Cl2; then MCPBA; (d) TsOH, MeOH; (e) RCOOH, DCC, DMAP, CH2Cl2; (f) H2, Pd/C, tBuOH-H2O (6:1), NaHCO3
R = C3H7, C7H15, C15H31

Figure 9. Synthesis of PtdIns(3)P analogs via introduction of acyl groups at end of synthesis. The counterions are sodium in final PtdInsP_ns.

All symmetrical diacyl derivatives can be prepared from a single intermediate, as outlined in Figure 9 for the synthesis of dibutanoyl, dioctanoyl, and dipalmitoyl PtdIns(3)P. Briefly, the PtdInsP$_n$ precursor is coupled to the isopropylidene-protected glyceryl phosphite reagent shown, oxidized to the phosphate triester, and the PMB group is replaced by a phosphate ester. Finally, the ketal is hydrolyzed, the diol is diacylated, and the protecting groups hydrogenolyzed to give the PtdIns(3)P. This methodology has the advantage that all three diacyl products are accessed from a single phosphoinositide intermediate.

We have improved the stability and versatility of the phosphitylating agent by using the cyanoethoxy rather than the benzyloxy phosphites (51). This strategy, which can be used for preparing triester PtdInsP$_n$ analogs (hydrogenolysis of the cyanoethyl group) or the parent PtdInsP$_n$ compounds (by trialkylamine cleavage of the cyanoethyl phosphate prior to hydrogenolytic deprotection), is illustrated for the dipalmitoyl PtdIns(3,4,5)P$_3$ in Figure 10. The more stable diacylglyceryl phosphitylating reagent can be chromatographed and stored for several days, and the yields of subsequent manipulations are unaffected or improved.

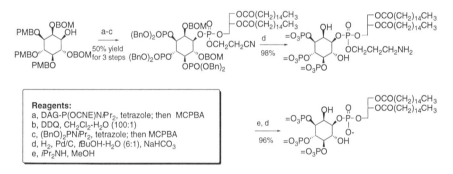

Reagents:
a, DAG-P(OCNE)N*i*Pr$_2$, tetrazole; then MCPBA
b, DDQ, CH$_2$Cl$_2$-H$_2$O (100:1)
c, (BnO)$_2$PN*i*Pr$_2$, tetrazole; then MCPBA
d, H$_2$, Pd/C, *t*BuOH-H$_2$O (6:1), NaHCO$_3$
e, *i*Pr$_2$NH, MeOH

Figure 10. Cyanoethylphosphotriester groups provide access to more stable and versatile intermediates.

Moreover, the P-1 aminopropyl phosphotriesters can be used to prepare photoactivatable or other affinity-labeled triesters. As illustrated in Figure 11 for PtdIns(3,4,5)P$_3$, the triester probe (far right) samples proteins that interact at the interface of the lipid bilayer and the charged water-solvated head group. This contrasts with the acyl photoprobe, which labels only hydrophobic regions binding the acyl chains, or the soluble BZDC-Ins(1,3,4,5)P$_4$ mimic of PtdIns(3,4,5)P$_3$, which labels sites proximal to the phosphorylated inositol.

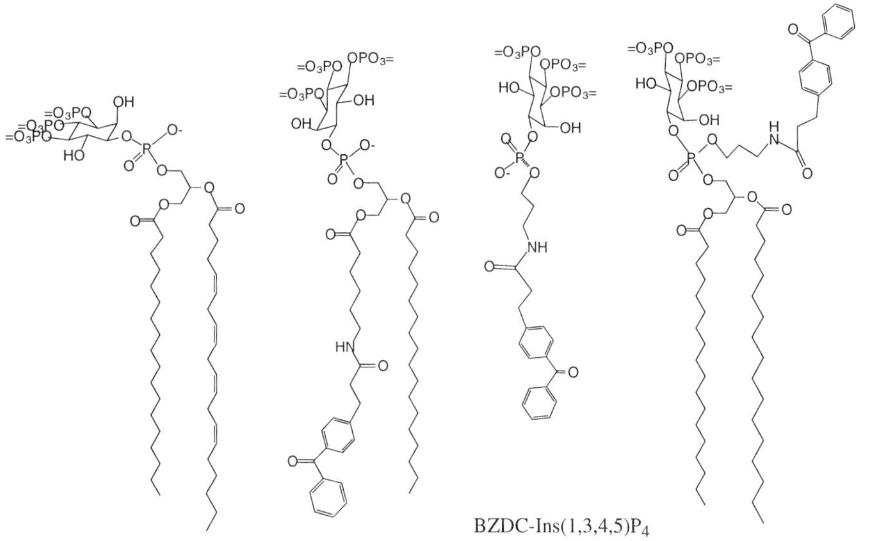

BZDC-Ins(1,3,4,5)P$_4$

PtdIns(3,4,5)P$_3$ *acyl*-BZDC-PtdIns(3,4,5)P$_3$ *triester*-BZDC-PtdIns(3,4,5)P$_3$

Figure 11. PtdIns(3,4,5)P$_3$ and three photoactivatable analogs that probe different environments for potential binding interactions.

Aminoacyl analogs have been converted to biotinylated, fluorophoric, and photoaffinity derivatives of the *sn-1-O*-(6-aminohexanoyl) affinity probes (Figure 12) *(49,65)*. These probes are valuable members of the treasure chest of biochemical and biophysical probes useful for studying PtdInsP$_n$ - protein interactions *in vitro*.

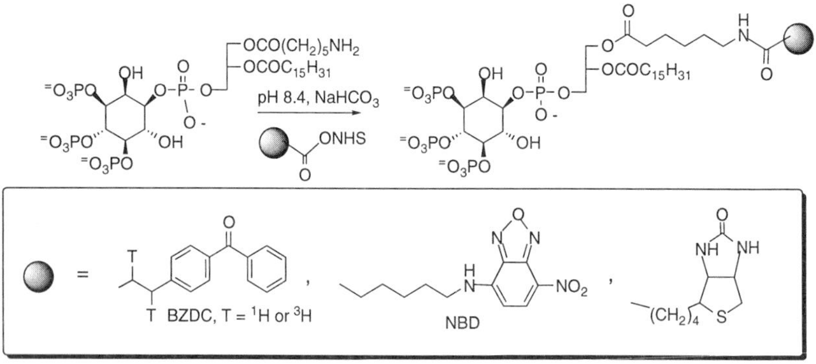

Figure 12. Coupling of aminoacyl PtdInsP$_n$ analogs to reporter groups, shown here for acyl-modified photoaffinity (BZDC), fluorescent (NBD), and biotinylated PtdIns(3,4,5)P$_3$ derivatives.

Finally, the best experimental control to distinguish ligand-specific binding from low affinity hydrophobic - electrostatic interactions is the enantiomer of the endogenous polyphosphoinositide. Thus, we synthesized the mirror image of the L-α-phosphatidyl

D-*myo*-inositol 4,5-bisphosphate ($D^{Ins}L^{DAG}$) to explore which proteins showed acyl, head group, or no specificity (Figure 13). In addition, we prepared two "twisted sisters," e.g., the diastereomers in which either the stereochemistry of head group only or the acyl chain only were inverted relative to the natural $D^{Ins}L^{DAG}$ isomer (J. Chen, B. Mehrotra, G. D. Prestwich, submitted for publication). Activation of PLD and activity as substrates for PI 3-kinase are being examined for these four stereoisomers (A. Morris, unpublished results). The corresponding four stereoisomers of PtdIns(3,4,5)P$_3$ have also been synthesized.

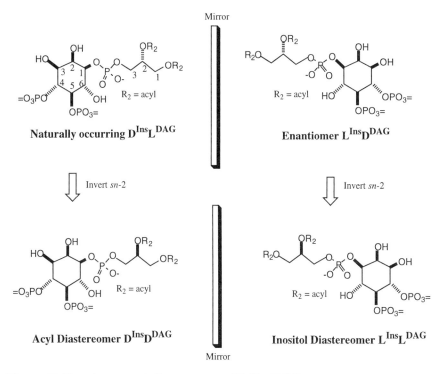

Figure 13. Enantiomers and diastereomers of PtdIns(4,5)P$_2$

New Biochemical Results using PtdInsP$_n$ Photoaffinity Labeling

The PtdInsP$_n$ photoaffinity labels have been extensively employed for identification of novel binding proteins, for elucidation of the PtdInsP$_n$-binding domains, and for establishing the selectivity for a particular PtdInsP$_n$. Examples of collaborative projects for which the biochemical studies have been conducted in my laboratories at The University at Stony Brook or The University of Utah are summarized below.

Synaptotagmin C2B Domains. Synaptotagmin II (Syt II), a synaptic vesicle protein involved as a mediator of exocytosis and vesicle recycling, contains two protein kinase C (PKC) homology regions, known as the C2A and C2B domains. The former acts as a Ca^{2+} sensor, while the latter binds inositol polyphosphates. Photoaffinity analogs of Ins(1,4,5)P$_3$, Ins(1,3,4,5)P$_4$, and InsP$_6$ containing the BZDC photophore were used to label GST fusion proteins of the Syt II-C2A and C2B domains. The P-2-linked

[^{3}H]BZDC-InsP$_6$ showed efficient, InsP$_6$-displaceable labeling of the GST-Syt II-C2B. The rank order of photocovalent modification paralleled the order of competitive displacement: InsP$_6$ (P-2-linked) > Ins(1,3,4,5)P$_4$ > Ins(1,4,5)P$_3$. When mixtures of the 32-amino acid basic peptide corresponding to the essential IP$_n$ binding region of the Syt II-C2B domain and GST-Syt II-C2B were labeled by a stoichiometric amount of P-2-linked [^{3}H]BZDC-InsP$_6$, the two polypeptides showed equivalent affinity for the photolabel. Although the CD spectrum of this 32-mer at two pH values showed a random coil, the photoaffinity analog of InsP$_6$ appeared to induce a binding-compatible structure in the short peptide *(62)*.

Assembly Proteins AP-2 and AP-3. Both AP-2 and AP-3 (aka, AP-180) are important in the clathrin coating of vesicles during endocytosis, and both soluble InsP$_n$ and PtdInsP$_n$ have been shown as ligands mediating this process. For both assembly proteins, PtdIns(3,4,5)P$_3$ was the most effective inhibitor of clathrin coating *(41,66)*. We had originally identified AP-2 by affinity purification and photoaffinity labeling with Ins(1,3,4,5)P$_4$ affinity reagents *(67)*. Recently, we have shown with a palette of PtdInsP$_n$ photoprobes that it is the α subunit of AP-2 that is labeled, with certain probes also labeling the μ subunit. In other studies, we have confirmed by photoaffinity labeling with a soluble BZDC-Ins(1,3,4,5)P$_4$ analog that the N-terminal third of AP-180 contains the PtdInsP$_n$ binding site (A. A. Profit, J. Chen, Q.-M. Gu, A. Chaudhary, B. Mehrotra, E. M. Lafer, G. D. Prestwich, unpublished results). However, the physiological importance and function of the PtdInsP$_n$ and InsP$_n$ binding is still unknown, as are the true effector ligands in a biological context.

Profilin. Profilin is a widely and highly expressed 14 kDa protein that binds actin monomers, poly(L-proline) and polyphosphoinositol lipids and participates in regulating actin filament dynamics that are essential for cell motility *(68)*. Human profilin I was covalently modified using three [^{3}H]BZDC-containing photoaffinity analogs of PtdIns(4,5,)P$_2$. The P-1-tethered Ins(1,4,5)P$_3$ analog showed efficient and specific photocovalent modification of profilin I, competitively displaceable by PtdIns(4,5)P$_2$ analogs but not InsP$_3$. The acyl-modified PtdIns(4,5)P$_2$ analog showed little protein labeling while the head-group modified PtdIns(4,5)P$_2$ phosphotriester showed labeling of both monomeric and oligomeric profilin I. [^{3}H]BZDC-Ins(1,4,5)P$_3$-labeled profilin I showed attachment to the N-terminal 13 amino acid fragment, and the covalent modification of Ala-1 of profilin I was demonstrated by Edman degradation. The PtdIns(4,5,)P$_2$ binding site includes a bisphosphate interaction with a base-rich motif in the C-terminal helix and contact between the lipid moiety of PtdIns(4,5,)P$_2$ with a hydrophobic inner patch of the N-terminal helix of profilin, providing the first direct evidence for a site of interaction of the lipid moiety of a phosphoinositide bisphosphate with specific amino acid residues of profilin (A. Chaudhary, W. Witke, D.J. Kwiatkowski, G.D. Prestwich, submitted for publication).

Golgi Coatomer. Coatomer, a complex of coat proteins (COPs) involved in the formation of specific Golgi intercisternal transport vesicles, consists of seven subunits *(69)*. Coatomer binds Ins(1,3,4,5)P$_4$ and InsP$_6$ with subnanomolar affinity. Phosphoinositide and subunit selectivity were examined using P-1-linked photoaffinity analogs of the soluble inositol polyphosphates Ins(1,4,5)P$_3$ and Ins(1,3,4,5)P$_4$, a P-2-linked analog of InsP$_6$, and P-1-linked phosphotriester analogs of the polyphosphoinositides PtdIns(3,4,5)P$_3$, PtdIns(4,5)P$_2$, and PtdIns(3,4)P$_2$. Labeling of Golgi coatomer COPI complex showed a highly-selective Ins(1,3,4,5)P$_4$-displaceable photocovalent modification of the αCOP subunit by [^{3}H]BZDC-Ins(1,3,4,5)P$_4$ probe, while the P-2-linked [^{3}H]BZDC-InsP$_6$ probe labeled six of the seven subunits. Most importantly, [^{3}H]BZDC-triester-PtdIns(3,4,5)P$_3$, the polyphosphoinositide analog with the same phosphorylation pattern as Ins(1,3,4,5)P_4, also showed completely specific, PtdIns(3,4,5)P_3-displaceable labeling of αCOP. Labeling by the PtdIns(4,5)P_2 and PtdIns(3,4)P_2 photoaffinity probes showed no discrimination based on PtdInsP$_n$ ligand. These data suggest the critical importance of the D-3 and D-5 phosphates in regulation of the recruitment of membranes during vesicle budding in signal transduction, and thus implicate PtdIns 3-kinase in this process (A. Chaudhary, J. Chen, A. A. Profit, O. Thum, L. Jeyakumar, Y. Qi, S. Fleischer, G. D. Prestwich, submitted for publication).

PI Transfer Protein. Labeling of a 35-kD recombinant mammalian PI-TP with a variety of photoaffinity probes was conducted to test the hypothesis that PI-TP bound principally to the acyl chains, thereby presenting the PtdIns to successive phosphorylation by 4-kinase and a 5-kinase to PtdIns(4,5)P_2. The acyl-linked PtdInsP$_n$ photoprobes showed strong labeling of PI-TP, while the soluble InsP$_n$ photoprobes showed the weakest labeling. The triester probes showed the most clearly ligand-specific labeling, with the highest affinity ligand apparently being PtdIns(3,4,5)P_3 (A Chaudhary, S. Cockcroft, and G. D. Prestwich, unpublished results). This unexpected result suggests that, in addition to the importance of the diacylglyceryl moiety for binding, PI-TP may be regulated by 3-phosphorylated phosphoinositides at a separate site from that involved in the canonical role of exchanging PC for PI in membranes.

Acknowledgments

We thank the National Institutes of Health (NS 29632 to G.D.P.) for primary financial support. We also thank The University of Utah, NEN Life Science Products, and our collaborators named herein for additional funding, radioligands, proteins, advice, and inspiration.

Literature Cited

1. Janmey, P. *Chem. & Biol.* **1995**, *2*, 61-65.
2. Decamilli, P.; Emr, S. D.; McPherson, P. S.; Novick, P. *Science* **1996**, *271*, 1533-1539.
3. Rothman, J. E. *Protein Sci.* **1996**, *5*, 185-194.
4. Schekman, R.; Orci, L. *Science* **1996**, *271*, 1526-1533.
5. Cockcroft, S. *Curr. Opin. Hematol.* **1996**, *3*, 48-54.
6. Berridge, M. J. *Annu. Rev. Biochem.* **1987**, *56*, 159-193.
7. Berridge, M. J. *Nature* **1993**, *361*, 315-325.
8. Berridge, M. J. *Molec. Cell. Endocrinol.* **1994**, *98*, 119-124.
9. Nishizuka, Y. *Science* **1992**, *258*, 607-614.
10. Saraste, M.; Hyvönen, M. *Curr. Opin. Struct. Biol.* **1995**, *5*, 403-408.
11. Rebecchi, M.; Peterson, A.; McLaughlin, S. *Biochemistry* **1992**, *31*, 12742-12747.
12. Hyvönen, M.; Macias, M. J.; Nilges, M.; Oschkinat, H.; Saraste, M.; Wilmanns, M. *EMBO J.* **1995**, *14*, 4676-4685.
13. Ferguson, K. M.; Lemmon, M. A.; Schlessinger, J.; Sigler, P. B. *Cell* **1995**, *79*, 199-209.
14. Ferguson, K. M.; Lemmon, M. A.; Schlessinger, J.; Sigler, P. B. *Cell* **1995**, *83*, 1037-1046.
15. Lambrechts, A.; Vandamme, J.; Goethals, M.; Vandekerckhove, J.; Ampe, C. *Eur. J. Biochem.* **1995**, *230*, 281-286.
16. Morris, A. J.; Engebrecht, J. A.; Frohman, M. A. *Trends Pharmacol. Sci.* **1996**, *17*, 182-185.

17. Hammond, S. M.; Jenco, J. M.; Nakashima, S.; Cadwallader, K.; Gu, Q. M.; Cook, S.; Nozawa, Y.; Prestwich, G. D.; Frohman, M. A.; Morris, A. J. *J. Biol. Chem.* **1997**, *272*, 3860-3868.
18. Stephens, L.; Cooke, F. T.; Walters, R.; Jackson, T.; Volinia, S.; Gout, I.; Waterfield, M. D.; Hawkins, P. T. *Curr. Biol.* **1994**, *4*, 203-214.
19. Stephens, L. R.; Jackson, T. R.; Hawkins, P. T. *Biochim. Biophys. Acta* **1993**, *1179*, 27-75.
20. Stephens, L.; Radenberg, T.; Thiel, U.; Vogel, G.; Khoo, K. H.; Dell, A.; Jackson, T. R.; Hawkins, P. T.; Mayr, G. W. *J. Biol. Chem.* **1993**, *268*, 4009-4015.
21. Toker, A.; Cantley, L. C. *Nature* **1997**, *387*, 673-676.
22. Carpenter, C. L.; Cantley, L. C. *Curr. Opin. Struct. Biol.* **1996**, *8*, 153-158.
23. Carpenter, C. L.; Auger, K. R.; Chanudhuri, M.; Yoakim, M.; Schaffhausen, B.; Shoelson, S.; Cantley, L. C. *J. Biol. Chem.* **1993**, *268*, 9478-9483.
24. Zvelebil, M. J.; MacDougall, L.; Leevers, S.; Volinia, S.; Vanhaesebroeck, B.; Gout, I.; Panayotou, G.; Domin, J.; Stein, R.; Pages, F.; Koga, H.; Salim, K.; Linacre, J.; Das, P.; Panaretou, C.; Wetzker, R.; Waterfield, M. *Philos. Trans R Soc. Lond. [Biol.]* **1996**, *351*, 217-223.
25. MacDougall, L. K.; Domin, J.; Waterfield, M. D. *Curr. Biol.* **1995**, *5*, 1404-1415.
26. Vanhaesebroeck, B.; Leevers, S.; Panayotou, G.; Waterfield, M. *TIBS* **1997**, *22*, 267-272.
27. Eberle, M.; Traynor-Kaplan, A. E.; Sklar, L. A.; Norgauer, J. *J. Biol. Chem.* **1990**, *265*, 16725-16728.
28. Traynor-Kaplan, A. E.; Thompson, B. L.; Harris, A. L.; Taylor, P.; Omann, G. M.; Sklar, L. A. *J. Biol. Chem.* **1989**, *264*, 15668-15673.
29. Burgering, B. M. T.; Coffer, P. J. *Nature* **1995**, *376*, 599-602.
30. Franke, T. F.; Yang, S.-I.; Chan, T. O.; Datta, K.; Kazlauskas, A.; Morrison, D. K.; Kaplan, D. R.; Tsichlis, P. N. *Cell* **1995**, *81*, 727-736.
31. Toker, A.; Meyer, M.; Reddy, K. K.; Falck, J. R.; Aneja, R.; Aneja, S.; Parra, A.; Burns, D. J.; Ballas, L. M.; Cantley, L. C. *J. Biol. Chem.* **1994**, *269*, 32358-32367.
32. Derman, M. P.; Toker, A.; Hartwig, J. H.; Spokes, K.; Falck, J. R.; Chen, C. S.; Cantley, L. C.; Cantley, L. G. *J. Biol. Chem.* **1997**, *272*, 6465-6470.
33. Zhang, J.; Falck, J. R.; Reddy, K. K.; Abrams, C. S.; Zhao, W.; Rittenhouse, S. E. *J. Biol. Chem.* **1995**, *270*, 22807-22810.
34. Toker, A.; Bachelot, C.; Chen, C. S.; Falck, J. R.; Hartwig, J. H.; Cantley, L. C.; Kovacsovics, T. J. *J. Biol. Chem.* **1995**, *270*, 29525-29531.
35. Klarlund, J. K.; Guilherme, A.; Holik, J. J.; Virbasius, J. V.; Chawla, A.; Czech, M. P. *Science* **1997**, *275*, 1927-1930.
36. Palmer, R. H.; Dekker, L. V.; Woscholski, R.; Le Good, J. A.; Gigg, R.; Parker, P. J. *J. Biol. Chem.* **1995**, *270*, 22412-22416.
37. Hammonds-Odie, L. P.; Jackson, T. R.; Profit, A. A.; Blader, I. J.; Turck, C.; Prestwich, G. D.; Theibert, A. B. *J. Biol. Chem.* **1996**, *271*, 18859-18868.
38. Theibert, A. B.; Estevez, V. A.; Ferris, C. D.; Danoff, S. K.; Barrow, R. K.; Prestwich, G. D.; S.H. Snyder *Proc. Natl. Acad. Sci. USA* **1991**, *88*, 3165-3169.
39. Theibert, A. B.; Estevez, V. A.; Mourey, R. J.; Marecek, J. F.; Barrow, R. K.; G.D. Prestwich; Snyder, S. H. *J. Biol. Chem.* **1992**, *267*, 9071-9079.
40. Estevez, V. A.; Prestwich, G. D. *J. Am. Chem. Soc.* **1991**, *113*, 9885-9887.
41. Hao, W. H.; Tan, Z.; Prasad, K.; Reddy, K. K.; Chen, J.; Prestwich, G. D.; Falck, J. R.; Shears, S. B.; Lafer, E. M. *J. Biol. Chem.* **1997**, *272*, 6393-6398.
42. Lu, P. J.; Shieh, W. R.; Rhee, S. G.; Yin, H. L.; Chen, C. S. *Biochemistry* **1996**, *35*, 14027-14034.
43. Schiavo, G.; Gu, Q.-M.; Prestwich, G. D.; Sollner, T. H.; Rothman, J. E. *Proc. Natl. Acad. Sci. USA* **1996**, *93*, 13327-13332.

44. Franke, T. F.; Kaplan, D. R.; Cantley, L. C.; Toker, A. *Science* **1997**, *275*, 665-668.
45. Whiteford, C. C.; Brearley, C. A.; Ulug, E. T. *Biochem. J.* **1997**, *323*, 597-601.
46. Liscovitch, M.; Cantley, L. C. *Cell* **1995**, *81*, 659-662.
47. Cunningham, E.; Thomas, G. M. H.; Ball, A.; Hiles, I.; Cockcroft, S. *Curr. Biol.* **1995**, *5*, 775-783.
48. Ferrier, R. J.; Middleton, S. *Chem. Rev.* **1993**, *93*, 2779-2831.
49. Prestwich, G. D. *Acc. Chem. Res.* **1996**, *29*, 503-513.
50. Chen, J.; Profit, A. A.; Prestwich, G. D. *J. Org. Chem.* **1996**, *61*, 6305-6312.
51. Gu, Q. M.; Prestwich, G. D. *J. Org. Chem.* **1996**, *61*, 8642-8647.
52. Thum, O.; Chen, J.; Prestwich, G. D. *Tetrahedron Lett.* **1996**, *37*, 9017-9020.
53. Falck, J. R.; Reddy, K. K.; Ye, J. H.; Saady, M.; Mioskowski, C.; Shears, S. B.; Tan, Z.; Safrany, S. *J. Am. Chem. Soc.* **1995**, *117*, 12172-12175.
54. Reddy, K. K.; Saady, M.; Falck, J. R.; Whited, G. *J. Org. Chem.* **1995**, *60*, 3385-3390.
55. Gou, D.-M.; Chen, C.-S. *J. Chem. Soc. Chem. Commun.* **1994**, 2125-2126.
56. Wang, D.-S.; Chen, C.-S. *J. Org. Chem.* **1996**, *61*, 5905-5910.
57. Bruzik, K. S.; Kubiak, R. J. *Tetrahedron Lett.* **1995**, *36*, 2415-2418.
58. Watanabe, Y.; Hirofuji, H.; Ozaki, S. *Tetrahedron Lett.* **1994**, *35*, 123-124.
59. Watanabe, Y.; Tomioka, M.; Ozaki, S. *Tetrahedron* **1995**, *51*, 8969-8976.
60. Glaser, M.; Wanaski, S.; Buser, C. A.; Boguslavsky, V.; Rashidzada, W.; Morris, A.; Rebecchi, M.; Scarlata, S.; Runnels; Prestwich, G. D.; Chen, J.; Aderem, A.; Ahn, J.; McLaughlin, S. *Nature* **1996**, submitted.
61. Tall, E.; Dormán, G.; Garcia, P.; Runnels, L.; Shah, S.; Chen, J.; Profit, A.; Gu, Q. M.; Chaudhary, A.; Prestwich, G. D.; Rebecchi, M. J. *Biochemistry* **1997**, *36*, 7239-7248.
62. Mehrotra, B.; Elliott, J. T.; Chen, J.; Olszewski, J. D.; Profit, A. A.; Chaudhary, A.; Fukuda, M.; Mikoshiba, K.; Prestwich, G. D. *J. Biol. Chem.* **1997**, *272*, 4237-4244.
63. Dormán, G.; Chen, J.; Prestwich, G. D. *Tetrahedron Lett.* **1995**, *36*, 8719-8722.
64. Martin, S. F.; Josey, J. A.; Wong, Y.-L.; Dean, D. W. *J. Org. Chem.* **1994**, *59*, 4805-4820.
65. Prestwich, G. D.; Dormán, G.; Elliott, J. T.; Marecak, D. M.; Chaudhary, A. *Photochem. Photobiol.* **1997**, *65*, 222-234.
66. Gaidarov, I.; Chen, Q.; Falck, J. R.; Reddy, K. K.; Keen, J. H. *J. Biol. Chem.* **1996**, *271*, 20922-20929.
67. Voglmaier, S. M.; Keen, J. H.; Murphy, J. E.; Ferris, C. D.; Prestwich, G. D.; Snyder, S. H.; Theibert, A. B. *Biochem. Biophys. Res. Commun.* **1992**, *187*, 158-163.
68. Sohn, R. H.; Chen, J.; Koblan, K. S.; Bray, P. F.; Goldschmidt-Clermont, P. J. *J. Biol. Chem.* **1995**, *270*, 21114-21120.
69. Bednarek, S. Y.; Orci, L.; Schekman, R. *Tr. Cell Biol.* **1996**, *6*, 468-473.

Chapter 3

Differential Recognition of Phosphoinositides by Actin Regulating Proteins and Its Physiological Implications

Pei-Jung Lu[1], Da-Sheng Wang[1], Keng-Mean Lin[2], Helen L. Yin[2], and Ching-Shih Chen[1,3]

[1]Division of Medicinal Chemistry and Pharmaceutics, College of Pharmacy, University of Kentucky, Lexington, KY 40536–0082
[2]Department of Physiology, University of Texas Southwestern Medical Center, Dallas, TX 75235–9094

In an effort to understand the intricate relationship between phosphoinositide 3-kinase (PI 3-kinase) and actin assembly, we examined the differential interaction between phosphoinositides and actin-binding proteins (ABPs). The affinities for inositol lipids of three important ABPs (profilin, gelsolin, and CapG) are assessed by gel filtration and fluorescent titration. These analyses indicate that profilin displays preferential binding to the D-3 phosphoinositides over PtdIns(4,5)P$_2$, whereas gelsolin and CapG favor PtdIns(4,5)P$_2$. Also noteworthy is that the binding of gelsolin and CapG to PtdIns(4,5)P$_2$ is Ca^{2+}-dependent. The binding affinities in the presence of Ca^{2+} are 7.5- and 2.4-fold higher than that without Ca^{2+} for gelsolin and CapG, respectively. Furthermore, the present data support the notion that these ABPs provide a link between actin dynamics and PtdIns(4,5)P$_2$-dependent signaling pathways. The regulatory effect of ABPs on actin assembly is altered by phosphoinositides with potencies conforming to the respective binding affinities. Meanwhile, ABPs exert an inhibitory effect on the activity of PtdIns(4,5)P$_2$-utilizing enzymes in part by controlling PtdIns(4,5)P$_2$ availability. This inhibition may represent a negative feedback control of PI 3-kinase. Taken together, a working model correlating PI 3-kinase activation and the regulation of actin assembly by profilin and gelsolin is proposed.

In many biological responses, cells respond to external stimuli by exhibiting a diverse repertoire of motile activities such as crawling, protruding, secreting, retracting, engulfing, and ruffling. These activities result from the rapid, coordinated remodeling of the same cytoskeleton that provides these cells with their shapes. Rapid reorganization of the actin cytoskeleton is usually associated with a marked increase in actin filament turnover, characterized by the selective polymerization of actin into specific regions, concurrent with the selective depolymerization of other regions. This actin remodeling has been shown to be regulated by a host of ABPs and membrane phosphoinositides in a dynamic and interactive manner (*1-3*). Evidence suggests that the functional role of phosphoinositides in actin regulation is multi-faceted. For

---

[3]Corresponding author.

example, phosphoinositides provide physical anchors for the attachment of ABPs to membranes, and this binding sequesters ABPs from actin monomers (4,5). The unique interactions between ABPs and phosphoinositides, especially PtdIns(4,5)P$_2$, have been well documented (1,2,6). The ABPs known to be regulated by PtdIns(4,5)P$_2$ include actin monomer-binding proteins [profilin (4,5,8) and cofilin (9,10)], actin severing and/or capping proteins [gelsolin (9,10), CapG (11, 12), and CapZ (13)], and other ABPs [α-actinin (14) and vinculin (15)]. At a resting state, these ABPs partition between the plasma membrane and the cytosol, in response to variation in PtdIns(4,5)P$_2$ levels. For example, in vivo evidence shows that under growth conditions where plasma membrane levels of PtdIns(4,5)P$_2$ were depleted, profilin was found to translocate from the membrane to cytosol, however, it translocated back to membrane when plasma membrane levels of PtdIns(4,5)P$_2$ levels were replenished (5).

The intricate relationship between phosphoinositides and actin in cells has been the focus of many in vivo investigations. Several lines of data dispute the premise that PtdIns(4,5)P$_2$ solely plays a regulatory role in the actin cytoskeleton. For example, N-formyl-peptide induced actin polymerization in neutrophils correlated with the intracellular levels of PtdIns(3,4,5)P$_3$, rather than with PtdIns(4,5)P$_2$ concentrations (16). It was reported that the cellular event that most closely correlated with PI 3-kinase activation was actin filament rearrangement (17,18). Also, cells with mutations that abrogate PI 3-kinase binding sites in the platelet-derived growth factor (PDGF) β-receptor failed to undergo actin rearrangement or other mobility responses such as membrane ruffling and chemotaxis in response to PDGF (19-21). In addition, actin rearrangement in fibroblasts mediated by PDGF was blocked by wortmannin (22), a PI 3-kinase inhibitor. Moreover, the small GTP-binding protein Rho that regulated cytoskeletal reorganization in growth factor-treated cells was found to activate PI 3-kinase in platelet extracts (23). Rac, another small GTP-binding protein that was implicated in membrane ruffling was suggested to be a major effector protein for the PI 3-kinase signaling pathway (24). All these findings have prompted a notion that activation of PI 3-kinase is also crucial to actin reorganization.

It has been shown that the output signal of PI 3-kinase activation is transient accumulations of PtdIns(3,4)P$_2$ and PtdIns(3,4,5)P$_3$ in the cell (25,26). These D-3 phosphoinositides are not susceptible to phospholipase hydrolysis, suggesting that they do not serve as precursors to inositol phosphate secondary messengers (27). Although the critical role of PI 3-kinase in many cellular functions has been demonstrated (18,28-31), the molecular targets of its lipid products remain unclear. Thus, to understand the mode of mechanism for D-3 phosphoinositides in actin reorganization, we examine the interfacial interactions of various phosphoinositides with three ubiquitous ABPs, namely, profilin, gelsolin and CapG. The phosphoinositides that were examined included PtdIns(3,4)P$_2$, PtdIns(3,4,5)P$_3$ and PtdIns(4,5)P$_2$. Here, we report that these ABPs display differential recognition among these phosphoinositides. This finding has implications concerning the regulation of signaling pathway and the functional role of PI 3-kinase in actin rearrangement. To illustrate these correlations, the characteristic features of profilin, gelsolin and CapG are summarized below.

Profilin forms a stable 1 : 1 complex with actin monomers with K$_d$ between 0.5 and 10 mM. It is well documented that this ABP modulates the rate of actin polymerization by affecting actin dynamics, in which several distinct mechanisms may be involved (2,3,32). However, evidence suggests that profilin preferentially binds to PtdIns(4,5)P$_2$ in the plasma membrane in resting cells (4,5,7). The putative utility of this PtdIns(4,5)P$_2$-profilin association is two-fold. First, profilin is sequestered from actin monomers, thus inhibiting actin-profilin complex formation (2,33). Second, this binding protects PtdIns(4,5)P$_2$ from being hydrolyzed by PLC-γ_1, which implicates profilin in the regulation of Ins(1,4,5)P$_3$ production (*34*). Inhibition is overcome when PLC-γ_1 is phosphorylated by a receptor tyrosine kinase, leading to the hydrolysis of PtdIns(4,5)P$_2$ accompanied by the dissociation of profilin from the plasma membrane (*35*).

Gelsolin exhibits multiple functions in the regulation of actin assembly, namely, severing and capping of actin filaments (*36*). These activities were intrinsically modulated by pH, Ca^{2+} and phosphoinositides (*9,10,37,38*). The concerted change in these factors during cell activation supports the notion that the discriminative regulation of gelsolin function by receptor-mediated signaling can lead to cytoskeleton rearrangement. Substantial data have implicated gelsolin in physiological functions beyond actin reorganization. For instance, PtdIns(4,5)P$_2$-gelsolin binding has been reported to affect the activities of PLC (*39,40*), PLD (*41*), DNase I (*42*), and PI 3-kinase (*43*). Moreover, gelsolin has been suggested to play a key role as a tumor suppressor in human urinary bladder carcinogenesis (*44*). Also, a recent study indicates that gelsolin inhibited the activation of CPP32 proteases, which protect cells from apoptosis (*45*).

CapG, a member of the gelsolin family, is present in both the nucleus and the cytoplasm while other members of the family are exclusively cytoplasmic (*46*). It is noteworthy that nuclear CapG is predominately present in the phosphorylated form. CapG binds to the barbed end of actin filaments, however, without the severing activity.

Experimental Procedures

Materials. 1-*O*-(1,2-Di-*O*-palmitoyl-*sn*-glycero-3-phospho)- D-*myo*-inositol 3,4,5-trisphosphate [PtdIns(3,4,5)P3], 1-*O*-(1,2-Di-*O*-octanoyl-*sn*-glycero-3-phospho)- D-*myo*-inositol 3,4,5-trisphosphate [di-C$_8$-PtdIns(3,4,5)P$_3$], and 1-*O*-(1,2-di-*O*-palmitoyl-*sn*-glycero-3-phospho)-D-*myo*-inositol 3,4-bisphosphate, [PtdIns(3,4)P$_2$] were synthesized as previously reported (*46,47*). The purity of these synthetic phosphoinositides was confirmed by ^{1}H and ^{31}P NMR and FAB mass spectrometry, in which no trace amounts of isomeric impurity could be detected. PtdIns(4,5)P$_2$, phosphatidylinositol (PtdIns), phosphatidyl-serine (PtdSer), phosphatidylcholine (PtdCho), phosphatidylethanolamine (PtdEA), and 1,2-dipalmitoyl-*sn*-glycerol were purchased from Calbiochem or Sigma. PLC-γ_1 and PI 3-kinase were obtained from Drs. Sue Goo Rhee (NIH) and Lewis Cantley (Harvard Medical School), respectively.

Homogenous micelles containing various phospholipids were prepared according to the procedure of Goldschmidt-Clermont et al. (*34*) by sonicating in distilled water in a Model 1210 Bransonic Ultrasonic Cleaner for 5 min at room temperature.

Poly-L-proline (M_r 12,000 - 15,000; Sigma) was coupled to CNBr-activated Sepharose (Sigma) according to a modification of the method developed by Lindberg et al. (*49*). Poly-L-proline (2 g) was stirred in 100 mL of distilled water for 2 days, and added to 40 mL of 0.1 M KHCO3, pH 8.3, containing 5 g of CNBr-activated Sepharose. The mixture was incubated at 4°C for 20 h with gentle shaking. The matrix beads were separated by filtration, washed with 0.1 M KHCO$_3$, pH 8.3, and added to the same buffer containing 0.2 M ethanolamine to remove the remaining active groups on the gel. The affinity matrix was stored in a high salt buffer (Buffer A) containing 10 mM Tris, pH 7.8, 0.1 M NaCl, 0.1 M glycine, and 1 mM dithiothreitol (DTT). After each use, the gel was washed with 8 M urea for 6 h to remove bound material.

Profilin preparation. Poly(L-proline) was coupled to CNBr-activated Sepharose according to a modification of a published procedure (*49*). Profilin was purified from outdated human platelets by affinity chromatography on poly(L-proline)-Sepharose based on the method of Kaiser et al (50). In brief, the cytosolic fraction of human platelets was diluted with buffer A that contained 10 mM Tris, pH 7.8, 0.1M NaCl, 0.1 M glycine and 1 mM dithiothreitol, at a 1:1 ratio, and applied onto a poly(L-proline)-Sepharose column (1 x 15 cm) pre-equilibrated with the same buffer. After the column had been washed with 10 bed volume of buffer A, actin and profilin were eluted by 4 and 8M urea, respectively. The homogeneity of affinity-purified protein was identified by a single band on 12.5% of SDS-PAGE. The profilin-containing fractions were collected and dialyzed against G buffer containing 2 mM Tris/HCl, pH7.2, 0.1 mM ATP, 0.1 mM $MgCl_2$, 0.1 mM $CaCl_2$, and 0.5 mM DTT for 48h. The profilin was concentrated by ultrafiltration and used for assays. The profilin concentration was determined by measuring the UV absorbance at 280 nm using an extinction coefficient of 0.015 mM^{-1} cm^{-1} (51).

Gelsolin and CapG preparation. The expression vectors for gelsolin and CapG have been described previously (52-54). Briefly, the full length gelsolin expression vector (encompassing the entire human plasma gelsolin coding sequence) was contracted by ligating gelsolin cDNA to pET3a via the BamH I site. Recombinant proteins were expressed in bacteria and purified using sequential anion and cation exchange chromatography (52-54). Protein purity was assessed by 5-20 % SDS PAGE and protein concentration was determined by the Bradford method. Both purified gelsolin and CapG were stored at -80°C in buffer B containing 25 mM Hepes, pH 7.0, 80 mM KCl, 3 mM EGTA , and 0.5 mM DTT.

Gel Filtration Assay for the ABP Binding to Lipid Micelles. This assay was similar to that described for the binding of profilin (34) and gelsolin (53) to PtdIns(4,5)P_2. For comparison, phosphoinositide-containing micelles were prepared in distilled water as originally described. For example, profilin (14 mM), was incubated with different amounts of micellar phospholipids for 30 min at 25°C. One hundred mL of the mixture was chromatographed at room temperature on a Sephacryl S-200 column (1 x 15 cm) equilibrated with 10 mM Tris/HCl pH 7.5, containing 75 mM KCl and 1 mM DTT. The column was eluted with the same buffer at 0.5 mL/min. and 0.5 mL fractions were collected. Protein assays were performed by the method of Bradford with bovine serum albumin as standard. Because phospholipids interfere with the dye-binding assay, protein concentrations were given in arbitrary units. The fraction of bound profilin was calculated as the difference between the total amount applied onto the column and the amount in the entire included peak of free protein (34). Free phospholipid concentrations were calculated as the difference between the total concentration and the amount of profilin-associated lipid. Accordingly, the apparent dissociation constants were determined by using Equation 1:

$$K_d = [protein]_{free} \times [lipid]_{free} / [protein-lipid] \qquad (Eq. 1)$$

These experiments were repeated with PtdIns(3,4)$_2$ and PtdIns(4,5)P_2. Determination of the binding affinity between gelsolin or CapG and phosphoinositides was carried out in a similar fashion (37).

Fluorescence Spectroscopy. Fluorescence spectra were recorded at 30°C with a Hitachi F-2000 spectrophotometer. Human profilin, for example, contains two N-terminal tryptophan residues at the position 3 and 31. The fluorescence excitation wavelength was set at 292 nm as described by Raghunathanet et al (55). The buffer used for the fluorescence experiments consisted of 10 mM Tris/HCl, 150 mM KCl, 1 mM EDTA, and 5 mM β-mercaptoethanol, pH 7.5. Individual phosphoinositides, either the water-insoluble di-C_{16} derivatives, or the water soluble di-C_8 counterparts, were gradually introduced into 800 mL of the buffer containing various concentrations of ABPs. The total volume of the phosphoinositides added was less than 10 mL. Within the concentration range of phosphoinositides used in this study, the bulk solution remained clear. Moreover, no appreciable spectra were noted when individual phosphoinositides at the indicated concentrations were incubated alone. Changes in the fluorescence intensity were used as a measure of the binding affinity of the protein-lipid complex according to equation 2:

$$1/\{[1 - (\Delta F/\Delta Fmax)] \times K_a\} = \{[phosphoinositide]_{total}/$$
$$(\Delta F/\Delta Fmax)\} - p[profilin]_{total} \qquad (Eq. 2)$$

where K_a denotes the association constant and p represents the number of binding sites.

Actin polymerization. Purified G-Actin (from rabbit muscle) was labeled with pyreneiodoacetamide (*56*) and stored in G buffer containing 0.5 mM ATP, 0.5 mM DTT, 0.1 mM $CaCl_2$, and 10 mM Tris/HCl, pH 7.5. Actin polymerization was analyzed by using a fluorescence assay described by Lee et al (*57*). The amount of actin was measured at 30°C from the fluorescence of pyrene-labeled actin with an excitation and emission wavelengthes of 365 nm and 407 nm, respectively. The experiments were performed in G buffer and were initiated by adding 100 mM KCl and 2 mM $MgCl_2$ to the assay mixture.

PLC Assay. The assay was carried out by measuring the formation of $Ins(1,4,5)P_3$ from $PtdIns(4,5)P_2$ hydrolysis by PLC-γ_1 according to the procedure of Goldschmidt-Clermont (*34,35*). The reaction mixture consisted of 50 mM Hepes, pH 7.0, $PtdIns(4,5)P_2$ vesicles containing 15 nCi of $[^3H]PtdIns(4,5)P_2$, 10 µM cold $PtdIns(4,5)P_2$, 30 µM PtdEA, and 30 µM PtdSer with or without 10 µM $PtdIns(3,4)P_2$ or $PtdIns(3,4,5)P_3$, varying amounts of profilin, 2 mM $MgCl_2$, 120 mM KCl, 10 mM NaCl, 2 mM EGTA and 1 µM free from $CaCl_2$, in a total volume of 45 µL. The reaction was initiated by adding 5 ng of PLC-γ_1 in 5 µL of 50 mM Hepes, pH 7.0, containing 0.1 mg/mL BSA. After incubating at 30°C for 10 min, the reaction was stopped by the addition to the mixture of 200 µL of 10% trichloroacetic acid and 100 µL of 10 mg/mL BSA. The sample stood on ice for 3 min and was centrifuged at 10,000 x g for 5 min. The $[^3H]Ins(1,4,5)P_3$ in the supernatant was transferred into 20 mL scintillation vial and was measured by liquid scintillation spectrometry.

PI 3-kinase Assay. $PtdIns(4,5)P_2$ (10 µg) and PtdSer (40 µg) were dried and suspended in 100 µL of 30 mM Hepes, pH 7.5, containing 1 mM EDTA and 1mM EGTA, sonicated in a water bath-type sonicator for 5 min, and mixed vigorously with a votex mixer before assays. Purified PI 3-kinase from rat liver (2 µL) and various amounts of individual ABPs (3 µL) were incubated in 25 mM Hepes buffer, pH 7.5, containing 125 mM ATP, 10 µCi $[\gamma\text{-}^{32}P]ATP$, and 6.2 mM $MgCl_2$, in a final volume of 80 µL. The reaction was initiated by adding 20 µL of the phospholipid substrate described above, incubated at 37°C for 10 min, and stopped by adding 5 µL of 1 M EDTA and 25 µL of 5 M HCl, followed by 160 µL of chloroform-methanol (1 : 1; v/v). The mixture was thoroughly mixed by vortexing for 30 sec., and the two phases were separated by centrifugation at 6,000 g for 5 min. The organic layer was dried by a steam of N_2, spotted onto 1% oxalic acid-treated t.l.c. plates, and then developed with n-propanol : 2 M acetic acid (65 : 35) for overnight. After drying, spots were located by autoradiography and compared with standards. The autoradiograms were scanned by a photodyne image system and quantified using and NIH image program (version 1.59). For the examination of Ca^{2+} effect on gelsolin and CapG, Ca^{2+} was compensated at $[Ca^{2+}]_{free} = 25$ µM by using a $CaCl_2$/EGTA/EDTA system, which was calculated by a computer program developed by Wojciech Warchei (1990, version 2.1).

Results

Differential recognition phosphoinositides by ABPs. The interfacial interactions between ABPs and various phosphoinositides were examined by two different methods. In the gel filtration-based analysis, individual ABPs were incubated with micelles containing varying amounts of the pure phosphoinositide, and were chromatographed on a short-path Sephacryl S-200 column. Figure 1 illustrates the representative elution profiles of profilin in the presence of increasing micellar PtdIns(3,4,5)P$_3$.

As shown, the micelle-bound profilin was eluted in the void volume and was well separated from the free profilin. PtdIns(3,4,5)P$_3$ altered the elution of profilin in a dose dependent manner. Accordingly, the apparent dissociation constant was estimated from the data of individual experiments using Eq. 1.

Further evidence that ABPs exhibited discriminative binding to phosphoinositides was provided by fluorescence spectroscopy. The fluorescence spectral change reflects the change in microenvironments surrounding tryptophan residues. With profilin as an example, Figure 2 shows that all three phosphoinositides quenched the tryptophan fluorescence in a dose-dependent and saturable manner.

It is worth mentioning that addition of other phospholipids such as PtdCho, PtdSer, PtdIns, PtdEA, or related inositol phosphates at comparable concentrations did not cause appreciable change in profilin fluorescence (data not shown). This tryptophan titration allowed the determination of K$_d$ values according to the linear relationship between $1/[1 - (\Delta F/\Delta F_{max})]$ and [phosphoinositide]$_{total}/(\Delta F/\Delta F_{max})$, as depicted in Eq. 2. However, this method could not be applied to study gelsolin binding because full-length gelsolin signal fluctuated and could not reach a steady level (*37*).

Table 1 summarizes the K$_d$ values of these ABPs with different inositol lipids. The K$_d$ values determined by fluorescence titration were consistent with those estimated by gel filtration in spite of the difference in their basic principles.

Despite largely shared structural motifs among the three phosphoinositides examined, the affinity with individual ABPs varied to a great extent. For profilin, the binding affinity with PtdIns(3,4)P$_2$ and PtdIns(3,4,5)P$_3$ was 10- and 5-fold, respectively, higher than that of PtdIns(4,5)P$_2$. In contrast, the relative binding affinity for gelsolin and CapG was in the order of PtdIns(4,5)P$_2$ > PtdIns(3,4,5)P$_3$ >> PtdIns(3,4)P$_2$ (data not shown). Our data revealed that Ca^{2+} played a key role in regulating the binding of gelsolin and CapG to phosphoinositide. On the basis of the gel filtration method, in the presence of 25 μM Ca^{2+}, the binding affinity increased by 7.5- and 2.4-fold for gelsolin and CapG, respectively. CD data indicate that both gelsolin and CapG underwent substantial conformational changes upon ligand binding (not shown), of which the biochemical significance is under investigation. According to the crystal structure of gelsolin, Burtnick et al. recently proposed that Ca^{2+} binding release the connections that joined the N- and C-terminal halves of gelsolin, enabling each half to bind actin independently (*59*). These domain shifts might provide a basis to account for the action of gelsolin in response to Ca^{2+} to sever, cap, or nucleate F-actin filaments.

Taken together, this differential phosphoinositide recognition by ABPs might bear important physiological implications for how ABPs are regulated by Ca^{2+} and inositol lipids during actin rearrangement and how ABPs affected PtdIns(4,5)P$_2$-dependent signaling pathways via substrate sequestration.

Effect on actin cytoskeletal rearrangement. Two types of experiments were used to assess the phosphoinositide effect on profilin-exerted inhibition of actin polymerization. In the competitive binding experiment (Figure 3A), individual

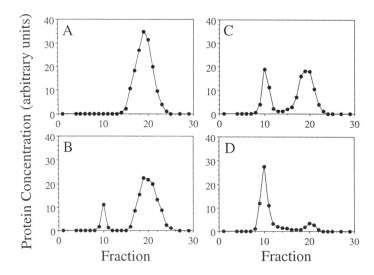

Figure 1. Gel filtration assay for profilin binding to di-C_{16}-PtdIns(3,4,5)P_3-containing micelles. (A) Elution profile (14 µM) alone at 25°C on a Sephacryl S-200 column (1 x 15 cm). (B-D) Elution profiles of mixtures of protein and varying amounts of micellar PtdIns(3,4,5)P_3 under the same conditions. The molar ratios of PtdIns(3,4,5)P_3 to profilin were 1.2:1, 2.5:1, and 4.9:1, respectively, for panels B-D. Due to the interference of phospholipids in the protein assay, the concentrations of eluted profilin are expressed in arbitrary units.

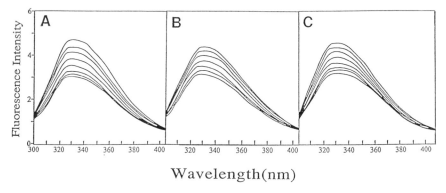

Wavelength(nm)

Figure 2. Tryptophan fluorescence emission spectra of profilin in the presence of varying amounts of (A) di-C_{16}-PtdIns(3,4)P_2, (B) di-C_{16}-PtdIns(3,4,5)P_3, and (C) di-C_{16}-PtdIns(4,5)P_2. Spectra were recorded with 17 µM profilin according to the method described in Experimental Procedures. Molar ratios of the phosphoinositides to profilin were (A) for PtdIns(3,4)P_2, (top to bottom): 0 : 1, 0.02 : 1, 0.06 : 1, 0.09 : 1, 0.26 : 1, 0.66 : 1, and 1.8 : 1; (B) for PtdIns(3,4,5)P_3 (top to bottom): 0 : 1, 0.14 : 1, 0.29 : 1, 0.43 : 1, 1 : 1, 1,6 : 1, and 2.7 : 1; (C) for PtdIns(4,5)P_2 (top to bottom): 0 : 1, 0.43 : 1, 0.85 : 1, 1.29 : 1, 2.1 : 1, 3.0 : 1, 3.9 : 1, and 5.6 : 1. The K_d values estimated according to Eq. 2 are 2 µM, 7 µM, and 36 µM for PtdIns(3,4)P_2, PtdIns(3,4,5)P_3 and PtdIns(4,5)P_2, respectively.

Table 1. Binding affinity of ABPs with various phosphoinositides (ref. *37, 58*)

	Apparent dissociation constants (K_d, μM)			
	Gel Filtration		Fluorescence Titration	
	- Ca^{2+}	+ Ca^2	- Ca^{2+}	+ Ca^2
Profilin				
PtdIns(4,5)P$_2$	11.0	ND	36.0	ND
PtdIns(3,4,5)P$_3$	5.7	ND	7.4	ND
PtdIns(3,4)P$_2$	1.1	ND	2.0	ND
Gelsolin				
PtdIns(4,5)P$_2$	305.4	40.2	ND	ND
CapG				
PtdIns(4,5)P$_2$	69.0	29.4	24.4	6.0

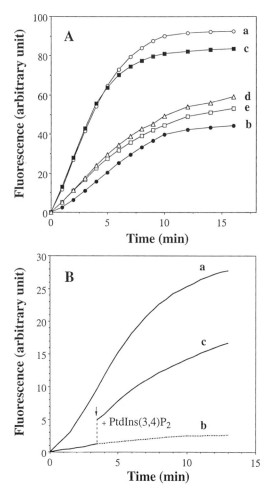

Figure 3. Effect of phosphoinositides on profilin-exerted inhibition of actin polymerization. In the competitive experiment (A), 10 μM pyrene-labeled actin was incubated in G buffer under various conditions: (a) control; (b) + 8 μM profilin; (c) + 8 μM profilin and 10 μM di-C_{16}-PtdIns(3,4)P_2; (d) + 8 μM profilin and 10 μM di-C_{16}-PtdIns(3,4,5)P_3; (e) + 8 μM profilin and 10 μM PtdIns(4,5)P_2 (Calbiochem). After 10 min incubation, actin polymerization was initiated by adding 100 mM KCl and 2 mM MgCl$_2$ to the assay mixture. All concentrations indicated above were based on the final volume of 1 mL. The extent of actin polymerization was indicated by the increase in fluorescence intensity. In the reversibility experiment (B), 2.4 μM pyrene-labeled actin and 5 μM unlabeled actin were incubated alone (curve a), or in the presence of 8 μM profilin (curve b) under actin polymerization conditions. The arrow indicates the addition of 32 μM di-C_{16}-PtdIns(3,4)P_2 to the mixture of actin and profilin at 3.5 min, and curve c represents the resulting time course

phosphoinositides (10 mM) competed with actin monomers (8 mM) for profilin binding (8 mM). Curves a and b depict the time course of actin polymerization of actin alone and in the presence of an equivalent amount of profilin, respectively.

As expected, the inhibition of actin polymerization by profilin was overcome by all three phosphoinositides examined, as a result of competitive profilin binding. The order of the compensatory effect was $PtdIns(3,4)P_2$ (curve c) > $PtdIns(3,4,5)P_2$ (curve d) > $PtdIns(4,5)P_2$ (curve e), which is consistent with their relative binding affinity with profilin. Correlated with this finding was the observation that $PtdIns(3,4,)P_2$ enhanced actin polymerization by facilitating the dissociation of actin-profilin complexes (Figure 3B). As shown, the rate of actin polymerization was suppressed in the presence of profilin (curve b), but significantly increased upon the addition of $PtdIns(3,4)P_2$ to the complex (curve c). $PtdIns(3,4)P_2$ was able to completely offset profilin inhibition after a few hours, as the extent of actin polymerization (curve c) reached the level of that of the control (curve a).

Concerning effect of phosphoinositide binding on gelsolin function, we found that the relative potency of phosphoinositides to inhibit gelsolin-mediated actin severing was in the order: $PtdIns(4,5)P_2$ > $PtdIns(3,4,5)P_3$ > $PtdIns(3,4)P_2$ (unpublished results), which conformed with the relative binding affinity. This inhibitory effect was attributable to the formation of phosphoinositide-gelsolin complexes.

Effect on $PtdIns(4,5)P_2$-utilizing signaling enzymes. The high affinity between ABPs and phosphoinositides might exert regulatory effects on enzymes that utilized $PtdIns(4,5)P_2$. This issue is addressed in the following studies.

Effect of D-3 phosphoinositides on profilin-mediated inhibition of PLC-γ_1. Profilin imposed a dose-dependent inhibition on PLC-γ_1-mediated $PtdIns(4,5)P_2$ hydrolysis (Figure 4A, curve a).

This inhibition has also been noted with gelsolin and CapG (*12,39*). Conceivably, competition for $PtdIns(4,5)P_2$ binding between ABPs and PLC-γ_1 inhibited the enzyme activity through substrate sequestration (*34,58*). This inhibition, however, could be overcome by $PtdIns(3,4,5)P_3$ or $PtdIns(3,4)P_2$ (Figure 4A, curve b and curve c, respectively). The reversal of PLC-γ_1 inhibition might be accounted for by two distinct mechanisms. First, $PtdIns(3,4,5)P_3$ and $PtdIns(3,4)P_2$ were able to stimulate PLC-γ_1 activity, which is evidenced by the relative PLC-γ_1 activities in the absence of profilin (Figure 4B, curves b and c). Secondly, this protective effect could be attributed to strong binding of profilin to the D-3 phosphoinositides. As shown, $PtdIns(3,4,5)P_3$ and $PtdIns(3,4)P_2$ at 10 µM were able to protect PLC-γ_1 from losing activity with profilin concentrations up to 1 µM (Figure 4A).

Effect of ABPs on PI 3-kinase. By the same token, sequestration of $PtdIns(4,5)P_2$ by ABPs could also cause PI 3-kinase inhibition. Figure 5 demonstrates the dose-dependent effect of profilin on PI 3-kinase activity. For example, at 0.3 : 1 and 0.6 : 1 molar ratios (profilin/$PtdIns(4,5)P_2$), 35% and 60%, respectively, of PI 3-kinase activity was inhibited.

Gelsolin and CapG also exerted PI 3-kinase inhibition, however, at lower potencies due to their lower binding affinity with $PtdIns(4,5)P_2$ in the absence of Ca^{2+}. For instance, the extents of inhibition were 15% and 12% for gelsolin and CapG, respectively, at a 0.3 : 1 molar ratio, vis-à-vis 35% for profilin. Nevertheless, Ca^{2+} could dramatically enhance such inhibitory ability, which connoted the Ca^{2+} effect on the $PtdIns(4,5)P_2$ binding shown in Table 1. Figure 6 illustrates the enhancement of gelsolin-mediated PI 3-kinase inhibition by Ca^{2+}. At the 0.7 : 1 molar ratio of gelsolin to $PtdIns(4,5)P_2$, 33% and 85% inhibition of $PtdIns(3,4,5)P_3$ production were noted without and in the presence of 25 µM free Ca^{2+}, respectively. Similar results were obtained concerning the Ca^{2+} effect on CapG (not shown). These results suggest a fine-tuning mechanism for the regulatory activity of gelsolin and CapG.

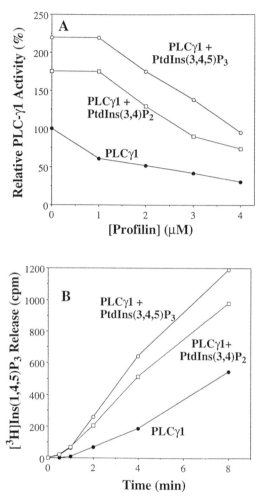

Figure 4. Effect of D-3 phosphoinositides on PLC-γ_1 activity. (A) Effect of di-C$_{16}$-PtdIns(3,4)P$_2$ (10 μM) and di-C$_{16}$-PtdIns(3,4,5)P$_3$ (10 μM) on profilin elicited PLC-γ_1 inhibition. (B) Stimulation of PLC-γ_1 by PtdIns(3,4)P$_2$ (10 μM) and PtdIns(3,4,5)P$_3$ (10 μM). PLC-γ_1-mediated PtdIns(4,5)P$_2$ hydrolysis was analyzed according to the method described under Experimental Procedures. PLC activity was analyzed in the presence of varying amounts of profilin under different conditions. Each data point represents the mean of three determinations.

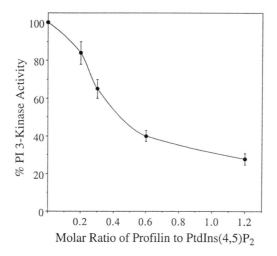

Figure 5. Dose-dependent inhibition of PI 3-kinase by profilin. Phosphorylation of PtdIns(4,5)P$_2$ by PI 3-kinase was analyzed in the presence of varying amounts of profilin as described under Experimental Procedures. Each data point represents the mean of three determinations.

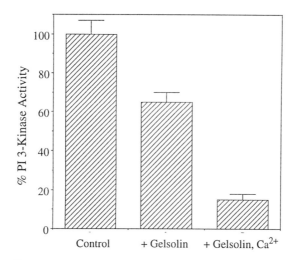

Figure 6. Effect of gelsolin on PI 3-kinase activity. PtdIns(4,5)P$_2$-containing lipid substrate was incubated with gelsolin at a molar ratio of 1 : 0.7, with or without 25 μM Ca^{2+}, for 20 min before purified PI 3-kinase was added to initiate the reaction. PI 3-kinase assay was analyzed according to the method described under Experimental Procedures. Each data point was the average of three independent experiments.

Results

The present study provides evidence that there exists an intricate relationship among membrane phosphoinositides, ABPs, and actin, of which the dynamic interactions provide a putative link between cytoskeletal reorganization and $PtdIns(4,5)P_2$ signaling. In quiescent cells, ABPs partition between the plasma membrane and cytosol, thus regulating monomeric actin pool and actin dynamics. The distributions of individual ABPs between membrane phosphoinositides and actin are dictated by their respective affinities and concentrations. For instance, in *Saccharomyces cerevisiae*, membrane-bound profilin accounted for 80% of total profilin, whereas the remainder was present in the cytosol in an actin-bound form (5). Although the distribution ratio of profilin in mammalian cells remains unclear, evidence shows that profilin-$PtdIns(4,5)P_2$ complexes block access of signaling enzymes, such as PLC-γ_1 and PI 3-kinase, to $PtdIns(4,5)P_2$. Gelsolin and CapG, on the other hand, are responsible for capping the fast growing ends of actin filaments. In resting cells, some gelsolin caps actin filaments (10-30%), some is attached to the plasma membrane (<5%), while the bulk is cytosolic. Gelsolin does not bind $PtdIns(4,5)P_2$ because most of the $PtdIns(4,5)P_2$ is sequestered by profilin, which is more than 10 times as abundant (10 vs. 0.7 mM, respectively, for profilin and gelsolin in NIH 3T3 fibroblasts). The combination of gelsolin's capping of filament barbed ends and profilin's inactivation by $PtdIns(4,5)P_2$ prevents actin polymerization at the barbed end..

This dynamic equilibrium is perturbed in response to agonist stimulation in part due to the change in membrane phosphoinositide compositions. This premise is supported by several lines of *in vivo* evidence. For example, N-formyl peptide-induced actin assembly in neutrophils was accompanied by a rapid decrease in $PtdIns(4,5)P_2$ levels and transient accumulations of the D-3 phosphoinositides (16). In addition, activation of platelets led to a significant increase in the membrane association of profilin (7).

The present data suggest that $PtdIns(3,4)P_2$ and $PtdIns(3,4,5)P_3$ produced in response to agonist stimulation may play a role in the regulation of actin dynamics by controlling the locality of ABPs (Figure 7). Due to the higher affinity with profilin, the D-3 phosphoinositides spontaneously take over a portion of the profilin from $PtdIns(4,5)P_2$.

However, an important factor that warrants consideration is the relatively small total mass of $PtdIns(3,4)P_2$ and $PtdIns(3,4,5)P_3$ produced versus that of the classical phosphoinositides, which may differ by 1 order of magnitude. Thus, it is plausible that the displacement of $PtdIns(4,5)P_2$ from profilin is localized to a specific area of the plasma membrane. Nevertheless, profilin regulation by D-3 inositol lipids is consistent with growing evidence for a role of PI 3-kinase in cytoskeletal rearrangement.

In addition, there is substantial evidence implicating PI 3-kinase in Ca^{2+} signaling. The present data show that $PtdIns(3,4,5)P_3$ and $PtdIns(3,4)P_2$ stimulated PLC-γ_1, which, in concert with the dissociation of $PtdIns(4,5)P_2$-profilin complexes, should stimulate $Ins(1,4,5)P_3$ and DAG production. This premise is supported by a recent observation in NIH 3T3 cells that exposure to $PtdIns(3,4,5)P_3$ resulted in a transient increase in intracellular Ca^{2+} (60) which could be blocked by a PLC inhibitor. Thus, this D-3 phosphoinositide-mediated stimulation may represent a new mechanism, auxiliary to PTK-mediated phosphorylation, for the regulation of PLC-γ_1 activity. Moreover, we recently discovered that in platelets, $PtdIns(3,4,5)P_3$ induced Ca^{2+} influx, though the exact mechanism remains unclear (unpublished data).

Consequently, two factors may contribute to the increased association of gelsolin

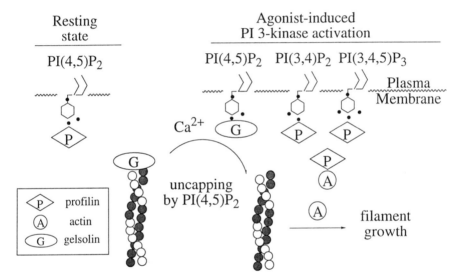

Figure 7. A proposed model for the coordinated regulation of actin filaments by gelsolin, profilin, and phosphoinositides.

and CapG with the plasma membrane during cell activation, despite a modest decrease in membrane PtdIns(4,5)P$_2$. First, PtdIns(3,4,5)P$_3$ and PtdIns(3,4)P$_2$ facilitate the dissociation of profilin-PtdIns(4,5)P$_2$ complexes, generating free PtdIns(4,5)P$_2$ for binding. Second, gelsolin and CapG binding to PtdIns(4,5)P$_2$ is enhanced significantly by Ca^{2+}, and thus can displace profilin from PtdIns(4,5)P$_2$. These combined effects allow the bulk of profilin to fall of membrane and capture actin monomers stored in the β-thymosin reservoir, and to drive actin assembly from the newly uncapped barbed ends. Multiple rounds of severing, uncapping and facilitated actin addition fuel the explosive amplification of filament growth observed in many physiological responses (1,2).

The association of ABPs with membrane PtdIns(4,5)P$_2$ raises a question concerning its potential impact on PtdIns(4,5)P$_2$-signaling enzymes. Our data suggests that the recruitment of ABPs to plasma membranes exerts an inhibition on PI 3-kinase, in part by controlling PtdIns(4,5)P$_2$ availability. This inhibition may represent a negative feedback control of PI 3-kinase.

In summary, the dynamic relationship between ABPs and membrane phosphoinositides provides a molecular basis for communications between actin cytoskeleton and PtdIns(4,5)P$_2$-dependent signaling pathways. The complexity of this issue, however, is beyond what is described here since there exist many other ABPs of different families in cells. These ABPs display different behaviors in response to membrane phosphoinositides and biological mediators. Presumably, differential regulation and cross-talk among these proteins may allow control to be exerted at different levels along the signaling process.

Acknowledgements: This work is supported by the National Institutes of Health grants R01 GM53448 (CSC) and R01 GM51112 (HLY).

Literature Cited

1. Stossel, T. P. *J. Biol. Chem.* **1989**, *264*, 18261-18264.
2. Sohn, R.; Goldschmidt-Clermont, P. *BioEssay* **1994**, *16*, 465-472.
3. Aderem A. *Trends Biochem. Sci.* **1992**, *17*, 438-443.
4. Lassing, I.; Lindberg, *Nature* **1985**,*314*, 472-474.
5. Ostrander, D. B.; Gorman, J. A.; Carman, G. M., *J. Biol. Chem.* **1995**, *270*, 27045-27050.
6. Janmey, P.A. *Annu. Rev.Physiol.* **1994**, 169-191.
7. Hartwig, J.; Chambers, K. A.; Hopcia, K. L.; Kwiatkowski, D. J. *J. Cell Biol.* **1989**, *109*, 1571-1579.
8. Yonezawa, N.; Homma, Y.; Yahara, I.; Sakai,H.; Nishida, E. *J. Biol. Chem.* **1991**, *266*, 17218-17221.
9. Janmey, P. A.; Stossel, T. P., *Nature* **1987**, *325*, 362-364.
10. Lamb, J. A.; Allen, P. G.; Tuan, B. Y.; Janmey, P. A. *J. Biol. Chem.* **1993**, *268*, 8999-9004.
11. Yu, F. X.; Johnston, P. A.;Sudhof, T. C.;Yin, H. L. *Science* **1990**, *250*, 1413-1415.
12. Sun, H. Q.; Kaiatkowska, K.; Wooten, D. C.; Yin, H. *J. Cell Biol.* **1995**, *129*, 147-156.
13. Schafer, D. A.; Jennings, P. B.; Cooper, J. A. *J. Cell Biol.* **1996**, *135*, 169-179.
14. Fukami, K; Furuhashi, K.; Inagaki, M.; Endo, T.; Hatano, S.; Takenawa, T. *Nature* **1992**, *359*, 150-152.
15. Gilmore, A. P.; Burridge, K. *Nature* **1996**, *381*, 531-535.
16. Eberle, M.; Traynor-Kaplan, A. E.; Sklar, L. A.; Norgauer, J. *J. Biol. Chem.* **1990**, *265* , 16725-16728.

17. Cantley, L. C.; Auger, K. R.; Carpenter, C.; Duckworth, B.; Graziani, A.; Kapeller, R.; Soltoff, S. *Cell* **1991**, *64*, 281-302.
18. Rittenhouse, S. E. *Blood* **1996**, *88*, 4401-4414.
19. Severinsson, L.; Ek, B.; Mellstrom, K.; Claesson-Welsh, L.; Heldin, C. H., *Mol. Cell Biol.* **1990**, *10*, 801-809.
20. Wennstrom, S.; Siegbahn, A.; Yokote, K.; Arvidsson, A. K.; Heldin, C. H.; Mori, S.; Claesson-Welsh, L. *Oncogene* **1994**, *9*, 651-660.
21. Kundra, V.; Escobedo, J. A.; Kazlauskas, A.; Kim, H. K.; Rhee, S. G.; Williams, L. T.; Zetter, B. R. *Nature* **1994**, *367*, 474-476.
22. Wymann, M.; Arcaro, A. *Biochem. J.* **1994**, *298*, 517-520.
23. Zhang, J.; King, W. G.; Dillon, S.; Hall, A.; Feig, L.; Rittenhouse, S. E. *J. Biol. Chem.* **1993**, *268*, 22251-22254.
24. Rodriguez-Viciana, P; Warne P. H.; Khwaja, A; Marte, B. M.; Pappin, D.; Das, P.; Waterfield, M. D.; Ridley, A.; Downward, J. *Cell* **1997**, *80*, 457-467.
25. Traynor-Kaplan, A. E.; Thompson, B. L.; Harris, A. L.; Taylor, P.; Omann, G. M.; Sklar, L. A. *J. Biol. Chem.* **1989**, *264*, 15668-15673.
26. Stephens, L. R.; Hughes, K. T.; Irvine, R. F. *Nature* **1991**, *351*, 33-39.
27. Serunian, L. A.; Haber, M. T.; Fukui, T;, Kim, J. W.; Rhee, S. G.; Lowenstein, J., M.; Cantley, L. C. *J. Biol. Chem.* **1989**, *264*, 17809-17815.
28. Kapeller, R.; Cantley, L. C. *Bioassays* **1994**, *16*, 565-576.
29. Stephens, L. R.; Jackson, T. R.; Hawkins, P. T. *Biochim. Biophys. Acta* **1993**, *1179*, 27-75.
30. Toker, A.; Cantley, L. C. *Nature* **1997**, *387*, 673-676.
31. Vanhaesebroeck, B.; Atein, R. C.; Waterfield, M. D. *Cancer Survey* **1996**, *27*, 249-270.
32. Theriot, J. A; Mitchison, T. J. *Cell* **1993**, *75*, 835-838.
33. Carlsson, L.; Nystrom, L. E.; Sundkvist, I.; Markey, F.; Lindberg, U. *J. Mol. Biol.* **1977**, *115*, 465-483.
34. Goldschmidt-Clermont, P. J.; Machesky, L. A.; Baldassare, J. J.; Pollard, T. D. *Science* **1990**, *246*, 1575-1578.
35. Goldschmidt-Clermont, P. J.; Kim, J. W.; Machesky, L. M.; Rhee, S. G.; Pollard, T. D. *Science* **1991**, *251*, 1231-1233.
36. Janmey, P. *Nature* **1993**, *364*, 675-676.
37. Lin, K. M.; Wenegieme, E.; Lu, P. J.; Chen, C. S.; Yin, H. L. *J. Biol. Chem.* **1997**, *272*, 20443-20450.
38. Janmey, P. *Annu. Rev. Physiol.* **1994**, *56*, 169-191.
39. Banno, Y.; Nakashima, T.; Kumada, T; Ebisawa, K.; Nonomura, Y.;Nozawa, Y. *J. Biol. Chem.* **1992**, *267*, 6488-6494.
40. Sun, H. Q.; Lin, K. M.; Yin, H. L. *J. Cell Biol.* **1997**, in press.
41. Steed, P. M.; Nadar, S.; Wennogle, L. P. *Biochemistry* **1996**, *35*, 5229-5237.
42. Davoodian, K.; Ritchings,B. W.; Ramphal, R.; Bubb, M. R. *Biochemistry* **1997**, *36*, 9637-9641.
43. Singh, S. S.; Chauhan, A.; Murakami, N.; Chauhan, V. P. *Biochemistry* **1996**, *96*, 16544-16549.
44. Tanaka, M.; Mullauer, L.; Ogiso, Y.; Fujita, H.; Shinohara, N.; Koyanagi, T.; Kuzumaki, N. *Cancer Research* **1995**, *55*, 3228-3232.
45. Ohtsu, M.; Sakai, N.; Fujita, H.; Kashiwagi, M.; Gasa, S.; Sakiyama, Y.; Kobayashi, K.; Kuzumaki, N. *EMBO J.* **1997**, *16*, 4650-4656.
46. Onoda, K.; Yin, H. L. *J. Biol. Chem.* **1993**, *268*, 4106-4112.
47. Gou, D.-M.; Chen, C.-S. *J. Chem. Soc. Chem. Commun.* **1994**, 2125-2126.
48. Wang, D.-S.; Chen, C.-S. *J. Org. Chem.* **1996**, *61*, 5905-5910.
49. Lindberg, U.; Schutt, C. E.; Hellston, E.; Tjader, A. C.; Hult, T. *Biochim. Biophys. Acta* **1988**, *967*, 391-400.

50. Kaiser, D. A.; Goldschmidt-Clermont, P. J.; Levine, B. A.; Pollard, T. D. *Cell Motil. Cytoskeleton* **1989**, *14*, 251-262.
51. Tseng, P. C.-H.; Runge, M. S.; Cooper, J. A.; Pollard, T. D. *J. Cell Biol.* **1984**, *98*, 214-221.
52. Kwiatkowski, D. P.; Stossel, T. P.; Orkin, S. H.; Colten, H. R.; Yin, H. L. *Nature* **1986**, *323*, 455-458.
53. Yu, F. X.; Sun, H. Q.; Janmey, P.; Yin, H. L. *J. Biol. Chem.* **1992**, *267*, 14616-14621.
54. Yu, F. X.; Zhou, D.; Janmey, P.; Yin, H. L. *J. Biol. Chem.* **1991**, *266*, 19269-19275.
55. Raghunathan, V.; Mowery, P.; Rozycki, M.; Lindberg, U.; Schutt, C. *FEBS Lett.* **1992**, *297*, 46-50.
56. Kouyama, I.; Mihashi, K. *Eur. J. Biochem.* **1981**, *114*, 33-38.
57. Lee, S.; Li, M.; Pollard, T. D. *Anal. Biochem.* **1988**, *168*, 148-155.
58. Lu, P.-J.; Shieh, W.-R., Rhee, S. G.; Yin, H. L.; Chen, C.-S. *Biochemistry* **1996**, 35, 14027-14034.
59. Burtnick, L. D.; Koepf, E. K.; Grimes, J.; Jones, E. Y.; Stuart, D. I.; McLaughlin, P. J.; Robinson, R. C. *Cell* **1997**, *90*, 661-670.
60. Bae, Y. S.; Cantley, L. G.; Chen, C.-S.; Kim, S.-R.; Kwon, K.-S.; Rhee, S. G. *J. Biol. Chem.* **1998**, *273*, 4465-4469.

Chapter 4

Studies on New Ins(1,4,5)P$_3$ Binding Proteins with Reference to the pH Domains

Takashi Kanematsu, Hiroshi Takeuchi, and Masato Hirata

Department of Biochemistry, Faculty of Dentistry, Kyushu University, Fukuoka 812-8582, Japan

Two inositol 1,4,5-trisphosphate (Ins(1,4,5)P$_3$) binding proteins, having molecular masses of 85 kDa and 130 kDa, were purified from rat brain using an Ins(1,4,5)P$_3$ affinity column; the former protein was found to be the δ_1-isozyme of phospholipase C (PLC-δ_1) and the latter protein was an unidentified novel protein. Subsequent molecular biology studies isolated the full-length cDNA for the 130 kDa-Ins(1,4,5)P$_3$ binding protein (p130) which encoded 1,096 amino acids. The predicted amino acid sequence of p130 had 38.2% homology to that of PLC-δ_1. The region of p130 responsible for Ins(1,4,5)P$_3$ binding was mapped in a pleckstrin homology (PH) domain of the molecule. The PH domain of p130 could also bind to Ins(1,4,5,6)P$_4$ with a similar affinity to Ins(1,4,5)P$_3$. The 85 kDa protein (PLC-δ_1) was also analyzed for the binding site by molecular biological, peptide synthetic chemical and immunological studies: the sequence 30-43 of PLC-δ_1 was primarily involved in binding, which was mapped in the N-terminal of the PH domain of the molecule. Experiments as to the effect of Ins(1,4,5)P$_3$ on PLC-δ_1 activity showed that Ins(1,4,5)P$_3$ at concentrations over 1 μM strongly inhibited PLC activity of PLC-δ_1. The PH domains derived from four different proteins, the N-terminal part of pleckstrin, RAC-protein kinase, diacylglycerol kinase and p130, were analyzed for the capability and specificity of binding of inositol phosphates and derivatives of inositol lipids. We concluded from these studies that inositol phosphates and/or inositol lipids might be common ligands for the PH domains, and therefore inositol phosphates/inositol lipids might be involved in more aspects of cellular functions than originally thought, because more than 90 proteins to date are known to include PH domain. Which ligands are physiologically relevant for the PH domain, would depend on binding affinities and their cellular abundance.

D-*myo*-Inositol 1,4,5-trisphosphate [Ins(1,4,5)P$_3$], a product of the receptor-activated hydrolysis of phosphatidylinositol 4,5-bisphosphate [PI(4,5)P$_2$], plays an important role as an intracellular second messenger by mobilizing Ca^{2+} from non-mitochondrial storage sites (*1*). Three types of proteins have been shown to interact with Ins(1,4,5)P$_3$: Ins(1,4,5)P$_3$ receptor on the endoplasmic reticulum involved in Ca^{2+} release (*2,3*) and two types of Ins(1,4,5)P$_3$ metabolizing enzymes known as Ins(1,4,5)P$_3$ 5-phosphatase and Ins(1,4,5)P$_3$ 3-kinase (*4*).

Our aim has been to elucidate the structural basis for the interaction of Ins(1,4,5)P$_3$ with these recognizing molecules. Based on these studies one can rationalize the chemical design of agonists or antagonists for receptor molecules, and for enzyme inhibitors. The two approaches useful for achieving this goal are studies on the recognizing proteins to clarify binding domains, as done with the Ins(1,4,5)P$_3$ receptor (*5,6*), and the other is studies on the ligand, Ins(1,4,5)P$_3$. Several analogues with modifications on the inositol ring or the phosphates have been synthesized and the role of the three phosphates and hydroxyl groups of Ins(1,4,5)P$_3$ in receptor binding and enzyme recognition has been elucidated (*7-10*).

In 1985, we described for the first time the chemical modification of Ins(1,4,5)P$_3$ (*11*). This analogue has the azidobenzoyl group at the C-2 position for photoaffinity labeling and causes an irreversible inactivation of the receptor protein for Ca^{2+} release, following photolysis. On the basis of these findings and the report (*12*) that biological activities of Ins(1,4,5)P$_3$ are related to two adjacent phosphates at C-4 and C-5 and that the phosphate at C-1 increases the affinity for its recognition by the receptor site, we attempted further chemical modifications of Ins(1,4,5)P$_3$ at the C-2 position and examined biologically the events related to recognition by the above-mentioned Ins(1,4,5)P$_3$-recognizing proteins (*13,14*).

The analogues we synthesized were designed to enable further functionalization for photoaffinity labeling (*15*) or Ins(1,4,5)P$_3$-immobilized resin (204-resin), the latter of which proved to be useful for purifying the well known Ins(1,4,5)P$_3$-interacting proteins (*16*). In this chapter, we describe the purification of novel Ins(1,4,5)P$_3$-binding proteins from rat brain cytosol using the Ins(1,4,5)P$_3$-immobilized resin, and mapping of binding sites of these proteins. The results obtained thus far allow us to propose that inositol compounds might be common ligands for the PH domains. To gain supportive evidence for the proposal, several PH domains derived from various proteins were examined for the capability and specificity of binding of inositol compounds. Structural features of the putative pockets of PH domains for accommodating inositol compounds are also discussed.

Purification of New Ins(1,4,5)P$_3$ Binding Proteins

Ins(1,4,5)P$_3$ Affinity Chromatography. Cytosol fractions of the rat brain homogenates prepared in buffer A [50mM NaCl, 10mM Hepes/NaOH (pH 8.0), 1mM EDTA, 2mM NaN$_3$ and 10mM 2-mercaptoethanol] were applied to an Ins(1,4,5)P$_3$ affinity column (*16,17*). After washing the column with initial buffer, a solution of 0.5 M-NaCl or 2 M-NaCl in buffer A was subsequently added to elute the adsorbed proteins. Each fraction was assayed for [^{3}H]Ins(1,4,5)P$_3$ binding and [^{3}H]Ins(1,4,5)P$_3$ metabolizing activities. As shown in Table I, the Ins(1,4,5)P$_3$ 5-phosphatase activity was eluted with a 0.5 M-NaCl solution, providing a 10-fold increase in specific activity. The Ins(1,4,5)P$_3$ 3-kinase activity was eluted in a 2M-NaCl solution with about a 50-fold purification, but the recovery was only 10%. These findings were comparable with our previous data (*16*). When measured the Ins(1,4,5)P$_3$ binding activity in each fraction obtained by Ins(1,4,5)P$_3$ affinity chromatography, we found remarkable Ins(1,4,5)P$_3$ binding activity in a 2 M-NaCl fraction: the specific binding of 110 pmol/mg protein at 1.3 nM [^{3}H]Ins(1,4,5)P$_3$ was obtained in a 2 M-NaCl eluate, while that in the cytosol fraction was 0.01 pmol/mg protein, indicating more than 10,000-fold purification. The total binding of the cytosol

Table I. Binding and Metabolizing Activities of [³H]Ins(1,4,5)P₃
on Ins(1,4,5)P₃ Affinity Chromatography

Fraction	Phosphatase		Kinase		Binding	
	Specific activity	Total activity	Specific activity	Total activity	Specific activity	Total activity
	nmol/mg/min	*nmol/min*	*nmol/mg/min*	*nmol/min*	*pmol/mg*	*pmol*
Cytosol[a]	44.9	1533	0.29	9.9	0.01	0.34
Flow-through	4.4	139	0.05	1.6	ND*	
0.5 M NaCl	447	1188	0.15	0.4	ND	
2 M NaCl	ND	ND	16.0	1	110	6.6

ND* not detected

[a] The cytosol obtained from tow rat brains was appleid to a 204 column. Data are the mean of duplicate determinations, with similar results from two other experiments.
SOURCE: Reprinted with permission from ref. 17. Copyright 1992

fraction was about one-twentieth of that of the partially purified sample (a 2M NaCl eluate), thereby indicating the presence of inhibitory factor(s) for Ins(1,4,5)P$_3$ binding in a crude cytosol fraction, which was removable by Ins(1,4,5)P$_3$ affinity chromatography (18).

Gel Filtration Chromatography. The 2 M NaCl eluate from the Ins(1,4,5)P$_3$ affinity column with a high [^{3}H]Ins(1,4,5)P$_3$ binding activity was further fractionated by the application to a HiLoad 26/60 Superdex 200 column. As shown in Figure 1a, the binding activity was divided into two peaks between the molecular weight markers of 200 kDa and 66 kDa. On the other hand, Ins(1,4,5)P$_3$ 3-kinase activity was eluted at 110 min (results not shown), indicating that the [^{3}H]Ins(1,4,5)P$_3$ binding activities are not due to the Ins(1,4,5)P$_3$ 3-kinase. Analyses by SDS-polyacrylamide gel electrophoresis of these fractions indicated that the proteins with an apparent molecular mass of 130 kDa or 85 kDa were most likely to be responsible for the respective peak of [^{3}H]Ins(1,4,5)P$_3$ binding activities.

Determination of Partial Amino Acid Sequence. To identify these two Ins(1,4,5)P$_3$ binding proteins, the purified samples of the 130 kDa- and 85 kDa-proteins were hydrolyzed with lysyl endopeptidase, and the peptides were separated by reversed-phase HPLC and sequenced using a protein sequencer. Three peptides derived from the 130 kDa protein were sequenced as follows: NTETFXNNGLADQICEDXXF (P-22), XPLXFMEGNQNTPXF (P-14) and AIESFAXNIXV (P-13). Sequence homology was examined using the SWISS-PROT protein database and NBRF protein database, but there was no similar protein. On the other hand, three peptides derived from the 85 kDa protein were also sequenced as follows: TIWQESRK, IIHHSGSMDQRQK and QGYRHVHLLSK, which were the same as those of PLC-δ$_1$, 50 to 57, 128 to 140 and 728 to 738 of amino acid sequence, respectively (19). Thus, the 130 kDa protein (p130) was likely to be a newly-identified Ins(1,4,5)P$_3$ binding protein, while the 85 kDa protein capable of binding Ins(1,4,5)P$_3$ was PLC-δ$_1$ (17).

Characterization of New Ins(1,4,5)P$_3$ Binding Proteins

Binding Specificity. [^{3}H]Ins(1,4,5)P$_3$ binding to both p130 and PLC-δ$_1$ was displaced by Ins(1,4,5)P$_3$ in a dose-dependent manner. Scatchard analyses of Ins(1,4,5)P$_3$ binding indicated that each protein had a single binding site with a K_d/B_{max} values of about 2.5 nM/2.5 nmol/mg protein or 5.2 nM/9.8 nmol/mg protein for p130 or PLC-δ$_1$, respectively (17). As shown in Figure 2, a series of inositol phosphates were examined for their potency to displace [^{3}H]Ins(1,4,5)P$_3$ bound to each protein. In the case of p130, Ins(1,4,5,6)P$_4$ was as potent as Ins(1,4,5)P$_3$. The order of potency was Ins(1,4,5)P$_3$ = Ins(1,4,5,6)P$_4$ > Ins(1,3,4,5)P$_4$ > Ins(1,3,4,6)P$_4$ = Ins(3,4,5,6)P$_4$ > Ins(1,3,4)P$_3$. On the other hand, in the case of PLC-δ$_1$, Ins(1,4,5,6)P$_4$ was a weaker ligand than Ins(1,4,5)P$_3$ but was equipotent to Ins(1,3,4,5)P$_4$. The order of potency was Ins(1,4,5)P$_3$ > Ins(1,4,5,6)P$_4$ = Ins(1,3,4,5)P$_4$ > Ins(1,3,4,6)P$_4$ = Ins(3,4,5,6)P$_4$ > Ins(1,3,4)P$_3$. These results indicated that p130 was an Ins(1,4,5)P$_3$/Ins(1,4,5,6)P$_4$ binding protein, whereas PLC-δ$_1$ was specific to Ins(1,4,5)P$_3$ (17,20).

PLC Activity. Since the 85 kDa protein was found to be PLC-δ$_1$, each fraction obtained from gel filtration chromatography was assayed for PLC activity using [^{3}H]PIP$_2$ as a substrate. As shown in Figure 3, the second peak with the binding activity showed PLC activity, whereas no activity was found in the first peak, confirming that the 85 kDa-protein was indeed PLC-δ$_1$.

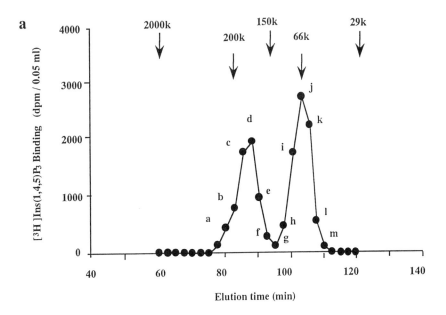

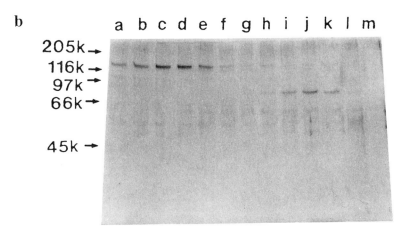

Figure 1. Gel-filtration Chromatography of Active Fractions Eluted from the Ins(1,4,5)P$_3$ Affinity Column (Adapted with permission from ref. 17. Copyright 1992 The American Society for Biochemistry and Molecular Biology).

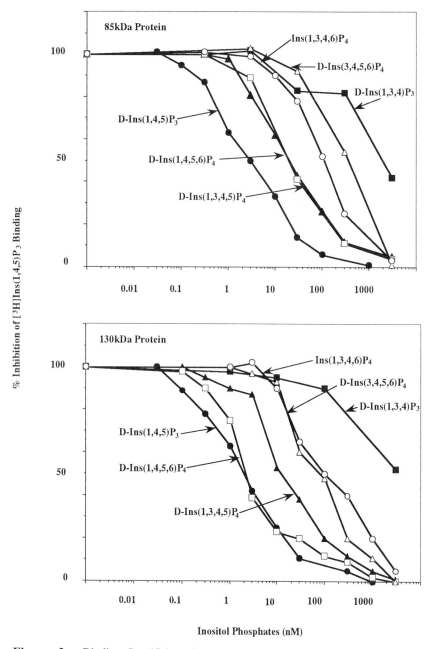

Figure 2. Binding Specificity of 130 kDa and 85 kDa Proteins to Inositol Phosphates (Adapted with permission from ref.17. Copyright 1992 The American Society for Biochemistry and Molecular Biology).

New Ins(1,4,5)P₃ Binding Proteins were Also Purified from Membrane Fraction of Rat Brain. We also isolated two [³H]Ins(1,4,5)P₃ binding proteins from the membrane fraction of rat brain extracted with 1 % Triton X-100, having the same molecular masses of 130 or 85 kDa on SDS-polyacrylamide gel electrophoresis (20). Partial amino acid sequence determination of these proteins revealed that the 130 kDa protein was a novel protein and 85 kDa protein was PLC-δ₁. Antibodies against p130 of a brain cytosol cross-reacted to the membranous 130 kDa protein. Specificity of the binding of membranous 130 kDa protein was the same as that of cytosolic p130. These results indicated that new Ins(1,4,5)P₃ binding proteins were present in both the cytosol and membrane fractions of rat brain.

Molecular Cloning of p130

Isolation of cDNA encoding p130. In order to know what p130 is, we isolated the cDNA clones (pcMT 1 to 3) encoding p130 from an enriched rat brain cDNA library (21). The restriction map was drawn from a series of digestion experiments using various kind of restriction enzymes, and all regions of pcMT3 with the longest cDNA insert were sequenced by the dideoxynucleotide chain termination method. The determined sequence (ACC #D45920) was 5,233 nucleotides which contained a single open reading frame, beginning at nucleotide 467 and ending at 3,754. The protein, therefore, was predicted to comprise 1,096 amino acids and to have a molecular weight of 122.8 kDa, identical to the apparent size as assessed by SDS polyacrylamide gel electrophoresis (17,20). The amino acid sequences of three lysylendopeptidase-cleaved peptides from the purified protein were identified in the deduced amino acid sequence (P-22 (NTETFXNNGLADQICEDXXF), positions 172-191; P-14 (XPLXFMEGNQNTPXF), 228-242 and P-13 (AIESFAXNIXV), 1024-1034), indicating that the clone we obtained was genuine for encoding p130. The hydropathy profile of deduced protein sequence did not show any significant hydrophobic segments shared with other transmembrane proteins. The determined nucleotide and amino acid sequences were subjected to the homology search, using the EMBL, GenBank, PIR and SWISS-PROT databases. p130 had an amino acid sequence (95-850) that was 38.2 % identical to that of PLC-δ₁ (1-756). PLC-δ₁ carries three known domains, the pleckstrin homology (PH) domain and the putative catalytic X and Y domains in the sequence (19,22). When sequence homology was examined with special reference to these domains, it was found that a PH domain and catalytic X and Y domains were localized in regions 110-222, 377-544, and 585-844 of p130 amino acid sequences with 35.2 %, 48.2 % and 45.8 % scores, respectively (Figure 4).

Expression of p130 in COS I Cells. The expression plasmid pcMT31 subcloned into pSG 5 vector, carrying the entire protein-coding sequence, was constructed and transfected into COS-1 cells. The vector alone was also transfected into COS-1 cells as a control. Transient expression of the p130 was detected by Western blotting analysis probed with anti-p130 antibodies (2F9) (20). A single band with an apparent molecular size of 130 kDa was recognized by 2F9 antibody in the extract from COS-1 cells transfected with pcMT31, but not in the extract with the vector alone. The positive band with the same molecular size was observed in a cytosolic fraction from rat brain (21).

The expressed protein in the cytosol fraction of COS-1 cells was examined for the ability to bind Ins(1,4,5)P₃. As shown in Figure 5a the cytosol fraction of COS-1 cells transfected with pcMT31 exhibited specific binding of [³H]Ins(1,4,5)P₃, whereas no significant [³H]Ins(1,4,5)P₃ binding was observed in the same amount of control cytosol. Binding specificity to recombinant p130 was examined using D-enantiomers

62

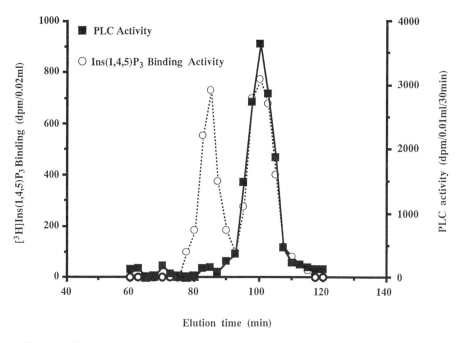

Figure 3. PLC Activity of Fractions Obtained from Gel Filtration Chromatography (Reproduced with permission from ref. 17. Copyright 1992 The American Society for Biochemistry and Molecular Biology).

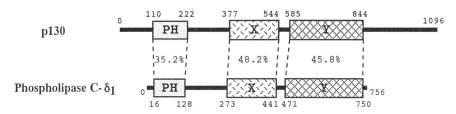

Figure 4. Comparison of Domain Structures of p130 and PLC-δ_1 (Reproduced with permission from ref. 21. Copyright 1996 The Biochemical Society and Portland Press).

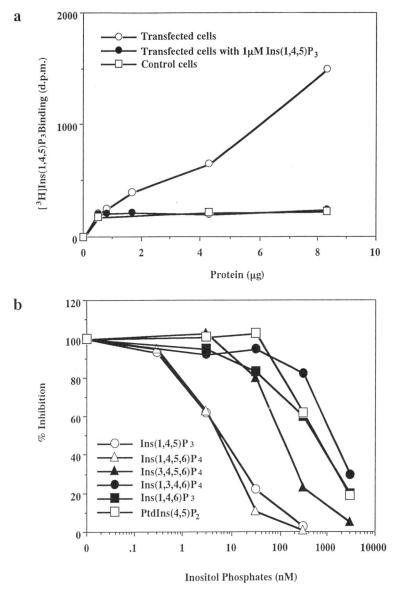

Figure 5. Ins(1,4,5)P$_3$ Binding and Binding Specificity of Recombinant p130 (Reproduced with permission from ref. 21. Copyright 1996 The Biochemical Society and Portland Press).

of several inositol polyphosphates and PIP$_2$ as inhibitors (Figure 5b). The binding specificity of the expressed protein was virtually identical to that of p130 isolated from brain (*17,20*); the binding to Ins(1,4,5,6)P$_4$ was observed with the same affinity as that to Ins(1,4,5)P$_3$.

Since p130 was found to be similar to PLC-δ_1, p130 is expected to have an intrinsic PLC activity; however, p130 isolated from rat brain did not show any PLC activity to PIP$_2$ (Figure 3). Recombinant p130 was also assayed for PLC activity. The cell extracts were assayed for PLC activity at free Ca^{2+} concentration of 10 μM and at pH of 7.2, using [^{3}H]PI(4,5)P$_2$ at 50 μM (containing 22,000 dpm [^{3}H]PIP$_2$/tube) as a substrate (*21*). The overexpressed- and control-cell extracts exhibited intrinsic PLC activities of 2,215±105, and 2,492±185 dpm [^{3}H]Ins(1,4,5)P$_3$ formed/0.5 mg protein of extract for 5 min (n=7), respectively. The enzyme assays were also performed at different free Ca^{2+} concentrations ranging from 0.1 μM to 100 μM; however, there was no difference in the activities between transfected-cell extract and control-cell extract. In addition, [^{3}H]PI at 100 μM (containing 44,000 dpm [^{3}H]PI/a tube) was also employed as a substrate for the assay of PLC activity at various free Ca^{2+} concentrations, but we could not observe any big differences between the two extracts. These results indicated that recombinant p130 had no intrinsic PLC activity to PIP$_2$ and PI.

These results clearly indicate that the p130 is a new Ins(1,4,5)P$_3$/Ins(1,4,5,6)P$_4$ binding protein homologous to PLC-δ_1, but has no detectable PLC catalytic activity under ordinary conditions.

Determination of Ins(1,4,5)P$_3$ Binding Site of p130 and PLC-δ_1

Localization of the Ins(1,4,5)P$_3$/Ins(1,4,5,6)P$_4$ Binding Site of p130.
To localize the region responsible for Ins(1,4,5)P$_3$ binding in p130, we prepared several groups of COS-1 cells, each of which was transfected with cDNA encoding either the full-length p130 (construct I) or its mutants truncated at various region (constructs II to VI). Figure 6 shows a schematic representation of these constructs, together with the binding results. Each recombinant product was examined for proper expression by an Western blotting analysis using two monoclonal antibodies against p130 (designated 4C1 and 2F9) (*23*). In every case, the apparent molecular mass of the band recognized by the antibodies coincided with the expected molecular sizes: 130-kDa (construct I), 105-kDa (construct II), 93-kDa (construct III), 80-kDa (construct IV), 93-kDa (construct V) and 108-kDa (construct VI).

Cytosol fractions were prepared from each group of COS-1 cells and assayed for the ability to bind [^{3}H]Ins(1,4,5)P$_3$. Cytosol fraction prepared from non-transfected cells, showed no [^{3}H]Ins(1,4,5)P$_3$ binding activity. On the other hand, construct I (the full-length p130) exhibited the highest activity of [^{3}H]Ins(1,4,5)P$_3$ binding (Figure 6). Removal of a portion at the amino-terminus (construct III, Δ1-232) completely eliminated [^{3}H]Ins(1,4,5)P$_3$ binding activity. Even when the concentration of [^{3}H]Ins(1,4,5)P$_3$ was increased 10-fold up to 13 nM, no binding to construct III was observed (data not shown), whereas construct II (Δ1-115) or construct IV (Δ1-94) showed the binding activity of [^{3}H]Ins(1,4,5)P$_3$, albeit the reduction of the values. Deletion of portions at the carboxyl-terminus, as observed with construct IV (Δ727-1096) and construct V (Δ851-1096), did not reduce the binding activity to zero. These results indicated that the residues 116-232 of p130, corresponding to the PH domain were essential for the binding of Ins(1,4,5)P$_3$. To gain conclusive evidence, recombinant PH domain (95-232) of p130 was expressed in bacteria as a His-tagged protein and then examined for [^{3}H]Ins(1,4,5)P$_3$ binding activity. Recombinant PH domain of p130 could bind [^{3}H]Ins(1,4,5)P$_3$, which was dose-dependently displaced by unlabeled Ins(1,4,5)P$_3$ (Figure 7). Scatchard analyses

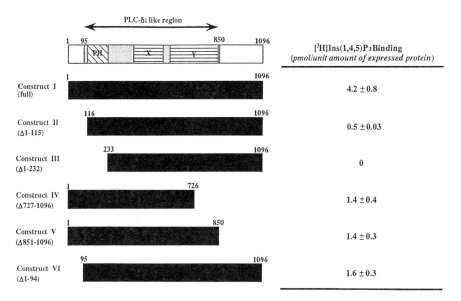

Figure 6. Schematic Representation of Construct I - VI and [³H]Ins(1,4,5)P₃ Binding Activity (Adapted with permission from ref. 23. Copyright 1996 The Biochemical Society and Portland Press).

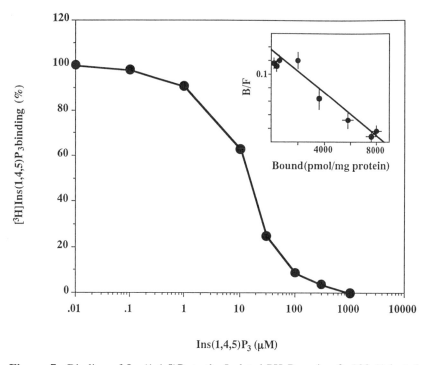

Figure 7. Binding of Ins(1,4,5)P$_3$ to the Isolated PH Domain of p130 (Adapted with permission from ref. 23. Copyright 1996 The Biochemical Society and Portland Press).

of three independent determinations revealed that the average K_d and B_{max} were 12.2 μM and 8350 pmol/mg protein, respectively. These data clearly indicated that the PH domain of p130 is the sole site responsible for binding of Ins(1,4,5)P$_3$. The ligand affinity, however, was about 3,500-fold lower than that seen with the whole molecule (compare Figure 2 with Figure 6).

Determination of the Ins(1,4,5)P$_3$ Binding Site of PLC-δ_1 and Effect of Ins(1,4,5)P$_3$ on PLC Activity. We, in collaboration with Dr. H. Yagisawa, Himeji Institute of Technology of Japan, determined the region involved in Ins(1,4,5)P$_3$ binding in PLC-δ_1 as follows (24,25). Recombinant PLC-δ_1 was expressed as a GST fusion protein (110 kDa) in *E. coli*, and cleaved by thrombin to produce PLC-δ_1 without GST. The procedure yielded a 76 kDa protein in addition to an expected 85 kDa protein. The former sample showed no binding activity of [³H]Ins(1,4,5)P$_3$, while the latter did. Sequence analysis from the N-terminus of each sample clarified that the 76 kDa protein lacks the first 60 amino acids of PLC-δ_1, indicating that the first 60 amino acids of PLC-δ_1 is primarily responsible for the binding of Ins(1,4,5)P$_3$. Of the first 60 amino acids, the sequence from 30 to 43 contains many basic amino acids which would be involved in recognizing anionic phosphates of Ins(1,4,5)P$_3$. Synthetic peptide comprising sequence 30-43 of PLC-δ_1, but not scramble peptide with the same amino acid composition showed the binding activity of [³H]Ins(1,4,5)P$_3$. Furthermore, polyclonal antibody raised against synthetic peptide PLC-δ_1(30-43) inhibited binding of [³H]Ins(1,4,5)P$_3$ to full-length PLC-δ_1. These results indicate that PLC-δ_1 (30-43) is primarily responsible for binding of Ins(1,4,5)P$_3$, and is mapped in N-terminal of the PH domain of PLC-δ_1 (24,25).

The effect of Ins(1,4,5)P$_3$ on PLC activity of purified PLC-δ_1 from rat brain was examined to explore the possibility of allosteric interaction since the site for binding Ins(1,4,5)P$_3$ (PH domian) is different from that for catalysis (17). Figure 8a shows the effect of various concentrations of Ins(1,4,5)P$_3$ on PLC activity at the substrate concentration of 50 μM. The activity was gradually inhibited by increasing the Ins(1,4,5)P$_3$ concentration up to 1 μM, and steeply inhibited when the concentration was greater than 1 μM. Such steep inhibition was observed also when the substrate concentration was varied from 12.5 to 100 μM in the same dependence, thereby confirming that the inhibition was not due to the competition with Ins(1,4,5)P$_3$ and PI(4,5)P$_2$ for the active site of the enzyme. Figure 8b shows the inhibition of 3 μM Ins(1,4,5)P$_3$ on PLC activity at various substrate concentrations. Three micromolar Ins(1,4,5)P$_3$ inhibited the PLC activity ranging from 12.5 to 133 μM, and the Lineweaver-Burk plot analysis (*inset* of Figure 8b) revealed that the inhibition was non-competitive, suggesting that Ins(1,4,5)P$_3$ did not compete with the substrate [PI(4,5)P$_2$] for the active site of the enzyme. The PLC-δ_1 PH domain is thought to play its role by binding to PI(4,5)P$_2$ present in the liposome (in *in vitro* experiments) or the plasma membrane (in intact cell level experiments) for processive catalysis (26,27). Therefore, the inhibition observed here might be caused by the competition between the enzyme and Ins(1,4,5)P$_3$ for binding to liposomes.

Binding Characterization of Several PH Domains

Binding Specificity of the PH Domain of p130. Ligand specificity of p130 was studied by determining IC$_{50}$ values for displacement of [³H]Ins(1,4,5)P$_3$ with a number of inositol phosphate derivatives (28) (Table II). COS-1 cytosol containing construct V (see Figure 6) was particularly useful for these studies because of the following reasons: it afforded the highest binding activity per unit of cytosolic protein, and exhibited the same range of affinity to Ins(1,4,5)P$_3$ as the native protein (23).

68

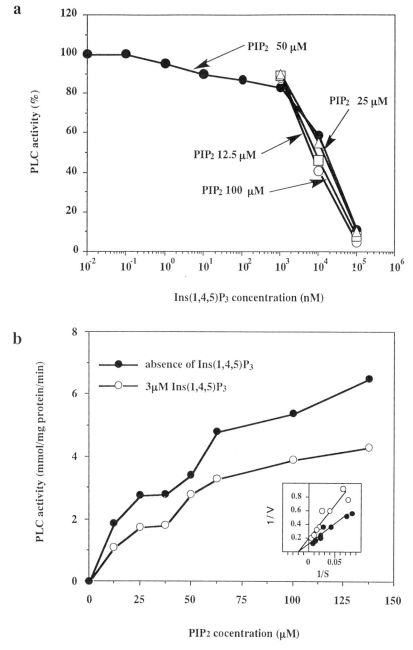

Figure 8. Effect of Ins(1,4,5)P$_3$ on PLC Activity for 85 kDa Protein (Adapted with permission from ref. 17. Copyright 1992 The American Society for Biochemistry and Molecular Biology).

Table II. Binding Specificity of Construct V and Purified p130

Compound	IC_{50}	
	Construct V (nM)	Purified p130 (nM)
Ins(1,4,5)P$_3$	4.3	4.5
Ins(1,4)P$_2$	>3000	nd
Ins(4,5)P$_2$	2000	nd
GroPIns(4,5)P$_2$	100	nd
DL-2-Deoxy-Ins(1,4,5)P$_3$	6.7	6.3
3-Deoxy-Ins(1,4,5)P$_3$	21.6	nd
2,3-Dideoxy-Ins(1,4,5)P$_3$	22.0	nd
#209[1]	3.9	4.3
#309[2]	20.7	nd
L-Ins(1,4,5)P$_3$	>3000	nd
DL-Ins(2,4,5)P$_3$	92	42
Ins(1,4,6)P$_3$	1000	nd
Ins(1,3,6)P$_3$	600	nd
Ins(1,3,4)P$_3$	>3000	>1000
Ins(1,5,6)P$_3$	2000	nd
DL-Ins(1,2,4,5)P$_4$	7.3	nd
Ins(1,3,4,5)P$_4$	11.3	21
Ins(1,4,5,6)P$_4$	4.1	3.1
Ins(3,4,5,6)P$_4$	110	84

[1] For detail structure see ref. 13
[2] For detail structure see ref. 51
SOURCE: Adapted with permission from ref. 23. Copyright 1996

We first examined the contributions made by the phosphate groups beginning with the 5-phosphate on inositol ring. As shown in Table II, at concentrations up to 3 μM, Ins(1,4)P$_2$ was unable to displace 1.3 nM [^{3}H]Ins(1,4,5)P$_3$, and Ins(1,4,6)P$_3$ was 250-fold weaker than Ins(1,4,5,6)P$_4$. The 4-phosphate is also an important determinant of specificity because Ins(1,5,6)P$_3$ was nearly 500-fold weaker than Ins(1,4,5,6)P$_4$. Ins(4,5)P$_2$ was about 500-fold less effective than Ins(1,4,5)P$_3$. It indicated that the 1-phosphate also makes an important contribution. The fact that the affinity for DL-Ins(2,4,5)P$_3$ was intermediate between that of Ins(4,5)P$_2$ and Ins(1,4,5)P$_3$ indicates that the phosphate at the 2-position can only partly substitute for the 1-phosphate. The intermediate affinity for DL-Ins(2,4,5)P$_3$ has also been observed with the native protein (Table II). The potency of GroPIns(4,5)P$_2$ was about 25-fold less than Ins(1,4,5)P$_3$. These results indicated that either a fully ionized 1-phosphate is important, and/or the glycerol moiety provides some steric hindrance. This result was unprecedented because for all the other PH domains examined to date, a free monoester 1-phosphate was not critical, so they bound Ins(1,4,5)P$_3$ with virtually the same affinity as GroPIns(4,5)P$_2$ (29-33). The fact that both Ins(1,4,5,6)P$_4$ and Ins(1,4,5)P$_3$ bound with equal affinity could be because neither the 6-phosphate nor the 6-OH have much impact upon specificity. Alternatively, it is possible that these two groups each interact with different amino-acids, perhaps by hydrogen-bonding, such that the absence of one interaction is quantitatively compensated by the presence of the other. Further data do indicate that this idea is feasible. For example, it appears that the 3-OH contributes to ligand specificity, presumably due to hydrogen bonding interaction, since the 3-deoxy analog displays 5-fold reduced affinity (Table II). This proposed contribution of the 3-OH is also supported by the 5-fold reduction in affinity brought about when it was substituted by a bulky aminobenzoyl group. However, when the 3-OH was replaced by another relatively bulky substituent, a phosphate group, there was only a 2.5-fold reduction in affinity. This result suggests that the 3-phosphate has its own interactions with p130 that can partly compensate for the loss of the 3-OH. Arguably the strongest argument that the 3- and/or 6-phosphates do interact with amino acids in p130 comes from our new observation that the affinity for Ins(3,4,5,6)P$_4$ was 20-fold higher than the affinity for Ins(4,5)P$_2$.

The contribution of the 2-OH was also studied. The analogue with a bulky aminobenzoyl substituent at the 2-position was similar in potency to Ins(1,4,5)P$_3$ (Table II). We also found that 2-deoxy-DL-Ins(1,4,5)P$_3$ and DL-Ins(1,2,4,5)P$_4$ both had less than a 2-fold lower affinity than Ins(1,4,5)P$_3$. Note, that the contribution from the L-enantiomers in these last two racemic mixtures was unlikely (since L-Ins(1,4,5)P$_3$ was not a significant ligand). Therefore, it seems that the 2-OH makes no contribution to ligand specificity. The latter conclusion is also supported by the observation that 3-deoxy-Ins(1,4,5)P$_3$ had a similar affinity to 2,3-dideoxy-Ins(1,4,5)P$_3$. The poor affinity for both Ins(1,4,6)P$_3$ and Ins(1,3,6)P$_3$ was informative, in part because when these structures are inverted and rotated, they both satisfy the requirement for correctly oriented 1-, 4- and 5-phosphates of Ins(1,4,5)P$_3$. However, in each case, the orientation of the OH groups fails to match that of Ins(1,4,5)P$_3$. Thus, although the OH groups of Ins(1,4,5)P$_3$ do not greatly contribute to specificity per se, if their orientation is switched between the axial and equatorial configuration, a steric constraint is probably introduced that substantially reduces ligand affinity.

Binding Specificity of Four Different PH Domains for Inositol Phosphate and Inositol Lipid Derivatives. The binding of PLC-δ_1 PH domain was relatively specific to Ins(1,4,5)P$_3$, while that of p130PH domain was to Ins(1,4,5)P$_3$ and Ins(1,4,5,6)P$_4$ with a similar affinity. On the basis of these observations, we have proposed that some of the PH domains derived from various proteins may specifically bind different inositol phosphates.

In order to gain the supportive evidence for the proposal, several PH domains derived from different proteins were assayed for the ability and specificity of binding inositol phosphates. The PH domains selected for this study have been derived from unrelated proteins with distinct cellular functions, N-terminal PH domain of pleckstrin (PleNPH), PH domain of the γ-subtype of RAC protein kinase (also known as Akt or PKB) (RACγPH), PH domain of δ-subtype of diacylglycerol (DGKδPH). p130PH was also included for comparison. Figure 9 shows the sequence alignments of the PH domains of proteins used here. The alignment is based on secondary structure elements from known three-dimensional structures of pleckstrin, β-spectrin, dynamin and PLC-δ_1. A conserved polypeptide fold allows prediction that the PH domains used in the present study form a similar three-dimensional structure: two orthogonal anti-parallel β-sheets are composed of three and four strands and a single α-helix caps one end of the β-barrel (29,31,34-39). The capability and specificity for binding inositol phosphate(s), however, cannot be predicted only from their sequence alignments.

All four PH domains could bind [^{3}H]Ins(1,4,5)P$_3$, which were displaced by unlabeled Ins(1,4,5)P3. Displacements experiments of [^{3}H]Ins(1,4,5)P$_3$ bound to each PH domain by several inositol phosphates have been performed, and the results using a inositol phosphates with the wide range of IC$_{50}$ values are shown in Table III. In the case of PleNPH, L-Ins(1,4,5)P$_3$ was as efficacious as its D-enantiomer, and the regioisomer Ins(1,3,4)P$_3$ had a similar potency. All examined InsP$_4$ regioisomers (Ins(1,3,4,5)P$_5$, Ins(1,4,5,6)P$_4$ and Ins(3,4,5,6)P$_4$), despite different dispositions of phosphates on the inositol ring, were more potent than the InsP$_3$ series, and each of InsP$_4$s had a similar affinity. Moreover, Ins(1,3,4,5,6)P$_5$ was even more potent than the InsP$_4$ series. All these data suggested non specific binding to inositol phosphates. A very limited specificity, however, could be due to a sterical constraint caused by the presence of the 2-phosphate because Ins(1,2,4,5,6)P$_5$ and Ins(1,2,3,4,5,6)P$_6$ were about 5-times less potent than Ins(1,3,4,5,6)P$_5$. In the case of RACγPH, Ins(1,4,5,6)P$_4$ and Ins(1,3,4,5,6)P$_5$ were most efficacious at displacing [^{3}H]Ins(1,4,5)P$_3$ closely followed by Ins(1,4,5)P$_3$ and InsP$_6$. There was a clear difference in the efficiency between the enantiomer and regioisomer of Ins(1,4,5)P$_3$; the same result was observed for InsP$_4$ and InsP$_5$. In all cases dispositions of four/five phosphates on the inositol ring made a clear difference in the affinity suggesting that these residues contribute towards specific binding of inositol phosphates by RACγPH. DGKδPH had the highest affinity for Ins(1,3,4,5,6)P$_5$ followed by Ins(1,4,5,6)P$_4$. Stereospecific and regioisomer specific binding among a series of InsP$_3$, InsP$_4$ and InsP$_5$ were observed. They, however, were less pronounced than affinities observed in RACγPH. Most striking differences in the affinity among stereo-isomers and regio-isomers of InsP$_3$ was observed in p130PH. L-Ins(1,4,5)P$_3$ or Ins(1,3,4)P$_3$ was 31-fold or >100-fold weaker than Ins(1,4,5)P$_3$, respectively. Ins(1,4,5,6)P$_4$ was as potent as Ins(1,4,5)P$_3$ at displacing [^{3}H]Ins(1,4,5)P$_3$ as shown in above sections.

To compare the affinity between inositol lipids and their polar heads, we used soluble derivatives of PI(4,5)P$_2$ and PI(3,4,5)P$_3$, GroPIns(4,5)P$_2$ and DiBzGroPIns(3,4,5)P$_3$, respectively; this allowed correct determination of inositol lipid concentrations that would otherwise be influenced by micelle formation. This, however, also limited the difference to the presence of only the glycerol moiety without lipid side-chains. In the case of PleNPH, both GroPIns(4,5)P$_2$ and DiBzGroPIns(3,4,5)P$_3$ had almost the same efficiency in the displacement as their respective polar head moieties, inositol tris- and tetrakis-phosphates (Table III, bottom part). Thus, the glycerol moieties did not seem to impose a sterical constraint for the

72

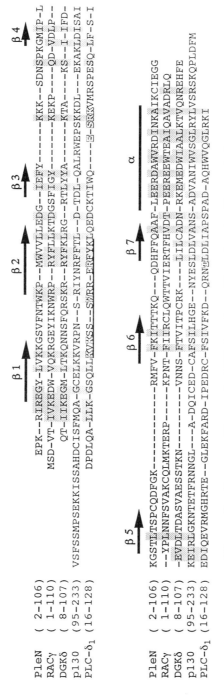

β 1

β 2

β 3

β 4

```
PleN    ( 2-106)    EPK--RIREGY-IVKKGSVFNTWKP--MWVVLLEDG--IEFY------KKK--SDNSPKGMIP-L
RACγ    ( 1-110)    MSD-VT-IVKEDW-VQKRGEYIKNWRP--RYFLLKTDGSFIGY------KEKP----QD-VDLP--
DGKδ    ( 8-107)    QT-IIKEGM-LTKQNNSFQRSKR--RYFKLRG--RTLYYA------KTA----KS--I-IFD-
p130    (95-233)    VSFSSMPSEKKISSAHDCISFMQA-GCELKKVRPN--S-RIYNRFFTL--D-TDL-QALRWEPSKKDL--EKAKLDISAI
PLC-δ₁  (16-128)    DPDLQA-LLK-GSQLLKVKSS--SWRR-ERFYKLQEDCKTIWQ-----E-SRKVMRSPESQ-LF-S-I
```

β 5

β 6

β 7

α

```
PleN    ( 2-106)    KGSTLTSPCQDFGK---------RMFV-FKITTKQ--QDHFFQAAF-LEERDAWVRDINKAIKCIEGG
RACγ    ( 1-110)    --YPLNNFSVAKCQLMKTERP------KPNT-FIIRCLQWTTVIERTFHVDT-PEEREEWTEAIQAVADRLQ
DGKδ    ( 8-107)    -EVDLTDASVAESSTKN-------VNNS-FTVITPCRK----LILCADN-RKEMEDWIAALKTVQNREHFE
p130    (95-233)    KEIRLGKNTETFRNNGL---A-DQICED-CAFSILHGE-NYESLDIVANS-ADVANIWVSGLRYLVSRSKQPLDFM
PLC-δ₁  (16-128)    EDIQEVRMGHRTE--GLEKFARD-IPEDRC-FSIVFKD--QRNtLDLIAPSPAD-AQHWVQGLRKI
```

Figure 9. Sequence Alignment of PH domains. (Reproduced with permission from ref. 28. Copyright 1997 Elsevir).

Table III Summary of data:IC_{50} values of inositol phosphates in the displacement of $[^3H]Ins(1,4,5)P_3$

	PleNPH	RACγPH	DGKδPH	p130PH
	(µM)	(µM)	(µM)	(µM)
Ins(1,4,5)P_3	3.6 (12)*	0.06 (1.7)	0.06 (5)	5.6 (1)
L-Ins(1,4,5)P_3	3.6 (12)	0.4 (11)	0.15 (12)	250 (45)
Ins(1,3,4)P_3	4.5 (15)	0.6 (17)	0.16 (13)	>1,000 (>180)
Ins(1,3,4,5)P_4	1.0 (3.3)	0.3 (8.3)	0.035 (2.9)	30 (5.4)
Ins(1,4,5,6)P_4	1.2 (4)	0.036 (1)	0.02 (1.7)	6.0 (1.1)
Ins(3,4,5,6)P_4	1.2 (4)	0.14 (3.9)	0.034 (2.8)	nd**
Ins(1,3,4,5,6)P_5	0.3 (1)	0.036 (1)	0.012 (1)	36 (6.4)
Ins(1,2,4,5,6)P_5	1.6 (5.3)	0.14 (3.9)	0.06 (5)	nd
Ins(1,2,3,4,5,6)P_6	1.8 (6)	0.06 (1.7)	0.055 (4.6)	38 (6.8)
GroPIns(4,5)P_2	7.0 (23)	2.5 (69)	0.6 (50)	45 (8)
DiBzGroPIns(3,4,5)P_3	1.1 3.7	0.45 (12)	0.05 (4.2)	22 (3.9)
GroPIns(4,5)P_2 /Ins(1,4,5)P_3	7/3.6 (1.9)	2.5/0.06 (42)	0.6/0.06 (10)	45/5.6 (8)
DiBzGroPIns(3,4,5)P_3 /Ins(1,3,4,5)P_4	1.1/1.0 (1.1)	0.45/0.3 (1.5)	0.05/0.035 (1.4)	22/30 (0.7)

*Number in *parenthesis* represents the value of IC_{50} relative to that seen for inositol phosphate which was the most potent.
** Not done.
Source: Reprinted with permission from ref. 28. Copyright 1997.

interaction, and therefore inositol lipids could have the same affinity as the corresponding inositol phosphate for binding to PleNPH. In other three cases, DiBzGroPIns(3,4,5)P_3 had similar affinity as Ins(1,3,4,5)P_4, while GroPIns(4,5)P_2 was weaker than Ins(1,4,5)P_3 by 8- to 42-fold. This difference could be caused by either a steric hindrance due to the glycerol moiety and/or requirement for the fully ionized 1-phosphate. An explanation for why such a difference is not seen between DiBzGroPIns(3,4,5)P_3 and Ins(1,3,4,5)P_4 could be that the absence of the 1-phosphate may be compensated by interactions of the 3-phosphate with the PH domains.

Direct comparison of four different PH domains described above showed that degree of specificity and identity of the preferred ligand varies, although they all could bind inositol phosphates. PleNPH demonstrated only very limited specificity, and it mainly bound inositol phosphates according to the number of phosphates on the inositol ring, *i.e.* more phosphate groups correlated with stronger binding. Thus, the negative charges of tris-, tetrakis- or pentakis-phosphates on the inositol ring appeared to provide interactions for binding. It indicated that binding was simply due to electrostatic interactions between negative charges of the phosphates and positive charges of basic amino acids. On the contrary, RACγPH and DGKδPH preferentially bound Ins(1,3,4,5,6)P_5 and to a slightly less extent Ins(1,4,5,6)P_4, whereas p130PH had an equal preference for Ins(1,4,5)P_3 and Ins(1,4,5,6)P_4. These three types of PH domains exhibited a clear discrimination in the recognition among regioisomers and enantiomers of InsP$_3$, InsP$_4$ and InsP$_5$. Analysis of these PH domains also provided insight into major determinants of their specificity. As with other PH domains characterized to date (*33,34,38,40-42*), the 4,5-vicinal phosphate pair was an essential determinant of ligand specificity while the 1-phosphate increased the affinity. This conclusion was supported by the findings that among inositol-bisphosphates only Ins(4,5)P_2 was recognized and Ins(1,4,5)P_3 was much more potent than Ins(4,5)P_2. Only the 3-phosphate positively contributed to the recognition by providing arms for salt-bridge formations. As shown in Table III, Ins(1,3,4,5)P_4 was a slightly stronger ligand than Ins(1,4,5)P_3 and in a similar way Ins(1,3,4,5,6)P_5 was more potent than Ins(1,4,5,6)P_4. Unlike DGKδPH, in cases of RACγPH and p130PH, the 3-phosphate appears to provide a structural constraint. This is illustrated by weaker binding of Ins(1,3,4,5)P_4 in comparison with Ins(1,4,5)P_3. Observation that Ins(1,3,4,5,6)P_5 was as potent as Ins(1,4,5,6)P_4 could be explained by compensation of the 3-phosphate charge with the 6-phosphate. Presence of the 6-phosphate seemed to be important in cases of RACγPH and DGKδPH since it was found that Ins(1,4,5,6)P_4 or Ins(1,3,4,5,6)P_5 was more potent than Ins(1,4,5)P_3 or Ins(1,3,4,5)P_4, respectively. Binding to p130PH, however, did not seem to be influenced by the 6-phosphate, since Ins(1,4,5)P_3 and Ins(1,4,5,6)P_4, as well as Ins(1,3,4,5)P_4 and Ins(1,3,4,5,6)P_5 bound to p130PH with equal affinity. Disposition of the 2-phosphate seemed to provide a steric hindrance for inositol phosphate binding to RACγPH, DGKδPH and PleNPH but had little effect on binding to p130PH, as observed with a whole molecule (Table II).

All these results suggested that detailed structures of the putative pockets for accommodating inositol phosphates were different, even if the overall three-dimensional structures of the PH domains are conserved (*29,31,34-39*). Both steric constraints that exclude additional phosphates and presence of basic amino acid(s) interacting with their negative charges, could contribute towards the binding specificity observed.

Ligand Binding Properties of The PH Domains and Their Physiological Significance. Together with specificity, determination of the PH domain affinities for their inositol phosphate/inositol lipid ligands provides a basis for consideration of their interactions *in vivo*. The IC_{50} values (the lowest was about 6 μM for p130PH, while those for PleNPH, RACγPH and DGKδPH were 0.3 μM, 0.036 μM and 0.012 μM, respectively) were within micromolar and submicromolar range. Surprisingly, the lowest affinity was determined for highly specific p130PH, and in comparison with the intact protein this was a substantial reduction. It is likely that folding and stability of the PH domain might be affected when proteins are expressed in isolation rather than as a part of the multidomain protein. Therefore, affinities measured for isolated PH domains could be lower than their affinities in the context of another intact protein. This, however, did not influence changes in the specificity, as described above (compare Table III with Table II).

The determination of physiologically relevant ligands does not depend only on relative binding affinities, but also on relative intracellular abundance of a ligand. Although free concentrations of inositol lipids are difficult to evaluate, measurements of their total intracellular concentrations have been reported. Total cellular levels of $PI(4,5)P_2$ and $PI(3,4,5)P_3$ have been estimated to be 30-160 μM and less than 0.1 μM respectively (*43,44*), but the local concentrations are likely to be higher. The concentration of $PI(4,5)P_2$ in the inner leaflet of neutrophils is estimated to be 5 mM decreasing to 3.5 mM upon chemotactic peptide-stimulation (*44*). The local concentration of $PI(3,4,5)P_3$ seems to be much lower, 5 μM in resting cells and 200 μM after stimulation (*44*). These estimates suggest that the affinity for this ligand should be considerably higher than for $PI(4,5)P_2$, in order to specifically recognize $PI(3,4,5)P_3$. This has not been observed with any of the PH domains tested with their soluble analogues or headgroups.

Abundance of different inositol phosphates also vary (1-100 μM) and is influenced further by agonist stimulation and differences in half-life. For example, inositol 5-phosphatase acting on $Ins(1,4,5)P_3/Ins(1,3,4,5)P_4$ is so active that their half-lives are within 10 sec, whereas those of $Ins(1,3,4)P_3/Ins(1,3,4,6)P_4$, whose concentration increases from 1-4 μM to 10-60 μM upon cell-stimulation, is about 10 min (*45*). Nonetheless, the PH domains of pleckstrin, RAC-PK and DGK showed preference and high affinity for $Ins(1,3,4,5,6)P_5$, and this inositol phosphate (as well as $Ins(1,2,3,4,5,6)P_6$) is assumed to be most abundant in cells (estimated to be 15-100 μM, see *Refs. 45,46* for a review). The function of this binding could be to inhibit interactions of the PH domains with inositol lipids affecting cellular localization of the PH domain-containing protein, or could directly influence protein function by changing the activity of the protein.

A role for inositol phosphates in determining subcellular localization by competing with other molecules binding to membrane inositol lipids has been previously discussed for PLC-δ_1 PH domain (*17,26,41*). Based on estimates of total inner-leaflet concentrations of $PI(4,5)P_2$, it is possible that concentrations of free $PI(4,5)P_2$, available for interactions with proteins, are also significantly higher than that of $Ins(1,4,5)P_3$ or other inositol phosphates. In this case, the ability to compete results from higher affinity for inositol phosphates.

There is also some evidence that binding of inositol lipids/inositol phosphates may not be only implicated in the regulation of membrane attachment, but could also directly influence the function of a PH domain-containing protein. For example, binding of $PI(4,5)P_2$ to dynamin results in changes of its GTPase activity (*40,47*). In case of RAC-PK (*48-50*), kinase activity of this protein is stimulated in the presence of the PH-domain ligand.

These considerations strongly suggest that inositol phosphates and/or inositol lipids might be common ligands for the PH domains and therefore, the function of proteins bearing PH domains might be modified through either the direct binding or the recruitment to the plasma membrane. Therefore inositol phosphates/inositol lipids might be involved in many more aspect of cellular function than originally thought, because more than 90 proteins to date are known to have PH domain. However, we should be careful not to conclude which ligand is physiologically relevant for the PH domain from their specificity in *in vitro* experiments, because there are more than 20 types of inositol compounds with varying cellular abundances, which dynamically change in response to cellular stimulation.

Acknowledgments

Authors wish to thank Dr. H. Yagisawa, Himeji Institute of Technology, Hyogo, Japan and Dr. Matilda Katan, Chester Beatty Laboratory, Institute of Cancer Research, London, UK for collaboration. The work was financially supported by a Grant-in-Aid for Scientific Research from the Ministry of Education, Science, Sports and Culture of Japan (to TK and MH), and by the ONO Medical Research Foundation, the Ryoichi Naito Foundation for Medical Research and the Fukuoka Cancer Society (to MH).

References Cited

1. Berridge, M. J; Irvine, R. F. *Nature* **1984,** *312,* 315-321.
2. Supattapone, S.; Worley, P. F.; Baraban, J. M.; Snyder, S. H. *J. Biol. Chem.* **1988,** *263,* 1530-1534.
3. Furuichi, T.; Yoshikawa, S.; Miyawaki, A.; Wada, K.; Maeda, N.; Mikoshiba, K. *Nature* **1989,** *342,* 32-38.
4. Shears, S. B. *Biochem. J.* **1989,** *260,* 313-324.
5. Mignery, G. A.; Südhof, T. C. *EMBO J.* **1990,** *9,* 3893-3898.
6. Miyawaki, A.; Furuichi, T.; Ryou, Y.; Yoshikawa, S.; Nakagawa, T.; Saitoh, T.; Mikoshiba, K. *Proc. Natl. Acad. Sci. USA* **1991,** *88,* 4911-4915.
7. Potter, B. V. L. in *"Trends in Receptor Research"* (Claassen, V. ed.) pp. 185-214, Elsevier Science Publishers B.V. 1993.
8. Billington, D. C. *"The Inositol Phosphates-Chemical synthesis and Biological Significance"*, VCH Publishers, Weinheim, 1993.
9. Wilcox, R. A.; Nahorski, S. R.; Sawer, D. A.; Liu, C.; Potter, B. V. L. *Carbohydr. Res.* **1992,** *234,* 237-246.
10. Ozaki, S.; Watanabe, Y.; Hirata, M.; Kanematsu, T. *Carbohydr. Res.* **1992,** *234,* 189-206.
11. Hirata, M.; Sasaguri, T.; Hamachi, T.; Hashimoto, T.; Kukita, M.; Koga, T. *Nature* **1985,** *317,* 723-725.
12. Irvine, R. F.; Brown, R. O.; Berridge, M. J. *Biochem. J.* **1984,** *221,* 269-272.
13. Hirata, M.; Watanabe, Y.; Ishimatsu, T.; Ikebe, T.; Kimura, Y.; Yamaguchi, K.; Ozaki, S.; Koga, T. *J. Biol. Chem.* **1989,** *264,* 20303-20308.
14. Hirata, M.; Yanaga, F.; Koga, T.; Ogasawara, T.; Watanabe, Y.; Ozaki, S. *J. Biol. Chem.* **1990,** *265,* 8404-8407.
15. Watanabe, Y.; Hirata, M.; Ogasawara, T.; Koga, T.; Ozaki, S. *Bioorg. Med. Chem. Lett.* **1991,** *1,* 399-402.
16. Hirata, M.; Watanabe, Y.; Ishimatsu, T.; Yanaga, F.; Koga, T.; Ozaki, S. *Biochem. Biophys. Res. Commun.* **1990,** *168,* 379-386.
17. Kanematsu, T.; Takeya, H.; Watanabe, Y.; Ozaki, S.; Yoshida, M.; Koga, T.; Iwanaga, S.; Hirata, M. *J. Biol. Chem.* **1992,** *267,* 6518-6525.
18. Hirata, M.; Yoshida, M.; Kanematsu, T.; Takeuchi, H. *Mol. Cel. Biochem.* in press.
19. Suh, P.-G.; Ryu, S. H.; Moon, K. H.; Suh, H. W.; Rhee, S. G. *Cell* **1988,** *54,* 161-169.
20. Yoshida, M.; Kanematsu, T.; Watanabe, Y.; Koga, T.; Ozaki, S.; Iwanaga, S.; Hirata, M. *J. Biochem. (Tokyo)* **1994,** *115,* 973-980.

21. Kanematsu, T.; Misumi, Y.; Watanabe, Y.; Ozaki, S.; Koga, T.; Iwanaga, S.; Ikehara, Y.; Hirata, M. *Biochem. J.* **1996**, *313*, 319-325.
22. Musacchio, A.; Gibson, T.; Rice, P.; Thompson, J.; Saraste, M. *Trends Biochem. Sci.* **1993**, *18*, 343-348.
23. Takeuchi, H.; Kanematsu, T.; Misumi, Y.; Yaakob, H. B.; Yagisawa, H.; Ikehara, Y.; Watanabe, Y.; Tan, Z.; Shears, S. B.; Hirata, M. *Biochem. J.* **1996**, *318*, 561-568.
24. Yagisawa, H.; Hirata, M.; Kanematsu, T.; Watanabe, Y.; Ozaki, S.; Sakuma, K.; Tanaka, H.; Yabuta, N.; Kamata, H.; Hirata, H.; Nojima, H. *J. Biol. Chem.* **1994**, *269*, 20179-20188.
25. Hirata, M.; Kanematsu, T.; Sakuma, K.; Koga, T.; Watanabe, Y.; Ozaki, S.; Yagisawa, H. *Biochem. Biophys. Res. Commun.* **1994**, *205*, 1563-1571.
26. Cifuentes, M. E.; Delaney, T.; Rebecchi, M. J. *J. Biol. Chem.* **1994**, *269*, 1945-1948.
27. Essen, L.-O.; Perisic, O.; Cheung, R.; Katan, M.; Williams, R. L. *Nature* **1996**, *380*, 595-602.
28. Takeuchi, H.; Kanematsu, T.; Misumi, Y.; Sakane, F.; Konishi, H.; Kikkawa, U.; Watanabe, Y.; Katan, M.; Hirata, M. *Biochim. Biophys. Acta* **1997**, *1359*, 275-285.
29. Ferguson, K. M.; Lemmon, M. A.; Schlessinger, J.; Sigler, P. B. *Cell* **1995**, *83*, 1037-1046.
30. Ferguson, K. M.; Lemmon, M. A.; Sigler, P. B.; Schlessinger, J. *Nature Struct. Biol. (London)* **1995**, *2*, 715-718.
31. Hyvönen, M.; Macias, M. J.; Nilges, M.; Oschkinat, H.; Saraste, M.; Wilmanns, M. *EMBO J.* **1995**, *14*, 4676-4685.
32. Lemmon, M. A.; Ferguson, K. M.; O'Brien, R.; Sigler, P. B.; Schlessinger, J., *Proc. Natl. Acad. Sci. USA* **1995**, *92*, 10472-10476.
33. Harlan, J. E.; Yoon, H. S.; Hajduk, P. J.; Fesik, S. W. *Biochemistry* **1995**, *34*, 9859-9864.
34. Ferguson, K. M.; Lemmon, M. A.; Schelessinger, J.; Sigler, P. B. *Cell* **1994**, *79*, 199-209.
35. Yoon, H. S.; Hajduk, P. J.; Petros, A. M.; Olejniczak, E. T.; Meadows, R. P.; Fesik, S. W. *Nature* **1994**, *369*, 672-675.
36. Macia, M. J.; Musacchio, A.; Ponstingl, H.; Nilges, M.; Saraste, M.; Oschkinat, H. *Nature* **1994**, *369*, 675-677.
37. Downing, A. K.; Driscoll, P. C.; Gout, I.; Salim, K.; Zvelebil, M. J.; Waterfield, M. D. *Curr. Biol.* **1994**, *4*, 884-891.
38. Harlan, J. E.; Hajduk, P. J.; Yoon, H. S.; Fesik, S. W. *Nature* **1994**, *371*, 168-170.
39. Fushman, D.; Cahill, S.; Lemmon, M. A.; Schelessinger, J.; Cowburn, D. *Proc. Natl. Acad. Sci. USA* **1995**, *92*, 816-820.
40. Zheng, J.; Cahill, S. M.; Lemmon, M. A.; Fushamn, D.; Schlessinger, J.; Cowburn, D. *J. Mol. Biol.* **1996**, *255*, 14-21.
41. Lemmon, M. A.; Ferguson, K. M.; Schlessinger, J. *Cell* **1996**, *85*, 621-624.
42. Garcia, P.; Gupta, R.; Shah, S.; Morris, A. J.; Rudge, S. A.; Scarlata, S.; Petrova, V.; McLaughlin, S.; Rebecchi, M. *Biochemistry* **1995**, *34*, 16228-16234.
43. Bunce, C. M.; French, P. J.; Allen, P.; Mountford, J. C.; Moor, B.; Greaves, M. F.; Michell, R. H.; Brown, G. *Biochem. J.* **1993**, *289*, 667-673.
44. Stephens, L. R.; Jackson, T. R.; Hawkins, P. T. *Biochim. Biophys. Acta* **1993**, *1179*, 27-75.

45. Shears, S. B. in *"myo-Inositol phosphate, phosphoinositides, and signal transduction"* Subcellular Biochemistry, eds. by Biswas, B. B.; Biswas, S., Plenum Press, New York, 1996, Vol. 26; pp. 187-226.
46. Shears, S. B. Adv. *Second Messenger Phosphorylation Res.* **1992**, *26*, 63-92.
47. Salim, K.; Bottomley, M. J.; Querfurth, E.; Zvelbil, M. J.; Gout, I.; Scaife, R.; Margolis, R. L.; Gigg, R.; Smith, C. I. E.; Driscoll, P. C.; Waterfield, M. D.; Panayotou, G. *EMBO J.* **1996,** *15*, 6241-6250.
48. Franke, T. F.; Kaplan, D. R.; Cantley, L. C.; Toker, A. *Science* **1997,** *275*, 665-668.
49. Klippel, A.; Kavanaugh, W.; Pot, D.; Williams, L. T. *Mol. Cell. Biol.* **1997,** *17*, 338-344.
50. Frech, M.; Andjelkovic, M.; Reddy, K. K.; Falck, J. R.; Hemmings, B. A. *J. Biol. Chem.* **1997,** *272*, 8474-8481.
51. Hirata, M.; Watanabe, Y.; Kanematsu, T.; Ozaki, S.; Koga, T. *Biochim. Biophys. Acta* **1995,** *1244,* 404-410.

MECHANISM AND STRUCTURE OF
PHOSPHATIDYLINOSITOL-SPECIFIC PHOSPHOLIPASES

Chapter 5

Structure and Mechanism of Ca^{2+}-Independent Phosphatidylinositol-Specific Phospholipases C

Dirk W. Heinz[1], Jürgen Wehland[2], and O. Hayes Griffith[3]

[1]Institut für Organische Chemie und Biochemie, Universität Freiburg, D-79104 Freiburg, Germany
[2]Gesellschaft für Biotechnologische Forschung, Mascheroder Weg 1, D-38124 Braunschweig, Germany
[3]Institute of Molecular Biology, Department of Chemistry, University of Oregon, Eugene, OR 97403

The three-dimensional structures of phosphatidylinositol-specific phospholipases C (PI-PLCs) from *Bacillus cereus* and the human pathogen *Listeria monocytogenes* were determined by X-ray crystallography, both in free form and in complex with the substrate-like inhibitor *myo*-inositol. Both enzymes share a very similar distorted $(\beta\alpha)_8$-barrel fold despite a moderate overall sequence identity of 24%. A high structural conservation is found for the active site where *myo*-inositol is recognized in a stereospecific fashion. Two histidine residues that are also conserved between prokaryotic and eukaryotic PI-PLCs act as a general base and a general acid during catalysis, while an arginine residue provides the electrostatic stabilization of the transition state. Based on the present crystal structures and sequence alignments, it is suggested that all Ca^{2+}-independent PI-PLCs known so far adopt the same fold and catalytic mechanism.

PI-PLCs catalyze the cleavage of the *sn*-3-phosphodiester bond in the membrane phospholipid, phosphatidylinositol (PI), to yield the products D-*myo*-inositol-1-phosphate (InsP) and diacylglycerol (DAG) (Figure 1). In higher eukaryots different PI-PLC isozymes (85 to 150 kDa) are key players in signal transduction cascades, controlling numerous important cellular events (*1,2*). These enzymes are strictly Ca^{2+}-dependent and show a clear preference towards the more highly phosphorylated PI-analogues, phosphatidylinositol 4-phosphate (PIP) and phosphatidylinositol 4,5-bisphosphate (PIP$_2$). The eukaryotic parasite *Trypanosoma brucei* also produces a much smaller PI-PLC (~39 kD) that is, however, Ca^{2+}-independent. It seems to be involved in the cleavage of glycosylphosphatidylinositol-anchors (GPI-anchors) of the parasite's variant surface glycoprotein (*3,4*), thereby defeating the host´s humoral immune system. Smaller PI-PLCs are also secreted by a variety of bacteria including the animal and human pathogens *Staphylococcus aureus* (*5*) and *Listeria monocytogenes* (*6,7*), and the insect pathogens *Bacillus thuringiensis* (*8*) and *Bacillus cereus* (*9*). These enzymes (~35 kDa) show sequence identities ranging from 24 to 98%. They are highly specific for PI and GPI as substrate, and do not require Ca^{2+} ions for catalysis. As known virulence factors, the PI-PLCs from *L. monocytogenes* and *S. aureus* are actively involved in the pathogenesis of these species (*5,10*).

Recently, the three-dimensional structures of the PI-PLCs from *B. cereus* (BPI-PLC) (*11*) and *L. monocytogenes* (LPI-PLC) (*12*), as well as a eukaryotic PI-PLC-δ1 (*13*), have been determined by x-ray crystallography. Here, we compare the structures of the bacterial PI-PLCs, both in complex with the inhibitor *myo*-inositol. The structures strongly support the previously proposed catalytic mechanism involving general base / general acid catalysis (*14,15*), that is also substantiated by data from site-directed mutagenesis (*16*). By using multiple sequence alignments, we also suggest that all known Ca^{2+}-independent PI-PLCs adopt a similar structure and catalytic mechanism.

Crystal Structures of Bacterial PI-PLCs

The structure of BPI-PLC solved at 2.6 Å resolution (*11*) (Figure 2) forms a single globular domain that folds as an irregular $(\beta\alpha)_8$-barrel (TIM-barrel) (*17*), where the parallel β-strands form an inner circular and closed barrel with α-helices located on the outside, connecting neighboring β-strands. In contrast to most members of the steadily increasing family of TIM-barrel containing proteins, the PI-PLC structure shows a number of unique differences. Only six instead of eight $(\beta\alpha)_8$-units are found in the barrel, with α-helices lacking between β-strands IV and V, and β-strands V and VI. Interestingly, the central eight-stranded β-barrel is not strictly closed due to the absence of main-chain hydrogen bonding interactions between β-strands V and VI. Another unique feature is the presence of an additional β-strand (Vb) that interacts with β-strand VI in an antiparallel fashion.

The structure of LPI-PLC solved at 2.0 Å resolution (*12*) (Figure 2) is very similar to the BPI-PLC structure. The β-barrel is not closed, as well, but an antiparallel β-strand similar to β-strand Vb in BPI-PLC is missing. Instead, two short β-strands (IIIb and IVb) are oriented antiparallel to β-strand V, producing an additional small antiparallel three-stranded β-sheet (Figure 2).

Active Site and Binding of *myo*-Inositol

Like in all TIM-barrel containing enzymes, the active site of bacterial PI-PLCs is located at the C-terminal end of the β-barrel. It forms a deep cleft that is lined by mainly polar and charged amino acids. The structures of BPI-PLC and LPI-PLC were also both determined in complex with the competitive substrate-like inhibitor *myo*-inositol at 2.6 Å resolution, thus providing essential insight into the binding of substrate to the active site. Upon complex formation the proteins did not show any significant conformational changes in the active site. *myo*-Inositol, that represents the polar headgroup of the natural substrate PI binds in an edge-on mode in the deepest well of the active site cleft, and hydrogen bonds via its hydroxyl groups in positions 2 to 5 (OH2 to OH5) with polar and charged side chains of the enzyme (Figure 3): OH2, which is the only axial hydroxyl group in *myo*-inositol, to atoms Nε2 of His32 and Nη2 of Arg69, OH3 to atom Oδ1 of Asp198, OH4 to atoms Nη1, Nη2 of Arg163 and Oδ2 of Asp198, and OH5 to atoms Nξ of Lys115 and Nη2 of Arg163. These residues are themselves firmly held in place by a network of hydrogen bonding and electrostatic interactions with oppositely charged amino acids located in their vicinity. In addition, there is a coplanar stacking interaction between the apolar side of the inositol ring and the side chain of Tyr200 (Figure 3). The high positional similarity of these side-chains in the free enzyme as well as in the complex with *myo*-inositol indicates the absence of any larger induced fit rearrangements in the active site during substrate binding. Not surprisingly, the hydroxyl groups in positions 1 and 6 of the inositol ring, that are linked to bulky groups in the natural substrates PI and GPI, are fully exposed to solvent, and do not form close contacts with the enzyme. Due to the highly specific

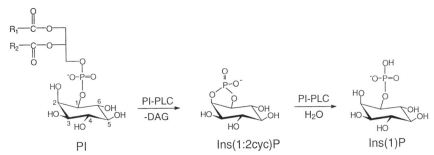

Figure 1. Catalytic reaction of PI-PLC.

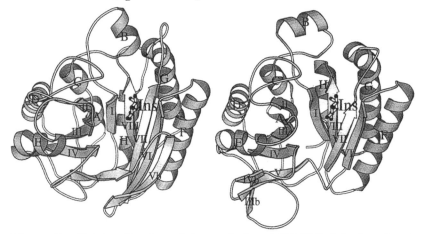

Figure 2. Side-by-side view of the topologies of BPI-PLC (left picture) and LPI-PLC (right picture) with bound *myo*-inositol (ball-and-stick, labeled Ins). β-Strands are symbolized by arrows and labeled with Roman numbers (I-VIII), α-helices with letters (A-G). Figure 2 and 3 were prepared by using MOLSCRIPT (*33*). Adapted from ref. 12.

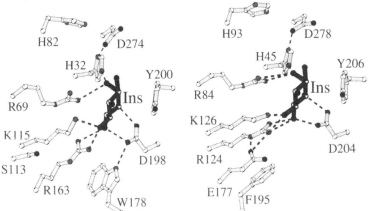

Figure 3. Side-by-side view of binding of *myo*-inositol (dark bonds, labeled Ins) to the active sites of BPI-PLC (left picture) and LPI-PLC (right picture). For clarity only the side-chains of active site residues (light grey bonds) are shown. Hydrogen bonds are indicated by dotted lines. Adapted from ref. 12.

interactions between enzyme and inhibitor it is very likely that the inositol head groups in the natural substrates PI and GPI bind to the active site pocket of PI-PLC analogously to *myo*-inositol. This is corroborated by the crystal structure of the complex between BPI-PLC and glucosaminyl($\alpha1\rightarrow6$)-D-*myo*-inositol, the central unit of GPI (*18*). In this structure, the *myo*-inositol group of glucosaminyl($\alpha1\rightarrow6$)-D-*myo*-inositol is virtually superimposable with the free *myo*-inositol in the active site.

In LPI-PLC *myo*-inositol interacts with the side chains of His45, Arg84, Arg124 and Asp204. Specifically, hydrogen bonds are formed between the 2-OH-group to atoms Nϵ2 of His45 and atoms Nη1 and Nη2 of Arg84, 3-OH to atom Oδ1 of Asp204, 4-OH to atoms Nη1 and Nη2 of Arg124 and atom Oδ2 of Asp204 and 5-OH to atoms Nη2 of Arg124 and atom Nξ of Lys126 (Figure 3). There is also a stacking interaction with Tyr206.

Most residues interacting with *myo*-inositol are not only conserved at protein sequence level, but also at the structure level where they are virtually superimposable (Figure 3). *Myo*-inositol binds in an identical orientation and location to the active site of the LPI-PLC as in that of BPI-PLC. The largest difference is found for an arginine residue whose side-chain interacts in both enzymes with the hydroxyls 4-OH and 5-OH of inositol but that originates from different β-strands. In BPI-PLC, Arg163 is located at the C-terminal end of β-strand IV, this residue is replaced by Glu177 in LPI-PLC. In LPI-PLC, the equivalent arginine (Arg124) is provided by β-strand III and is replaced by a serine (Ser113) in BPI-PLC. Interestingly, the side-chain of Arg124 is stabilized by two hydrogen bonds with the side-chain of Glu177 the C$_\alpha$-atom of which exactly superimposes on the C$_\alpha$-atom of Arg163 in BPI-PLC (Figure 3). The other major difference is found for the boundary between active site pocket and hydrophobic core of BPI-PLC, where the bulky side-chain of Trp178 forms a hydrogen bond to Asp198 that itself interacts with hydroxyls 3-OH and 4-OH of inositol. This tryptophane is lacking in LPI-PLC due to a large deletion and the vacated space is partially filled with the likewise aromatic side-chain of Phe195.

The rim of the active site in both PI-PLCs is formed by several loops as well as the short helix B. Some of these structural units show an unusual clustering of hydrophobic amino acids that are fully exposed to solvent. They belong to loops connecting α-helices B and C, β-strand II and α-helix D, β-strand VII and helix G, as well as helix B itself. These residues may be involved in contacting the lipophilic part of PI, and therefore possibly contribute to the interfacial activation observed for BPI-PLC (*15,19,20*).

Catalytic Mechanism and Substrate Specificity of Bacterial PI-PLCs

Based on biochemical data (*14,21*) that were strongly supported by the crystal structure of BPI-PLC (*11*), a catalytic mechanism involving a general acid and general base has been proposed for BPI-PLC. Briefly, this mechanism consists of two steps: a phosphotransfer followed by a phosphodiesterase reaction (Figure 4). First, the catalytic base abstracts a proton from the 2-OH-group of the inositol moiety of PI. The formed oxyanion then nucleophilically attacks the phosphorus atom belonging to the phosphodiester bond which leads to the formation of a pentacovalent transition state. This collapses into the products Ins(1:2cyc)P and DAG, the latter being protonated by the catalytic acid. In the second reaction the roles of both catalytic residues are reversed, and the phosphodiester bond in Ins(1:2cyc)P becomes hydrolyzed by an activated water molecule, resulting in the final product, InsP.

In BPI-PLC, His32 and His82 were identified as catalytic residues, and both are conserved in strikingly similar positions and orientations in LPI-PLC (His45 and His93) (Figure 3). Therefore, it is most likely that BPI-PLC and LPI-PLC utilize the same catalytic mechanism. The activity of the catalytic base (His32 in BPI-PLC; His45

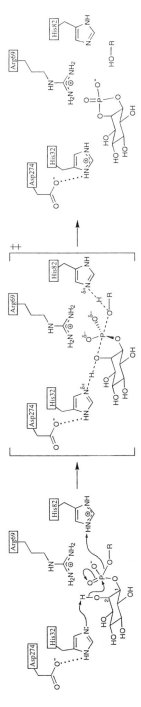

Figure 4. Proposed catalytic mechanism of BPI-PLC. In the first reaction the imidazole ring of His32 acts as a base while the protonated imidazole ring of His82 acts as an acid. The positively charged Arg69 stabilizes the negatively-charged pentacovalent transition state. The Ins(1:2cyc)P product of this phosphotransferase reaction can then act as a substrate in the second reaction, a slow hydrolysis to produce InsP. The mechanism of the second reaction is essentially the reverse of that shown above, but with water substituting for HO-R, and producing InsP.

in LPI-PLC) may be increased by an electrostatic interaction with the side-chain of an aspartate (Asp274 in BPI-PLC; Asp278 in LPI-PLC), that is ideally positioned to stabilize the imidazolium cation of the catalytic base formed during catalysis. Structurally, the arrangement of the aspartate, the catalytic base and the 2-OH-group of the inositol resemble very much the "catalytic triads" observed in numerous proteases and hydrolases.

The critical role of amino acids located in the active site of BPI-PLC in substrate binding and catalysis has been investigated by replacing them with other residues via site-directed mutagenesis, followed by enzyme kinetics and x-ray crystallography (16). In this study, the catalytic activity of the mutants was compared to wild-type BPI-PLC for two different substrates, the water soluble chromogenic substrate myo-inositol 1-(4-nitrophenyl phosphate) (NPIP) (K_m = 5 mM; 22,23) and the radiolabeled lipid ^{3}H-PI. In all cases, even conservative amino acid replacements led to a drastic reduction or complete abolishment of activity, confirming the exquisite specificity of interactions between enzyme and substrate (Figure 5).

The relative activities of the mutants at positions 32 and 82 show an important difference in both assays. Mutations of His32 (e.g. H32A) completely abolish the activity in both assays, whereas the H82L mutant retains 13% of activity in the NPIP-assay (Figure 5; 16). A similar observation has been made for RNase A (24) where His12 (analogous to His32 in BPI-PLC) and His119 (analogous to His82 in BPI-PLC) were replaced by alanine, and where activity of the mutants was assayed by using oligonucleotide substrates, as well as the chromogenic substrate, uridine 3'(4-nitrophenyl phosphate). For the H12A mutant, activity was abolished in both assays, whereas the H119A mutant still showed activity towards the chromogenic substrate. This could be explained by the fact that the 4-nitrophenoxy anion is a much better leaving group than an alkoxide anion, thus not requiring the assistance of His119 as a catalytic acid. This result provided a direct evidence for the role of His119 as the catalytic acid, and the same principle applies to His82 in BPI-PLC. Like His12 in RNase A, His32 in BPI-PLC most likely acts as the catalytic base.

In the crystal structure, Arg69 is ideally located to stabilize the pentacovalent negatively charged transition state during catalysis. The replacement of Arg[69] by other amino acids, leading to almost complete abolishment of activity even in case of the conservative replacement by a lysine (Figure 5; 16), confirmed its critical role in catalysis. More direct evidence is coming from a study where Arg69 in B. thuringiensis PI-PLC was mutated, as well (25). Thus, the R69K mutant showed a significant reduction in the stereoselectivity towards diastereomers of dipalmitoyl-PsI, suggesting a direct interaction of Arg69 with the phosphate moiety of the substrate. Interestingly, the role of Arg69 is fulfilled by a similarily positioned Ca^{2+}-ion in the structure of mammalian PI-PLC-δ1 (13,20,26). Mutation of Asp274 in BPI-PLC, that has no direct access to the active site pocket, led to a drastic loss in activity, confirming its supportive role as a member of the proposed catalytic triad consisting of Asp[274], His[32] and the 2-OH-group of the inositol head group (11).

The reduced activities caused by mutations at positions 115, 163, 178, 198 and 200 in BPI-PLC (Figure 5; 16) demonstrate the critical role of these residues in the specific binding of the substrate via its inositol headgroup. In contrast to the mammalian enzymes, bacterial PI-PLCs show no activity towards more highly phosphorylated PIs, PIP and PIP_2 that carry additional phosphate groups in positions 4 and 5 of the inositol ring (27). The close and specific interactions formed between the bacterial PI-PLCs and the free hydroxyls at these positions of myo-inositol rule out the bulky phosphate groups present in PIP or PIP_2.

Bacterial PI-PLCs, however, are able to specifically release GPI-anchored proteins from membranes (28). When compared with B. thuringiensis PI-PLC that differs by only 7 amino acid replacements from BPI-PLC, a greatly reduced catalytic

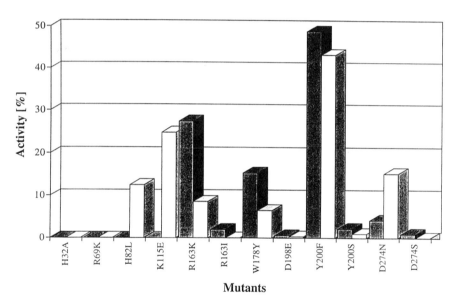

Figure 5. Bar diagram showing the catalytic activities (relative to wild-type BPI-PLC) of mutants of BPI-PLC using two different substrates. The two columns per mutant indicate the different substrates used in the activity assay: ^{3}H-PI (first column, dark grey) and NPIP (second column, light grey) (see text).

activity on different GPI-anchored proteins has been reported for LPI-PLC (*29*). For BPI-PLC, a 50-100 times stronger binding of GPI-anchors than that of PI has been reported (*30*). From the crystal structure of BPI-PLC in complex with glucosaminyl($\alpha1\rightarrow6$)-D-*myo*-inositol, the central fragment of all GPI-anchors, it was predicted that the remainder of the GPI-glycan makes interactions with a shallow groove extending from the active site pocket of BPI-PLC (*18*). A boundary of this groove is formed by β-strand Vb that is completely missing in LPI-PLC. The structural consequence is a large widening of the extended binding site that might be responsible for the significant reduction of interactions with the glycan part of GPI, resulting in the observed decrease in activity of LPI-PLC.

Structural Comparison of Ca^{2+}-Independent PI-PLCs

In previously published sequence alignments a sequence identity of approximately 24% was found between BPI-PLC and LPI-PLC (*6,7*). A superposition of both structures shows a strong overall positional similarity with the r.m.s. deviation of 1.46 Å for 228 common C_α-atom positions (using a 3.8 Å cutoff). The r.m.s. deviation for corresponding secondary structure elements is even lower: 0.74 Å for 48 common C_α-positions of residues belonging to β-strands I to VIII comprising the β-barrel (except β-strand V) and 1.0 Å for 70 common C_α-positions of residues belonging to α-helices A-H (excluding helix B). β-Strands V are shifted by more than 3 Å relative to each other, and α-helices B are tilted by approximately 70° (Figure 2). In contrast to the excellent conservation of the secondary structure elements, significant differences are found for loop regions where most of the 15 amino acid insertions and deletions are found. Further deviations between both structures are caused by additional β-strands that interact with β-strands of the β-barrel in an antiparallel fashion. They include β-strands IIIb, IVb that interact with the N-terminal half of β-strand V in LPI-PLC, and β-strand Vb that interacts with β-strand VI in BPI-PLC (Figure 2). In place of β-strand Vb, a large deletion of 14 residues is found in LPI-PLC, leading to an almost direct link between β-strands V and VI. This deletion causes an alteration of the boundary of the extended active site pocket of LPI-PLC as compared to BPI-PLC. In BPI-PLC, Trp178 is located in a short loop linking β-strands V and Vb, and defines part of the boundary of the inositol-binding pocket. It also makes a hydrogen bond with Asp198. Likewise, Asp180 forms a hydrogen bonding interaction with the side-chain of Tyr200 that makes a stacking interaction with the inositol ring.

Besides BPI-PLC and LPI-PLC, the amino acid sequence of two other bacterial PI-PLCs is known. The sequence of PI-PLC from *B. thuringiensis* (*8*) differs from BPI-PLC by only 7 amino acid substitutions that are located at the surface of the BPI-PLC structure, remote from the active site (*11*). PI-PLC from *S. aureus* shows a sequence identity of approximately 41% as compared with BPI-PLC, but only 23% as compared to *L. monocytogenes* (*5*). In PI-PLC from *S. aureus*, amino acids belonging to all β-strands found in BPI-PLC, including antiparallel β-strand Vb, are strikingly well conserved (Figure 6). The high conservation extends also to most α-helices. In addition, all residues involved in inositol binding and catalysis are conserved, as well. This indicates that BPI-PLC and PI-PLC from *S. aureus* are structurally much more related to each other than BPI-PLC and LPI-PLC, or LPI-PLC and PI-PLC from *S. aureus*, respectively.

When compared with bacterial PI-PLCs, GPI-PLC from *T. brucei* is of similar size and also does not depend on Ca^{2+}-ions for activity (*3*). A sequence alignment of the GPI-PLC with the bacterial PI-PLCs (Figure 6) shows conservation of the catalytic base and acid, as well as of a number of amino acids located in the active site, despite of sequence identity of less than 10%. The critical role of His34 (that aligns with His32 in BPI-PLC) has been confirmed by mutagenesis (Carrington, M., University

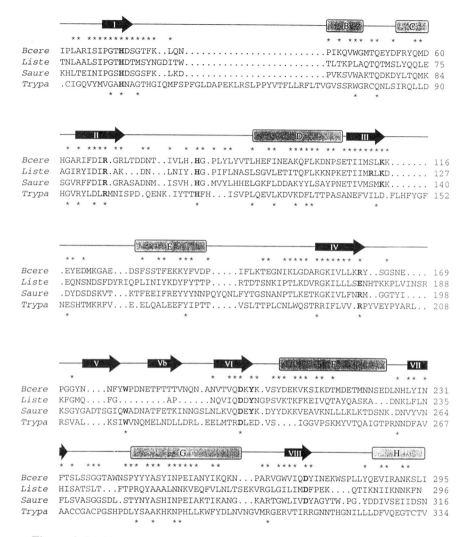

Figure 6. Multiple sequence alignment of prokaryotic PI-PLCs (from *B. cereus*, *L. monocytogenes* and *S. aureus*) and *T. brucei* GPI-PLC (top to bottom). The alignment of BPI-PLC and LPI-PLC is based on the crystal structures of both enzymes. The alignment of the other sequences with respect to these was based on an automatic sequence alignment that was manually adjusted. The secondary structure elements of BPI-PLC are shown above the sequences. α-helices (B-G) are symbolized as grey rectangles, β-strands (I-VIII) as black arrows. Amino acid identities between at least two bacterial PI-PLC are indicated by stars above the sequences. Amino acid identities between *T. brucei* GPI-PLC and at least two bacterial PI-PLCs are indicated by stars below the sequences. Common residues critical for catalysis and substrate binding are shown as bold letters.

of Cambridge, personal communication, 1997). GPI-PLC behaves like a membrane protein, requiring detergents to be extracted from the membrane and to remain stable in solution (*31*). Its sequence, however, does not show extended arrays of hydrophobic residues typical for integral membrane proteins. In the sequence alignment (Figure 6) ,a large insertion is present in GPI-PLC, comprising residues 46 to 71, that is missing in the bacterial PI-PLCs as well as in the mammalian PI-PLCs (*20*). In this insertion, 65% of the residues are hydrophobic, and these residues are separated by mainly polar and charged amino acids. This arrangement of residues is typical for amphipathic α-helices, and modeling of the insertion as an α-helix reveals that most of the hydrophobic residues lie on one side of the helix, whereas the polar residues reside on the opposite site. We therefore propose that it is this insertion that is responsible for the behavior of GPI-PLC as a membrane protein, and that the hydrophobic side of the amphipathic helix faces outward, away from the body of the protein, forming a large hydrophobic patch on the exterior of the protein that might partially penetrate into one leaflet of the membrane, similarly as observed in the structure of the monotopic membrane protein prostaglandin H_2 synthase-1 (*32*). In this structure, three amphipathic helices on the outside of the protein form a huge hydrophobic surface that directly penetrates into the membrane. Under the assumption that GPI-PLC folds similarly to BPI-PLC, the putative helix in GPI-PLC would be located between β-strand I and helix B on the outside of the protein, in an area extending from the active site pocket that most likely interacts with the membrane (*11,20*).

Acknowledments: We are most grateful to M. Ryan, C. Gässler, J. Moser, Dr. B. Gerstel, Dr. J. Meyer, Dr. T. Liu, Prof. T. Chakraborty, Dr. M. Carrington and Prof. G. E. Schulz for their help and support at various stages of the project.

Literature Cited

1. Berridge, M. J.; Irvine, R. F. *Nature* **1989**, *341*, 197-205.
2. Rhee, S. G.; Bae, Y. S. *J. Biol. Chem.* **1997**, *272*, 15045-15048.
3. Carrington, M.; Walters, D.; Webb, H. *Cell Biol. Int. Rep.* **1991**, *15*, 1101-1114.
4. Webb, H.; Carnall, N.; Carrington, M. *Brazilian J. Med. Biol. Res.* **1994**, *27*, 349-356.
5. Daugherty, S.; Low, M. G. *Infect. Immun.* **1993**, *61*, 5078-5089.
6. Leimeister-Wächter, M.; Domann, E.; Chakraborty, T. *Mol. Microbiol.* **1991**, *5*, 361-366.
7. Mengaud, J.; Braun-Breton, C.; Cossart, P. *Mol. Microbiol.* **1991**, *5*, 367-372.
8. Kupke, T.; Lechner, M.; Kaim, G.; Götz, F. *Eur. J. Biochem.* **1989**, *185*, 151-155.
9. Kuppe, A.; Evans, L. M.; McMillen, D. A.; Griffith, O. H. *J. Bacteriol.* **1989**, *171*, 6077-6083.
10. Camilli, A.; Goldfine, H.; Portnoy, D. A. *J. Exp. Med.* **1991,** *173*, 751-754.
11. Heinz, D. W.; Ryan, M.; Bullock, T.; Griffith, O. H. *EMBO J.* **1995**, *14*, 3855-3863.
12. Moser, J.; Gerstel, B.; Meyer, J.E.W.; Chakraborty, T.; Wehland, J.; Heinz, D.W. *J. Mol. Biol.* **1997**, *273*, 269-282.
13. Essen, L.-O.; Perisic, O.; Cheung, R.; Katan, M.; Williams, R. L. *Nature* **1996**, *380*, 595-602.
14. Volwerk, J. J.; Shashidhar, M. S.; Kuppe, A.; Griffith, O. H. *Biochemistry* **1990**, *29*, 8056-8062.

15. Lewis, K. A.; Garigapati, V. R.; Zhou, C.; Roberts, M. F. *Biochemistry* **1993**, *32*, 8836-8841.

16. Gässler, C. S.; Ryan, M.; Liu, T.; Griffith, O. H.; Heinz, D. W. *Biochemistry* **1997**, *36*, 12802-12813.

17. Reardon, D.; Farber, G. K. *FASEB J.* **1995**, *9*, 497-503.

18. Heinz, D. W.; Ryan, M.; Smith, M.; Weaver, L.; Keana, J.; Griffith, O. H. *Biochemistry* **1996**, *35*, 9496-9504.

19. Hendrickson, H. S.; Hendrickson, E. K.; Johnson, J. L.; Khan, T. H.; Chial, H. J. *Biochemistry* **1992**, *31*, 12169-12172.

20. Heinz, D. W.; Essen, L.-O.; Williams, R. L. *J. Mol. Biol.* **1998**, *275*, 635-650.

21. Lin, G.; Bennett, F.; Tsai, M.-D. *Biochemistry* **1990**, *29*, 2747-2757.

22. Shashidhar, M. S.; Volwerk, J. J.; Griffith, O. H.; Keana, J. F. W. *Chem. Phys. Lipids* **1991**, *60*, 101-110.

23. Ryan, M.; Smith, M. P.; Vinod, T. K.; Lau, W. L.; Keana, J. F. W.; Griffith, O. H. *J. Med. Chem.* **1996**, *39*, 4366-4376.

24. Thompson, J. E.; Raines, R. T. *J. Am. Chem. Soc.* **1994**, *116*, 5467-5468.

25. Hondal, R. J.; Riddle, S. R.; Kravchuk, A. V.; Zhao, Z.; Liao, H.; Bruzik, K. S.; Tsai, M.-D. *Biochemistry* **1997**, *36*, 6633-6642.

26. Essen, L.-O.; Perisic, O.; Katan, M.; Wu, Y.; Roberts, M. F.; Williams, R. L. *Biochemistry* **1997**, *36*, 1704-1718.

27. Bruzik, K. S.; Tsai, M.-D. *Bioorg. Med. Chem.* **1994**, *2*, 49-72.

28. Ikezawa, H. *Cell. Biol. Intl. Rep.* **1991**, *15*, 1115-1131.

29. Gandhi, A. J.; Perussia, B.; Goldfine, H. *J. Bacteriol.* **1993**, *175*, 8014-8017.

30. Stieger, S.; Brodbeck, U. *Biochimie* **1991**, *73*, 1179-1186.

31. Mensa-Wilmot, K.; Morris, J. C.; al-Qahtani, A.; Englund P. *Meth. Enzymol.* **1995**, *250*, 641-655.

32. Picot, D.; Loll, P. J.; Garavito, M. *Nature* **1994**, *367*, 243-249.

33. Kraulis, P. J. *J. Appl. Cryst.* **1991**, *24*, 946-950.

Chapter 6

Characterization of the Histidine Residues of *B. cereus* Phosphatidylinositol-Specific Phospholipase C by NMR

Tun Liu, Margret Ryan, Frederick W. Dahlquist, and O. Hayes Griffith

Institute of Molecular Biology and Department of Chemistry, University of Oregon, Eugene, OR 97403

Histidine residues are known to play important roles in the catalytic mechanism of the mammalian and bacterial phosphatidylinositol-specific phospholipase C (PI-PLC). Progress on the characterization of the histidines of *B. cereus* PI-PLC is reviewed. All six histidines have been resolved in 2D ^{1}H-^{13}C correlated NMR spectra and identified with the aid of site-directed mutagenesis. The pK_a values of the solvent-exposed histidines have been determined for the native enzyme. The tautomeric states for three of the histidines have been assigned. Preliminary data are presented for inhibitor binding effects on active site histidines. An unusual low-field resonance (16 ppm) in the ^{1}H NMR spectrum has recently been detected. It is identified as the proton donated by His32 to the hydrogen bond between Asp274 and His32, which together with the hydroxyl group of the substrate form a catalytic "triad". These data provide insights into the catalytic mechanism of *B. cereus* PI-PLC.

Phosphatidylinositol-specific phospholipases C (PI-PLC) are soluble membrane-associated proteins that play a central role in the early events of signal transduction in mammalian cells (*1*). The mammalian PI-PLCs are large multidomain proteins ranging in size from 85 kD to about 150 kD, and the activities are Ca^{2+}-dependent. In contrast, PI-PLCs excreted by bacteria such as *Bacillus cereus* and the essentially identical *B. thuringiensis* are much smaller (35 kD) and the activity is Ca^{2+}-independent (*2*). It was established by ^{31}P NMR and product analysis that the *B. cereus* PI-PLC exhibits sequential phosphotransferase and cyclic phosphodiesterase activities (*3, 4*). The enzyme initially cleaves phosphatidylinositol (PI) into two parts, a lipid-soluble component, diacylglycerol and the water-soluble component *myo*-inositol 1,2-cyclic phosphate, I(1:2cyc)P (Figure 1). In a much slower reaction, the latter can be converted to *myo*-inositol 1-phosphate. It has been proposed that the reaction proceeds in a two-step, ribonuclease-like mechanism (*3*). This focuses attention on histidine residues because of their importance in the ribonuclease acid-base catalysis. Recently, the first crystal structures of both bacterial and mammalian PI-PLCs have become available (*5, 6*). Two histidines are located at the active site of both the

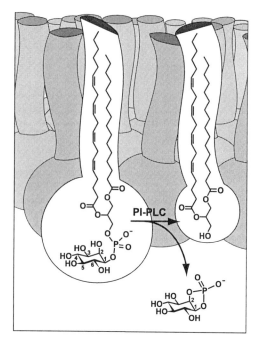

Figure 1. Cleavage of phosphatidylinositol (PI) catalyzed by *B. cereus* PI-PLC. In the phosphotransferase reaction the enzyme cleaves PI into two components, diacylglycerol and *myo*-inositol 1,2-cyclic phosphate, I(1:2cyc)P. In a second and much slower reaction (not shown), the enzyme cleaves I(1:2cyc)P to form *myo*-inositol 1-phosphate.

mammalian and bacterial enzymes, and these histidines are essential for activity as shown by site-directed mutagenesis (5, 6). Remarkably, the bacterial enzyme and the catalytic domain of the mammalian PI-PLC have similar structures, suggesting that studies of the smaller bacterial enzymes will aid in understanding the broader family of PI-PLCs. In order to correlate the structure with function, NMR studies have been initiated on the bacterial enzyme, with a special emphasis on histidines. In this paper we review the progress made thus far, including unpublished observations on inhibitor and substrate binding, and conclude with a summary of the mechanistic implications.

Resolution of All Six Histidines in the NMR Spectrum of *B. cereus* PI-PLC

Histidines are often involved in general acid-base catalyzed enzymatic reactions. The pK_a value of the imidazole group lies within the physiological range, and thus it can act as both a general acid and a general base. Among the early triumphs of histidine characterization by protein NMR was the resolution of the peaks from the $C^{\epsilon 1}$-H and $C^{\delta 2}$-H protons in small proteins such as bovine pancreatic ribonuclease A, chicken egg white lysozyme, and staphylococcal nuclease (7) (see Figure 2 for labeling conventions). The 500 MHz proton NMR spectrum (1D ^{1}H NMR) of *B. cereus* PI-PLC is shown in Figure 3. Considering the size of the protein (298 residues, 35 kD), it is not surprising that the peaks are broad and not resolved throughout the spectrum. Spin-echo experiments were performed with the aim of filtering out broad components in the spectrum and singling out the $C^{\epsilon 1}$-H and $C^{\delta 2}$-H resonances, and two peaks were resolved. One of them was identified as His32 based on the observation that this peak was absent in the H32L mutant. A second observation consistent with this assignment is that this peak was shifted upon addition of *myo*-inositol to wild-type PI-PLC. In the crystal structure (5) *myo*-inositol is within hydrogen-bonding distance to only one histidine, His32. The weak peaks that were observed at pH 8 become broader as the pH is decreased, and also merge with other strong peaks, making it difficult to follow the signals (data not shown).

Recognizing the limits of 1D NMR spectroscopy, a different strategy was chosen. Isolation of the histidine signals from the bulk of the protein resonances can be achieved by selectively labeling the histidine residues. The histidine residues in *B. cereus* PI-PLC were selectively labeled with $^{13}C^{\epsilon 1}$-histidine (97 – 98%) by producing recombinant PI-PLC in histidine auxotroph *E. coli* GM31, grown in a synthetic medium (8). 2D ^{1}H-^{13}C heteronuclear single quantum coherence (HSQC) NMR experiments (9) were performed. Experimental conditions were chosen to detect only the protons attached to the ^{13}C atoms. The result was a single peak for each histidine, and the strong coupling between $^1H^{\epsilon 1}$ and $^{13}C^{\epsilon 1}$ ($^1J_{HC}$ = 220 Hz) provided excellent sensitivity. For the NMR experiments, the enzyme concentrations were typically 0.2 to 0.3 mM in 20 mM Tris-d$_{11}$-maleate-d$_2$ buffer with 10% D$_2$O. The pH values of the experiments were not corrected for isotope effect. All NMR spectra reported here were acquired on a General Electric Omega 500 MHz spectrometer equipped with an 8 mm Nalorac triple-resonance gradient probe. Both ^{1}H and ^{13}C chemical shifts were referenced to an external standard, 2,2-dimethyl-2-silapentane-5-sulfonate (DSS) at 0 ppm. Six resonances are present and well resolved in the ^{1}H-^{13}C HSQC spectra for the wild type PI-PLC, each corresponding to one histidine residue, subsequently assigned by site-directed mutagenesis.

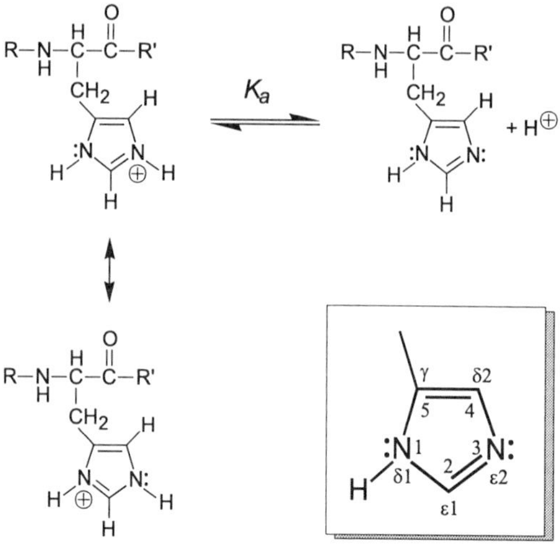

Figure 2. Histidine ionization and common conventions for identifying the imidazole ring positions. Only one of the two possible tautomers ($N^{\delta 1}$-H or $N^{\varepsilon 2}$-H) of the neutral histidine is shown. R and R' refer to adjacent groups in the polypeptide chain.

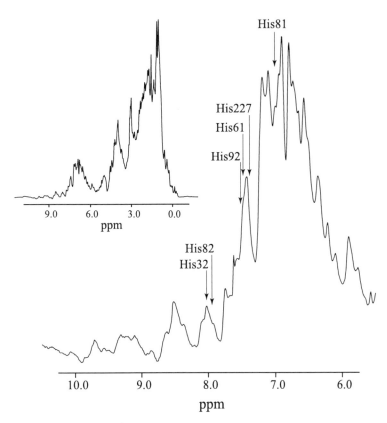

Figure 3. 1D 500 MHz ¹H NMR spectrum of *B. cereus* PI-PLC acquired in 30 mM potassium phosphate buffer with 99.9% D₂O at pH 8.0 and 30°C. Indicated are the positions corresponding to Cᵉ¹-H proton resonances of the six histidines as found in the 2D ¹H-¹³C HSQC NMR spectrum (Figure 4).

Identification of the Histidines in the ^{1}H-^{13}C HSQC NMR Spectrum by Site-Directed Mutagenesis

Five site-specific alanine mutants were generated for the individual histidine residues to allow resonance assignments in the ^{1}H-^{13}C HSQC NMR spectrum. One histidine was replaced in each mutant. The mutant proteins were then selectively labeled with ^{13}C$^{\epsilon 1}$-histidine and a ^{1}H-^{13}C HSQC spectrum was acquired for each mutant. The chemical shifts for the remaining five histidines in the mutants are nearly the same as in the wild type, indicating that the mutations did not cause any substantial alterations in the three-dimensional structure of the protein. Small effects were observed on the shifts in ^{1}H dimension of His32 and His81 in the H82A mutant, and on His82 in the H81A mutant, which is not surprising since these residues are relatively close together in the crystal structure (5). Since there are no changes, or at most very small changes, in the chemical shifts of the mutants, the assignments were straightforward and are summarized in Figure 4 along with a representative example of how the assignment was made for one of the histidines, His32.

pH Titration Curves and pK_a Determination

The histidine residues of wild-type *B. cereus* PI-PLC have been titrated at 25°C over a pH range of 4.0 – 9.0 in 20 mM Tris-d$_{11}$-maleate-d$_2$. Ionic strength was kept constant throughout by adding aliquots of 20 mM Tris-d$_{11}$/10% D$_2$O or 20 mM maleic acid-d$_2$/10% D$_2$O to change the pH. A ^{1}H-^{13}C HSQC spectrum was acquired at each pH value. The ^{1}H and ^{13}C chemical shifts of the histidines *vs.* pH are shown in Figure 5. The most important histidines in the catalytic mechanism are His32 and His82. As can be seen in the left panel of Figure 5, the ^{1}H chemical shifts for His32 and His82 are similar and cross at two points. Fortunately, their ^{13}C chemical shifts are very different (right panel, Figure 5), which removes any ambiguity in the assignment. The other histidines have sufficiently different chemical shifts that the assignments are unambiguous. For His61 and His227 the ^{1}H and ^{13}C chemical shifts cross or may cross, respectively, at pH 7.6. As the pH decreases, one resonance shifts downfield while the chemical shift of the other remains unchanged. To determine the continuity of the titration curves from the crossover point at pH 7.6, several additional ^{1}H-^{13}C HSQC spectra were acquired for the H61A mutant between pH 6.8 – 7.2, and the results further support the assignments shown in Figure 5.

These titration data were fit by non-linear regression to a one-proton titration curve to obtain the pK_a values of the histidines. The resulting pK_a values are summarized as follows: His32, pK_a = 7.6; His82, pK_a = 6.9; His92, pK_a = 5.4; His227, pK_a = 6.9. The other two histidines, His61 and His81, did not titrate over the range examined, suggesting that these residues are not solvent-accessible.

Studies of Uniformly ^{15}N-Labeled PI-PLC

Another way to study the histidines is to use uniformly ^{15}N-labeled protein. Both 1D ^{15}N NMR spectra and 2D ^{1}H-^{15}N heteronuclear multiple quantum coherence (HMQC) spectra were collected for ^{15}N-labeled PI-PLC. The 1D ^{15}N spectrum was acquired at 25 °C in 10 mM potassium phosphate buffer, pH 8.5, with 10% D$_2$O. The ^{15}N chemical shifts were referenced to liquid ammonia using the ^{15}N/^{1}H zero frequency ratio of 0.101329118 as determined by Wishart and coworkers (10). Two broad and weak peaks are observed around 170 ppm. These two signals originate from the N$^{\delta 1}$ and N$^{\epsilon 2}$ ring nitrogens of the histidines (data not shown). The 2D HMQC spectrum is shown in Figure 6B. In this experiment, the delay was set to observe two-bond

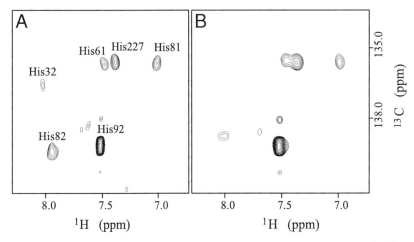

Figure 4. Assignment of the histidine NMR resonances. (A) The 2D ^{1}H-^{13}C HSQC NMR spectrum of wild type, ^{13}C$^{\epsilon 1}$-histidine-labeled PI-PLC. (B) An example of the spectral assignment: the corresponding spectrum of the mutant enzyme with His32 replaced by alanine. The missing peak is assigned to His32. Weak peaks above and below the His92 signal are data truncation artifacts from the strong His92 resonance. Both spectra were collected at pH 8.0 and 25°C. (Adapted from reference 8 with permission).

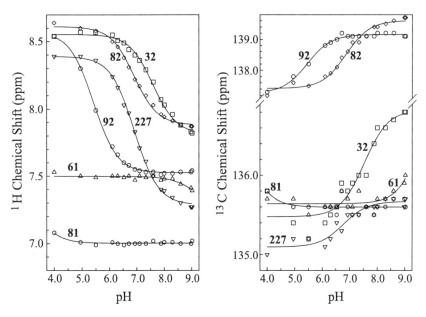

Figure 5. The pH titration curves of the six histidines of *B. cereus* PI-PLC derived from the ^{1}H-^{13}C HSQC NMR peaks of wild-type enzyme. The pH values were measured at 25°C and not corrected for isotope effect. (From reference 8, with permission).

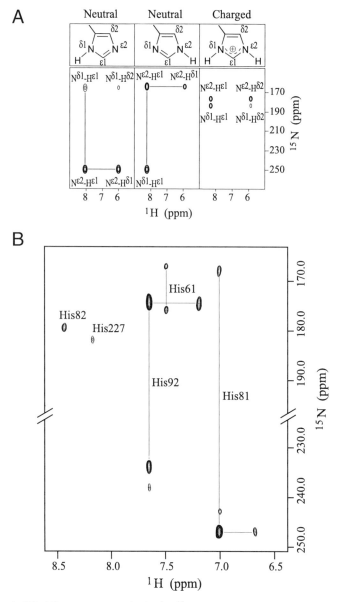

Figure 6. Histidine resonances in the ^{1}H-^{15}N HMQC spectrum. (A) Schematic diagram of the different tautomeric forms that a histidine residue can adopt and their characteristic chemical shifts and spectral patterns. (Adapted from reference *11*, with permission). (B) The ^{1}H-^{15}N HMQC spectrum of uniformly ^{15}N-labeled *B. cereus* PI-PLC acquired at 25°C and pH 6.3. Histidine residues are identified and cross-peak patterns indicated.

coupling between $N^{\delta 1}$-$H^{\epsilon 1}$, $N^{\epsilon 2}$- $H^{\epsilon 1}$, and $N^{\epsilon 2}$-$H^{\delta 2}$ and possibly three-bond coupling between $N^{\delta 1}$- $H^{\delta 2}$. The spectrum was acquired at 25 °C in 20 mM Tris-d_{11}-maleate-d_2 buffer, the same buffer as used in the ^{1}H-^{13}C HSQC experiments. Five of the six histidines were detected. Only the His32 signal was too weak to be observed. Based on the ^{1}H$^{\epsilon 1}$ chemical shifts in the ^{1}H-^{13}C HSQC spectrum and assignments determined by site-directed mutagenesis, the peaks in the HMQC spectrum were assigned to the corresponding histidines.

By combining ^{15}N chemical shift data of model compounds with J-coupling information for histidine, Pelton et al. (*11*) have constructed the different cross-peak patterns expected in ^{1}H-^{15}N HMQC spectra for the three possible protonation states of histidine (Figure 6A). Interpreting the PI-PLC data in terms of these patterns for the tautomeric states, His92 is in the neutral form at pH 6.3 and the proton resides on the $N^{\epsilon 2}$ nitrogen. His81 is also in the neutral form, but the tautomer present is the one with a proton on the $N^{\delta 1}$ nitrogen. His61 is in the acid form. Only one of the two nitrogens was observable for both His82 and His227. From the pH titration data, it is known that both His61 and His81 do not titrate over the pH range of 4.0 – 9.0. Taken together, it can be concluded that His61 is protonated until pH > 8.5 whereas His81 stays neutral until pH < 4.0.

Correlation of NMR, Crystal Structure, and Enzyme Activity

Crystal structures of *B. cereus* PI-PLC have been solved for the free enzyme, and for the enzyme complexed with *myo*-inositol or with glucosaminyl(α1→6)-D-*myo*-inositol, a central portion of glycosylphosphatidylinositol (GPI) membrane protein anchors (*5, 12*). The crystal structures reveal that *B. cereus* PI-PLC consists of a single globular domain, and folds as an irregular $(\alpha,\beta)_8$ barrel (for a review see *13*). A ribbon diagram of PI-PLC complexed with *myo*-inositol is shown in Figure 7. The positions of all six histidines are indicated. From the crystallographic and NMR data, the histidines can be divided into three groups. The first group consists of the two histidines located at the active site, His32 and His82. Histidines that do not titrate belong to the second group. These are His61 and His81. It is immediately apparent from the structure why these two histidines do not titrate, they are shielded from solvent in the protein structure by other residues. Histidines of the third group, His92 and His227, are located at some distance from the active site and are exposed at the protein surface. In agreement with the crystallographic data members of the first and third groups show normal titration curves. Thus, the NMR data and crystal structure data are consistent for all six histidines.

Enzyme activities were measured for all five of the histidine mutants constructed for the NMR assignments. Histidines 32 and 82 are required for activity (*5*). In a ^{3}H-PI assay, the residual activities compared to wild type are about 1×10^{-4} % for H32A, and less than 7×10^{-4} % for H82A. The activity of H81A is reduced to 16% of wild type, which is reasonable considering its proximity to the essential His82. In contrast, mutants H92A and H61A that are distant from the active site retain essentially full activity (*8, 14*).

Effect of Ligand Binding on ^{1}H-^{13}C HSQC NMR of Active Site Histidines

In order to fully understand the role of His32 and His82 in the catalytic mechanism, detailed X-ray crystallographic and NMR data are needed for complexes of the enzyme with substrate analogs or inhibitors that fully represent the structure of the natural substrate, i. e. they would consist of the inositol head group, a bridging phosphate group, the glycerol moiety, and *sn*-1 and *sn*-2 linked lipid chains. This information is not yet available. However, progress is being made. The chemical shifts of the histidines in the presence of 7.5 mM glucosaminyl(α1→6)-D-*myo*-inositol have been measured at pH 7.4 (*8*). In the ^{1}H-^{13}C HSQC spectrum, only the resonance of His32 was shifted (upfield by 0.3 – 0.4 ppm in the ^{1}H dimension). The NMR peaks of the other five histidines were unaffected by the presence of the inhibitor. Here we report some additional unpublished observations (Figures 8–10).

When 25 mM *myo*-inositol is added to H61A, only His32 is shifted upfield, and all the other histidines remain at the same chemical shifts. The chemical shift changes for His32 are 0.27 ppm at pH 7.1, 0.2 ppm at pH 8.2 (Figure 8), and 0.1 ppm at pH 8.5. The changes refer to the ^{1}H dimension of the spectra, which has higher resolution than the ^{13}C dimension. Using reported IC_{50} values (*15*) and assuming competitive inhibition, glucosaminyl(α1→6)-D-*myo*-inositol and *myo*-inositol K_I values are calculated to be 1.9 mM and 9.5 mM at pH 7.0, respectively. These K_I values suggest that NMR data were collected at saturating concentrations for these weak inhibitors. The results for *myo*-inositol are consistent with the NMR data recorded with glucosaminyl(α1→6)-D-*myo*-inositol, and with the crystallography data because, of the two active site histidines, only His32 forms a hydrogen bond with *myo*-inositol.

Binding of the water-soluble synthetic substrate, *myo*-inositol 1-(4-nitrophenyl hydrogen phosphate) [NPIP], was selected to explore the effect of the phosphate group on the active site histidines. The enzyme is specific for the D-isomer while the L-isomer has no effect on activity. In a racemic mixture the K_M for D-NPIP is about 5 mM, the same as for enantiomerically pure D-NPIP (*16*). For this experiment, the inactive mutant H32A was used so that the substrate would not be cleaved. Upon addition of 7.8 mM racemic NPIP, His82 was shifted downfield by about 0.05 ppm at pH 6.15 and 0.22 ppm at pH 7.05 compared to wild type (Figure 9). Since these pH values are close to the pK_a of 6.9 for His82, the chemical shift is very sensitive to pH changes and an internal control is needed to distinguish specific interactions from nonspecific effects of solvent pH. Fortunately, His227 can serve as a control. His227 also has a pK_a value of 6.9, is solvent-exposed and far from the active site. The chemical shift change for His227 upon addition of NPIP was only 0.01 ppm at pH 6.15 and 0.02 ppm at pH 7.05. Thus, we conclude that the chemical shift change for His82 is the result of a specific interaction with NPIP. This change is likely due to the electrostatic effect of the substrate's phosphate group.

Short-chain phosphonate inhibitors are useful for studying the interactions between PI-PLC and its substrate because of their structural similarity to the natural substrate and their relatively high-affinity binding compared to analogs that lack the lipid portion of the substrate (*15*). Shown in Figure 10 are the ^{1}H -^{13}C HSQC spectra for the wild type PI-PLC in the absence and presence of 13.4 mM racemic phosphonate inhibitor diC$_6$CH$_2$PI at pH 8.3. The principal observation is that the peaks of His32 and His82 are *both* shifted upon binding of the inhibitor. Taking into consideration a slight pH difference for the two spectra, both His32 and His82 are shifted downfield about 0.05 – 0.07 ppm. Although the changes are small, they are specific since the chemical shifts of the other four histidines are unaffected. The amount of inhibitor required to produce the observed effects on His32 and His82 is substantially greater than expected from the K_I of the inhibitor (14 *m*M, translating into 7 *m*M for the D-enantiomer, *15*). The reason for this insensitivity is not known, but with a critical micelle concentration of about 14 mM, it can not be ruled out that the micellar form of diC$_6$CH$_2$PI is responsible for the observed chemical shift changes. His32 was shifted downfield upon binding of this inhibitor, the opposite direction observed for *myo*-inositol binding. The implication is that there are competing effects on the chemical shift of His32 by the inositol group and the phosphate group. It is

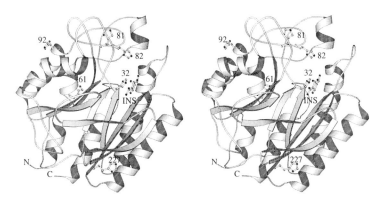

Figure 7. Ribbon diagram of PI-PLC showing the locations of the six histidines of *B. cereus* PI-PLC. *Myo*-inositol (INS) is bound at the active site. This figure was prepared from the crystallographic data of Heinz et al. (5) using MOLSCRIPT (29).

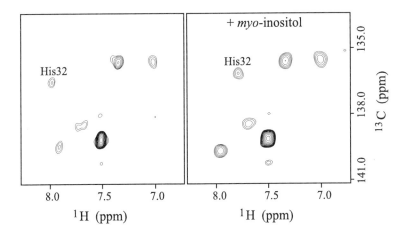

Figure 8. The ^1H-^{13}C HSQC NMR spectra of the H61A mutant of *B. cereus* PI-PLC in the absence and presence of 25 mM *myo*-inositol at 25°C and pH 8.2.

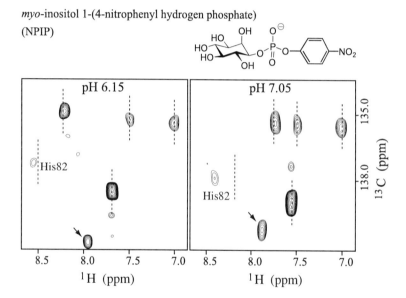

Figure 9. The ¹H-¹³C HSQC NMR spectra of the H32A mutant of *B. cereus* PI-PLC at 25°C in the presence of a chromogenic water-soluble substrate, 7.8 mM racemic NPIP (*30, 31*), at pH 6.15 and 7.05. The dashed vertical lines indicate the proton peak positions of wild type in the absence of NPIP (see Figure 5). The arrow points to an unidentified peak attributed to an impurity in the NPIP preparation.

worth noting that the resonance of His82 was not affected by the *myo*-inositol group, but was shifted only when the substrate or inhibitor bearing a phosphate or phosphonate group was present.

It is interesting to compare these results with the NMR results of the more extensively studied RNase A. RNase A has four histidines. Two of these (His12, pK_a = 5.8 and His119, pK_a = 6.2) are close together at the active site in an arrangement comparable to that found in *B. cereus* PI-PLC. His105 is distant from the active site, exposed to the solvent and has a pK_a of 6.7, and His48, the most buried of the histidines, has a pK_a of about 5.8 to 6.4 (7). These values are for the unliganded enzyme. Binding of dianionic mononucleotides resembling the product of catalysis induces downfield shifts of the His12 and His119 peaks, and an increase in the pK_a of His12 and His119. In contrast, binding of a monoanionic dinucleotide phosphonate more closely resembling the substrate has very little effect on the pK_a values of the two histidines, and resulting chemical shift changes are small (17, 18, 19). The reason for this is unknown, but it is consistent with the insensitivity we observed for *B. cereus* PI-PLC to the phosphonate substrate analog (Figure 10).

A Low Field Peak in the ^{1}H NMR Spectrum

The $C^{\varepsilon 1}$ protons of the histidines have been the focus of attention because they provide the critical data for determining the pK_a values. However, the ring N-H protons can, under certain circumstances, also provide valuable information. Normally, the ring N-H protons of histidines are not observable because of their rapid exchange with solvent on the NMR time scale. However, there are some interesting cases in the literature where these N-H proton resonances can be observed in proteins. Their signals are recorded in the low field region, i. e. with chemical shifts greater than 11 ppm, therefore resolved from backbone and side chain N-H protons and aromatic protons which have ^{1}H chemical shifts of 5 – 10 ppm. Serine proteases and ribonucleases exhibit such low field resonances from ring N-H protons of histidines. In chymotrypsin, for example, the resonance for a ring N-H proton of His57, a member of the catalytic triad, occurs at about 18 ppm at low pH and shifts to about 15 ppm at high pH (20, 21, 22). The low field resonance for RNase A is observed at 14 ppm, and is assigned to one of the active site histidines (23, 24). These unusual low field peaks in serine proteases have been attributed to short, strong hydrogen bonds or low barrier hydrogen bonds (LBHB) (25). This bond results in a downfield shift of the ^{1}H resonance and a slowed exchange rate with solvent, permitting the observation of the NMR peak. We have recently observed a single, broad resonance at 16.3 ppm in the ^{1}H NMR spectrum of *B. cereus* PI-PLC acquired at pH 6.3 and 25°C (Figure 11), the first time such a low field peak has been observed in a phospholipase. This peak is present in the wild type but not the H32A mutant. The position of this resonance is pH dependent with an inflection point at pH 8.0 at 5.8°C, comparable to the pK_a of 7.6 for His32 at 25°C. The crystal structure shows a short hydrogen bond (2.6 Å) between $N^{\delta 1}$ of His32 and a carboxyl oxygen of Asp274. From these observations we assign the low field resonance to the hydrogen bond between His32 and Asp274.

Catalytic Mechanism of *B. cereus* PI-PLC

Prior to the availability of the crystal structures, a general-acid-general-base catalysis was proposed for PI-PLC, based on the stable intermediate I(1:2cyc)P and the final product, *myo*-inositol 1-phosphate (26, 27, 3). The crystal structure of *B. cereus* PI-PLC bound to *myo*-inositol strengthened the concept by locating two histidines, His32 and His82, in positions to act as general base and acid, respectively, and these two

myo-inositol 1-[hydrogen [3,4-bis(hexyloxy)butyl]phosphonate]

(diC₆CH₂PI)

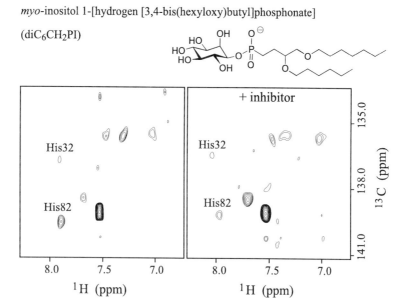

Figure 10. The ¹H-¹³C HSQC NMR spectra of *B. cereus* PI-PLC in the absence and presence of 13.4 mM short chain substrate analog inhibitor diC₆CH₂PI at 25°C at nearly matched pH values (pH 8.3 and pH 8.2, respectively).

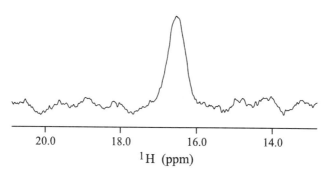

Figure 11. Low field line in the 500 MHz ¹H NMR spectrum of uniformly ¹⁵N-labeled *B. cereus* PI-PLC. The spectrum was acquired at pH 6.3 and 25°C with a spectral width of 16,129 Hz and 5,000 transients.

histidines are required for activity (*5*). His82 has been identified as the acid (i. e. proton donor) by demonstrating markedly different catalytic activities of the mutant H82L towards the hydrolysis of substrates that differ in their requirements for protonation of the leaving group. His82L is virtually inactive towards PI, where the leaving group without protonation would be an alkoxide ion (pK_a about 16) whereas NPIP is cleaved at a rate greater than 10^3-times that of PI (*14*). The leaving group of NPIP is a more favorable *p*-nitrophenolate anion ($pK_a = 7.1$).

The evidence for His32 acting as the general base in catalysis consists of its location at the active site where His32 is in a position to activate the catalytic 2-hydroxyl group of the substrate, and the observation that mutation of this residue results in an essentially inactive enzyme. Based on this information alone, one might predict that in the free enzyme His32 is deprotonated and His82 protonated. However, the NMR data clearly show that both His32 and His82 are protonated at a pH of about 6, which is optimal for catalysis (Figure 12). This apparent paradox can be resolved by assuming that a proton displacement takes place upon substrate binding, resulting in a deprotonated His32 that is hydrogen bonded to the proton of the 2-hydroxyl group of *myo*-inositol. The catalytic reaction then proceeds as diagrammed in Figure 12. His32 abstracts the 2-OH proton, facilitating the nucleophilic attack by the 2-oxygen on phosphorous at the same time the catalytic acid His82 protonates the leaving group. His32 could function to raise the Gibbs free energy of the bound substrate by requiring a proton displacement to form the Michaelis complex. The fact that His32 is a comparatively strong base also serves to lower the energy of the transition state. Both effects reduce the energy barrier and therefore increase the rate of reaction.

The catalytic histidines are part of a hydrogen bonding network involving Asp33, Asp67 and Arg69 in addition to Asp274. Figure 13A shows the actual positions of these residues, as found in the crystal structure of *B. cereus* PI-PLC complexed with *myo*-inositol (*5*). This arrangement is likely responsible for precise positioning of the imidazole ring in addition to influencing the pK_a values of the histidines. Figure 13B shows the proposed pentacovalent transition state, with His32 acting as the general base and His82 as the acid. A direct participation in catalysis has been proposed for Arg69 (*5, 28, 14*). Its guanidinium group is positioned close to the catalytic *myo*-inositol 2-hydroxyl group of the substrate and could function to stabilize the pentacovalent phosphorous of the transition state (Figure 13B). Based on the short distances between Asp274 and His32 and between His32 and the 2-OH group of the *myo*-inositol ring, a "catalytic triad" involving Asp274, His32, and the 2-OH group has been proposed, analogous to that in the serine proteases, except that the 2-OH group of the inositol moiety of the substrate takes the place of the serine hydroxyl group. Asp274 is required for activity. The D274A mutant retains only 3×10^{-4} % of the wild type activity (Liu et al., unpublished results). This is about the same level of residual activity found for active site mutants H32A and H82A. The observation of a low field resonance in the ^{1}H NMR spectrum of *B. cereus* PI-PLC provides direct evidence for an important hydrogen bond between Asp274 and His32. The functions of this hydrogen bond in the catalytic triad are evidently to orient His32 in the correct position to initiate catalysis, and to increase the basicity of this residue, facilitating the proton transfer reaction.

Acknowledgments

We are pleased to acknowledge our colleagues Drs. Dirk W. Heinz and G. Bruce Birrell for useful discussions. We thank Dr. S. Michael Strain for his expert assistance in NMR experiments. This work was supported by NIH Grant GM25698.

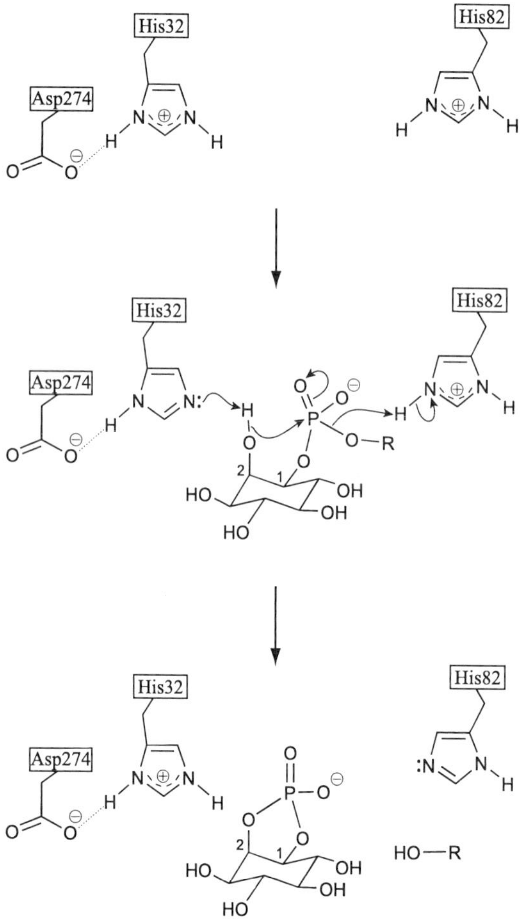

106

Figure 12. The proposed catalytic mechanism of *B. cereus* PI-PLC. Top: the state of protonation of the catalytic histidines prior to substrate binding. Center: the enzyme-induced electron and proton transfers occurring during catalysis. Bottom: state of protonation during product release.

A

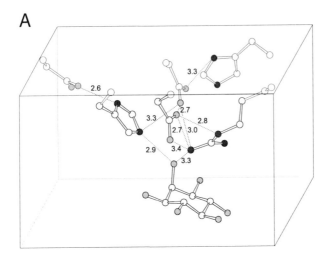

B

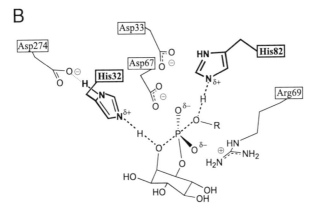

Figure 13. Transition state in the *B. cereus* PI-PLC phosphotransferase reaction. (A) Hydrogen bonding network (distances in C) at the active site involving two catalytic histidines and potentially assisting residues. The arrangement of amino acids is that found in the crystal structure of PI-PLC complexed with *myo*-inositol (Brookhaven National Laboratory Protein Data Bank entry code 1ptg) (5). Light gray and dark gray spheres represent oxygen and nitrogen atoms, respectively, carbon atoms are white. (B) The proposed pentacovalent transition state resulting in the formation of I(1:2cyc)P. The orientation of amino acid residues and *myo*-inositol is the same as in A.

Literature Cited

1. Rhee, S. G.; Bae, Y. S. *J. Biol. Chem.* **1997**, *272*, 15045-15048.
2. Bruzik, K. S.; Tsai, M.-D. *Bioorg. Med. Chem.* **1994**, *2*, 49-72.
3. Volwerk, J. J.; Shashidhar, M. S.; Kuppe A.; Griffith, O. H. *Biochemistry* **1990**, *29*, 8056-8062.
4. Ikezawa, H. *Cell Biol. Int. Rep.* **1991**, *15*, 1115-1131.
5. Heinz, D. W.; Ryan, M.; Bullock, T.; Griffith, O. H. *EMBO J.* **1995**, *14*, 3855-3863.
6. Essen, L.-O.; Perisic, O.; Cheung, R.; Katan, M.; Williams, R. L. *Nature* **1996**, *380*, 595-602.
7. Markley, J. L. *Acc. Chem. Res.* **1975**, *8*, 70-80.
8. Liu, T.; Ryan, M.; Dahlquist, F. W.; Griffith, O. H. *Protein Sci.* **1997**, *6*, 1937-1944.
9. Wider, G; Wüthrich, K. *J. Magn. Res. Ser. B* **1993**, *102*, 239-241.
10. Wishart, D. S.; Bigam, C. G.; Yao, J.; Abildgaard, F.; Dyson, H. J.; Oldfield, E.; Markley, J. L.; Sykes, B. D. *J. Biomol. NMR* **1995**, *6*, 135-140.
11. Pelton, J. G.; Torchia, D. A.; Meadow, N. D.; Roseman, S. *Protein Sci.* **1993**, *2*, 543-558.
12. Heinz, D. W.; Ryan, M.; Smith, M. P.; Weaver, L. H.; Keana, J. F. W.; Griffith, O. H. *Biochemistry* **1996**, *35*, 9496-9504.
13. Heinz, D. W.; Wehland, J.; Griffith, O. H. **1998**, this volume.
14. Gässler, C. S.; Ryan, M.; Liu, T.; Griffith, O. H.; Heinz, D. W. *Biochemistry* **1997**, *36*, 12802-12813.
15. Ryan, M.; Smith, M. P.; Vinod, T. K.; Lau, W. L.; Keana, J. F. W.; Griffith, O. H. *J. Med. Chem.* **1996**, *39*, 4366-4376.
16. Leigh, A. J.; Volwerk, J. J.; Griffith, O. H.; Keana, J. F. W. *Biochemistry* **1992**, *31*, 8978-8983.
17. Griffin, J. H.; Cohen, J. S.; Schechter, A. N. *Ann. N. Y. Acad. Sci.* **1973**, *222*, 693-708.
18. Haar, W.; Maurer, W.; Rüterjans, H. *Eur. J. Biochem.* **1974**, *44*, 201-211.
19. Jardetzky, O.; Roberts, G. C. K. *NMR in Molecular Biology* Academic Press: New York, NY, 1981; pp 333-343.
20. Robillard, G.; Shulman, R. G. *J. Mol. Biol.* **1972**, *71*, 507-511.
21. Markley, J. L. *Biochemistry* **1978**, *17*, 4648-4656.
22. Bachovchin, W. W. *Proc. Natl. Acad. Sci. USA* **1985**, *82*, 7948-7951.
23. Patel, D. J.; Woodward, C. K.; Bovey, F. A. *Proc. Nat. Acad. Sci.* **1972**, *2*, 543-558.
24. Griffin, J. H.; Cohen, J. S.; Schechter, A. N. *Biochemistry* **1973**, *12*, 2096-2099.
25. Frey, P. A.; Whitt,. S. A.; Tobin, J. B. *Science* **1994**, *264*, 1927-1930.
26. Lewis, K. A.; Venkata, R.; Garigapati, C. Z.; Roberts, M. F. *Biochemistry* **1993**, *32*, 8836-8841.
27. Lin, G. L.; Bennett, C. F.; Tsai, M. D. *Biochemistry* **1990**, *29*, 2747-2757.
28. Hondal, R. J.; Riddle, S. R.; Kravchuk, A. V.; Zhao, Z.; Liao, H.; Bruzik, K. S.; Tsai, M.-D. *Biochemistry* **1997**, *36*, 6633-6642.
29. Kraulis, P. J. *J. Appl. Cryst.* **1991**, *24*, 946-950.
30. Shashidhar, M. S.; Volwerk, J. J.; Griffith, O. H.; Keana, J. F. W. *Chem. Phys. Lipids* **1991**, *60*, 101-110.
31. Rukavishnikov, A. V.; Zaikova, T. O.; Griffith, O. H.; Keana, J. F. W. *Chem. Phys. Lipids* **1997**, *66*, 153-157.

Chapter 7

Mechanism of Phosphatidylinositol-Specific Phospholipase C Revealed by Protein Engineering and Phosphorothioate Analogs of Phosphatidylinositol

Robert J. Hondal[1], Zhong Zhao[1], Alexander V. Kravchuk[1], Hua Liao[2], Suzette R. Riddle[2], Karol S. Bruzik[2,5], and Ming-Daw Tsai[*,1-3,5]

[1]Department of Chemistry, [2]Ohio State Biochemistry Program, and [5]Department of Biochemistry, The Ohio State University, Columbus, OH 43210

The catalytic mechanism of phosphoinositide-specific phospholipase C was investigated by the combined use of protein engineering and application of phosphorothioate analogs of phosphatidylinositol as substrates. The results showed that three residues: His32, His82, and Arg69, each contribute to catalysis up to 10^5-fold rate acceleration factor. In addition, the carboxylic acid residue of Asp274 also contribute to catalysis by interacting with His32 to form a general base, whereas Asp33 forms a triad with Arg69 and His82 which function as general acid both activating the phosphate group for nucleophilic attack and assisting the leaving group. The overall mechanism can be described as involving a complex general acid-general base catalysis.

From an enzymologists' point of view, the mechanism of phosphatidylinositol-specific phospholipase C (PI-PLC, EC 3.14.10) is of great interest because of its close similarity to the mechanism of ribonuclease A, an enzyme which has been studied for more than 35 years (*1-3*). A general base-general acid (GB-GA) mechanism for ribonuclease A involving His12 and His119, respectively, has been widely accepted by most enzymologists. However, the "classical" GB-GA mechanism of ribonuclease A has been the subject of controversy in recent years because of the proposal by Breslow (*4*) in which he suggested that the role of His119 is to first protonate the nonbridging oxygen of the phosphate group to yield a "triester-like" species. The picture of catalysis was further complicated by Raines and coworkers (*5*) who have proposed a catalytic role for Lys41 *via* a hydrogen bonding to a nonbridging oxygen atom of the phosphate group. Gerlt and Gassman have termed this type of hydrogen bonding interaction a *low barrier hydrogen bond* (*6*). In summary, a complete catalytic site of RNase A is comprised of three elements, GB, GA, and the phosphate stabilizing residue.

PI-PLC catalyzes the conversion of phosphatidylinositol (PI) to 1-inositol phosphate (IP) in two distinct steps, *via* the formation of 1,2-cyclic phosphate (IcP) and its hydrolysis to inositol 1-phosphate as shown in Figure 1. The first transesterification step is ca. 10^3-fold faster than the second hydrolysis step (*7*). The

[4]Current address: Department of Medicinal Chemistry and Pharmacognosy, College of Pharmacy, University of Illinois at Chicago, Chicago, IL 60612
[5]Corresponding author.

results of our earlier stereochemical studies using phosphorothioate and oxygen-isotope labeled analogs (8,9) demonstrated that bacterial PI-PLC catalyzes two S_N2-type reactions resulting in the overall retention of configuration at the phosphorus atom, which immediately suggested a mechanism similar to that of ribonuclease A (10). Consistently, the recent x-ray structure of PI-PLC complexed with myo-inositol revealed the presence of His32 and His82 at the active site (11). These two histidine residues are almost superposable on the two histidines of the ribonuclease A active site (11), and are likely to perform the analogous functions. Both enzymes, PI-PLC and ribonuclease A, catalyze the conversion of a phosphodiester to a cyclic intermediate via intramolecular attack of a β-hydroxyl group, followed by slow hydrolysis of the five-membered cyclic product to a linear phosphomonoester. In the case of ribonuclease A, the first transphosphorylation step produces the 2',3'-cyclic nucleotide, which is released (12,13). Analogously to the hydrolysis of IcP, the cyclic phosphodiester is then only slowly hydrolyzed to form a 3'-phosphomonoester.

The initial use of site-directed mutagenesis to study the mechanism of PI-PLC (14,15) has largely confirmed the catalytic mechanism proposed based on the x-ray structure (11) and stereochemical results (8,9), and its similarity to that of ribonuclease A. The pair of active-site histidines was found absolutely essential for catalysis, and Arg69 was proposed to play the role of the third element analogous to that of Lys41 in ribonuclease A (5,11,14,15). Since both enzymes catalyze an intramolecular phosphate-transfer reaction, it is perhaps not surprising that they operate by using analogous chemical mechanisms. It is remarkable however, that both enzymes have evolved to utilize homologous catalytic machineries, when one is a phospholipase that works at a water-lipid interface, and the other is a hydrolytic nuclease.

The results of SDM obtained for PI-PLC allowed definition of the catalytically important residues, however by themselves they did not allow drawing detailed mechanistic inferences. A more precise picture of the mechanism was obtained by application of the combined approach using SDM, and kinetic and stereochemical analysis of phosphorothioate substrate analogs (14,16-18). The results obtained in this way for PI-PLC could also be significant to the mechanism of ribonuclease A, due to similar topology of the two active sites. This review summarizes our work on PI-PLC to date (14,16-18), and compares and contrasts the mechanisms of these two enzymes.

Site-Directed Mutagenesis

In addition to the presence of His32 and His82, the x-ray structure of PI-PLC-inositol complex revealed that each histidine is in close relationship with the carboxyl group of Asp274 and Asp33, respectively. It has been previously shown that His-Leu mutations completely abolished activities in mutants, indicating that these two histidines are essential for catalysis (11). It has been further demonstrated that the mammalian PI-PLC also uses homologous histidines for catalysis (19,20). Using site-directed mutagenesis, we have mutated the two active site histidines to alanine residues (18). The results of specific activity assay using ^{3}H-PI showed that His32 and His82 each contribute a factor of ca. 10^5 toward catalysis (Table 1). Likewise, the carboxylic residues of Asp274 and Asp33 also contribute significantly toward catalysis. We have found that the D274A mutant is 10^4-fold less active than WT PI-PLC, whereas the D274N mutant retains significant activity. The latter finding can be explained by the fact that asparagine is isosteric with aspartate and can maintain a similar hydrogen bonding pattern. Both the structure (11) and SDM data (14,15,18) indicate that Asp274 and His32 function together with the 2-OH group of inositol as a "catalytic triad" similar to those of serine proteases (21). These results are consistent with the proton-relay function of His32•••Asp274. In contrast to Asp274, the mutation of Asp33 to alanine decreased the catalytic rate only by a factor of 10^3, and unlike the D274N mutant, the D33N mutant had only slightly higher activity than D33A (14,18).

Finally, mutation of Arg69 to alanine caused ca. 10^5-fold decrease in activity with respect to WT, while mutation to lysine lowered the rate 10^3-fold (14). Thus, the results of SDM described so far indicated that His32, His82, and Arg69 each contribute a factor of 10^5 toward catalysis, however their specific function in the catalytic machinery remained somewhat unclear.

Use of Phosphorothioates to Study Active Site Interactions

Substitution of a nonbridging oxygen of a phosphate group with sulfur has been a widely used approach to study enzyme mechanisms, in view of the fact that phosphorothioate diesters bear a stereogenic phosphorus atom, and can be used to investigate steric course of the displacement reactions at phosphorus. These *phosphorothionates* are chemically less reactive than phosphates, and display also slower rates of enzymatic cleavage reaction (23). The rate decrease upon *O/S* modification is commonly referred to as the "thio-effect" (classified as the Type I thio-effect). In contrast to phosphorothionates, substitution of a bridging oxygen by a sulfur atom produces a *phosphorothiolate* which is chemically *more reactive* than the corresponding phosphate. The difference in reactivity obtained upon such modification is referred to as the Type II thio-effect. It is worth mentioning that despite rather large rate effects in enzymic reactions, the *O/S* modification can be regarded as a minor one from the structural point of view, allowing exactly the same binding modes of phosphorothioates as compared to phosphates. Thus, given the differential ability of oxygen and sulfur to act as hydrogen bond acceptors (24), the combination of SDM and phosphorothioate analogs offers a unique possibility of examining interactions of specific enzymic residues (such as e.g. Arg69) with specific sites of the substrate (such as the nonbridging/bridging oxygen/sulfur atom). In order to get further insights into the catalytic mechanism of PI-PLC we have employed both types phosphorothioate analogs of phosphatidylinositol to study reactions catalyzed by mutants altered at Arg69, Asp33 and His82 positions. The results obtained from this approach provide far more detailed insights into the enzyme mechanism than application of SDM or substrate alteration alone.

Role of Arg69 in Catalysis: Nonbridging Thio-Effects. Substitution of the nonbridging oxygen atom with sulfur results in formation of a pair of diastereomers of the phosphorothioate analog of PI, designated (R_p)-DPPsI and (S_p)-DPPsI (Figure 2). The R_p-isomer has been previously used in our investigation of the steric course of PI-PLC catalyzed formation of IcP (8). This isomer is turned over by the enzyme only slightly more slowly than the natural substrate ($k_o/k_s = 3$) (Table 2). The ^{31}P NMR time course of the reaction of the 1:1 mixture of both diastereomers with the WT PI-PLC revealed that this enzyme prefers the R_p-diastereomer over the S_p-DPPsI isomer by a factor of 1.6×10^5. To the best of our knowledge, this is the largest enzymic stereoselectivity observed to date between phosphorothioate diastereoisomers. Another important finding was that the R69K mutant showed 10^4-times lower stereoselectivity, $k_{Rp}/k_{Sp} = 16$ (14). This great relaxation of stereoselectivity is achieved by the mutant enzyme by maintaining the same rate of conversion of the S_p-isomer as the WT enzyme, while reducing activity toward the R_p-isomer by a factor of 10^4-fold. This result is consistent with the PI-PLC crystal structure showing that Arg69 is in the correct position to stabilize the negative charge of the phosphate group in the transition state (11). The extremely high stereoselectivity observed with the WT enzyme, and the great relaxation of stereoselectivity observed with the R69K mutant, is an unambiguous evidence for a strong direct interaction between the guanidinium side chain of Arg-69 and the *pro-S* oxygen of the phosphate group. Assuming tentatively that there is no hydrogen bonding stabilization between Arg69 and the sulfur atom in

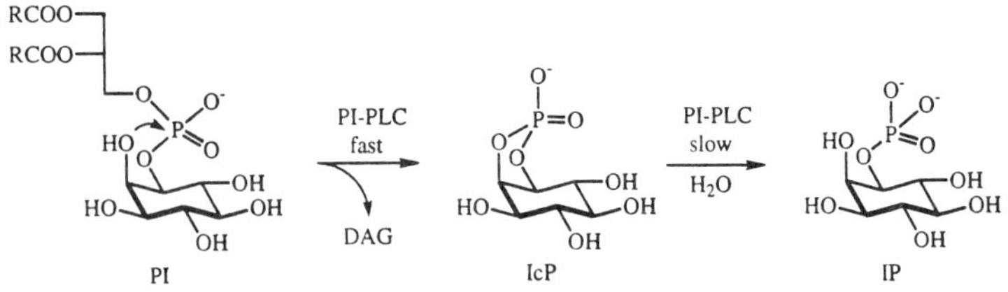

Figure 1. Reactions catalyzed by bacterial phosphatidylinositol-specific phospholipase C.

Table 1. Summary of Kinetic Data for WT and Mutant PI-PLC.

Enzyme	^{3}H-PI	DPsPI	
	Specific Activity (μmol min^{-1} mg^{-1})	V_{max} (μmol min^{-1} mg^{-1})	$K_{m,app}$ (mM)
WT	1300	53.5	0.18
H32A	0.0301	<0.0026	N.D.
H82A	0.013	0.1	0.17
D274A	0.060	<0.0028	N.D.
D274N	21.0	0.29	0.04
D33A	1.3	13.4	0.031
D33N	4.3	0.75	0.030
R69A	0.024	N.D	N.D.
R69K	1.0	0.03	0.08

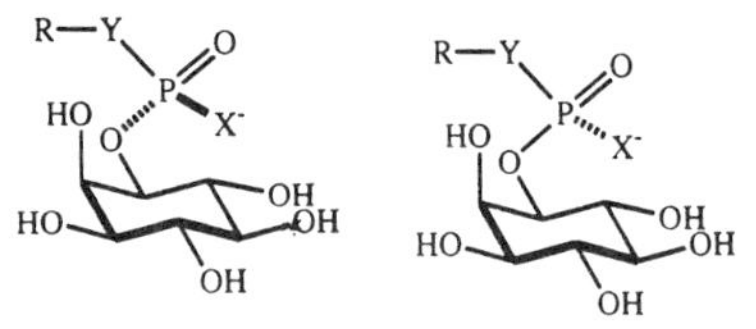

R = -DPG, X = S, Y =O	R_p-DPPsI	S_p-DPPsI
R = -Ph-4-NO$_2$, X = S, Y = O	S_p-NPIPs	R_p-NPIPs
R = DPG, X = O, Y = S	DPsPI	

Figure 2. Phosphorothioate analogs of phosphatidylinositol used in this study.

the reaction of the S_p-DPPsI, the determined stereoselectivity factor allows estimation of the strength of the Arg69-phosphate interaction at 7.0 kcal/mol. This assumption is validated by the fact that the observed stereoselectivity factor is very similar to the rate reduction upon R69A mutation. Consistently, in the reactions of R69K and R69A mutants, where such interactions are reduced or eliminated altogether, very low stereoselectivity is observed.

Similar rate effects as those observed with R_p/S_p-DPPsI were also found for the WT PI-PLC reaction with the phosphorothioate analogs of IcP (18, Table 2). Thus, the *trans*-isomer of IcPs exhibits a thio-effect $k_o/k_s = 5$, similar to that of R_p-DPPsI, while *cis*-IcPs could not be hydrolyzed even with a large amount of enzyme, allowing estimation of the lower limit of stereoselectivity at $k_o/k_s > 10^5$.

Although Asp33 is not implicated by the x-ray structure as interacting directly with the phosphate group, it could indirectly affect the phosphate protonation by way of interaction with Arg69. Examination of the rates of the cleavage of DPPsI isomers with the D33A and D33N mutants revealed unexpectedly that both had very low stereoselectivity, $k_{R_p}/k_{S_p} = 6$. The stereoselectivities of the DPPsI reactions with other active site mutants such as H32A and H82A could not be determined, owing to their very low turnover rates.

We believe that the mechanism that accounts for the above results involves the hydrogen bond formation/protonation of the *pro-S* oxygen atom of the phosphate group by Arg69 in the transition state, although the extent of proton transfer between Arg69 and the oxygen atom of the transition state remains an open question. The large energy of these interactions suggests formation of a low barrier hydrogen bond. It is well established that sulfur is a poor hydrogen bond acceptor, since no hydrogen bonds to sulfur atoms are found in the available crystal structures of nucleoside phosphorothioates (24). It is hence logical that replacement of oxygen by sulfur at the hydrogen bond acceptor site should result in a loss of an important stabilizing interaction, and consequently in a significant impairment of the catalytic efficiency of the enzyme. In contrast, in the mutant enzyme such as R69K, much of the stabilization is already absent, even in the case of the natural substrate, and consequently only a small rate difference between the two isomers is observed. In summary, the use of the diastereomers of the phosphorothioate analogs of PI with nonbridging sulfur unequivocally establishes the Arg69 residue as the phosphate activator function.

Thio-Effects with Phosphophosphorothionates Featuring Good Leaving Group.

Since the rate difference between the diastereomers of DPPsI is related to the free energy of the hydrogen bonding/protonation of the pro-S and pro-R oxygen atoms in the transition state, a lower stereoselectivity would be expected with substrates that require less phosphoryl group protonation/hydrogen bonding in the transition state. It has been proposed by Perreault and Anslyn (25) that the cleavage of phosphodiesters featuring leaving groups with low pKa, such as a stabilized phenoxy group, require less protonation of the nonbridging oxygen in the transition state. To examine this effect we have determined stereoselectivity of reactions of R_p- and S_p-diastereomers of inositol *p*-nitrophenyl phosphorothioate (NPIPs, Figure 2) with WT and mutant PI-PLCs, and the results are also shown in Table 2. In agreement with Anslyn's proposal (25), the stereoselectivity with this substrate is reduced over 100-fold for the WT PI-PLC. As might be anticipated, stereoselectivities are also low for the R69K, R69A, D33A, and D33N mutants, unexpectedly however, stereoselectivity is also reduced 10-fold with the H82A mutant.

The above results establish unequivocally, that the three residues, Arg69, Asp33 and His82 constitute a triad which participate in the activation of the phosphate group. The mechanistic significance of this finding is further explored in the following section by examining the reactivity of Asp33 and H82 mutants with the phosphorothioate analog bearing sulfur in the leaving group.

Role of His-82 and Asp-33: Bridging Thio-Effects. In this work we have used (2R)-1,2-dipalmitoyloxypropane-3-thiophospho-(1-D-*myo*-inositol) (DPsPI), which features a sulfur atom in place of a bridging oxygen atom of the leaving group (*16,18*). Since the cleavage of the P-S bond results in formation of the thiol product, the progress of the reaction can be conveniently monitored in a continuous fashion (Figure 3) (*26,27*). As might be expected, the reactivity of this analog with the WT PI-PLC was significantly lower than that observed with PI. A unique result obtained with this substrate analog is that both D33A and H82A mutants displayed several hundred fold higher activity towards DPsPI than would be predicted based on their activities with PI. We have termed this unique phenomenon the "reverse thio-effect", because for these mutants $k_s/k_o > 1$, an apparent anomaly (*16,18*). The slower turnover of DPsPI could be attributed to the loss or weakening of the leaving group stabilization by the general acid. The reversal of this effect in H82A and D33A mutants (the reverse thio-effect) is almost certainly related to the lower pK_a of the thiol (9.5) as compared to the OH group of diacylglycerol (14.2), and reflects the behavior of phosphorothiolates in the uncatalyzed reactions. The lower pK_a of the thiol allows development of more negative charge on the sulfur atom, thus stabilizing the transition state, and compensates for the loss of the general acid in these mutants. An analogous effect has been demonstrated for ribonuclease A using a low pK_a oxygen leaving group (*22*). The above results clearly established that the major function of both His82 and Asp33 is to interact with the oxygen atom of the leaving group. We have further corroborated this conclusion by showing that D33N mutant displays higher sensitivity than the WT PI-PLC to the pK_a of the alcohol in the reverse transesterification reaction of alcohols with IcP (*18,28*).

It seems useful to compare the results of our studies with DPsPI to those obtained for the cleavage reaction of phosphorothiolate analogs of RNA with the hammerhead ribozyme. Kuimelis and McLaughlin have found that *chemical* reactivity of the bridging 5'-phosphorothiolate RNA analog is 10^6 times greater than the corresponding oxo-substrate (*29*). The increased lability of the phosphorothiolate is due to two factors: (i) the pK_a of the thiol is much lower than the corresponding alcohol, and (ii) the P-S bond is much weaker than the corresponding P-O bond. However, it was found that the hammerhead ribozyme had nearly the same activity towards the 5'-phosphorothiolate RNA analog as it did towards RNA (*30,31*). This results is in concert with our results obtained using DPsPI. Even though the thiolate is chemically more reactive, DPsPI has a 20-fold lower reactivity than PI with the WT enzyme (comparing V_{max} for each substrate). The lower enzymatic activity of the WT PI-PLC towards DPsPI is due to partial compensation of the opposite rate effects: (i) lowering of the activity due to the loss of hydrogen bonding to the leaving group, and (ii) increase of the inherent reactivity due to lower pK_a of the leaving group, with the first effect slightly predominating.

While the reaction catalyzed by PI-PLC is metal independent, both the chemical and ribozyme mediated cleavage reactions of RNA are metal dependent and show the preference for thiophilic metal ions such as Cd^{2+} upon thio-substitution (*24*). For example, a RNA substrate, in which the bridging 3'-oxygen atom has been substituted with sulfur, is a poorer substrate for the ribozyme by a factor of 10^3 when Mg^{2+} is the metal cofactor. The replacement of Mg^{2+} with Mn^{2+} restores activity of the ribozyme towards the thio-analog, so that the ribozyme-catalyzed reaction is only 3-times slower ($k_o/k_s = 3$) (*31*). This effect was interpreted as an indication of existence of specific interactions between the bridging 3'-oxygen atom and the metal ion. In much the same way, we have demonstrated here the presence of interactions between

Table 2. Thio-Effects for WT and Mutant PI-PLC[a].

Enzyme	R_p-DPPsI	S_p-DPPsI	*trans*-IcPs	*cis*-IcPs	(R_p+S_p)NPIPs[a,b]
WT	3	1.6×10^5	5	$\geq 10^5$	10^3
R69K[a]		16			6
R69A[a]					9
D33N[a]		7			6
D33A[a]		5			6
H82A[a]					120

[a] thio-effects as defined by the rate ratio of R_p- vs S_p-isomer. Please, note the configurational assignment of NPIPs isomers is opposite to that of the DPPsI isomers due to difference in the ligands precedence; [b] determined at 10 mM concentration of each isomer.

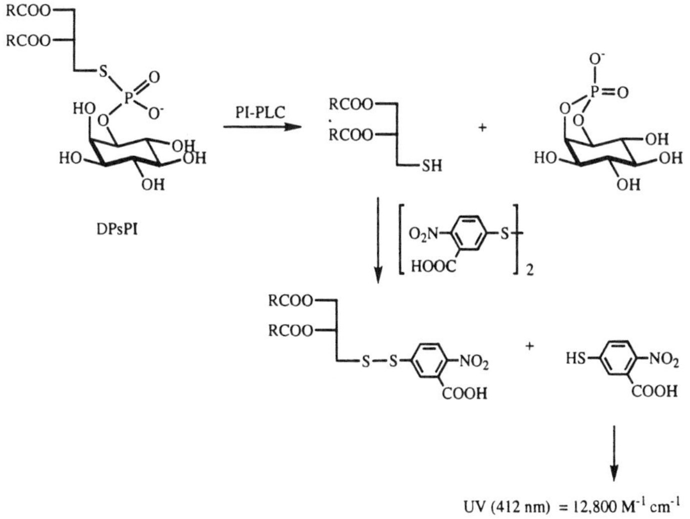

Figure 3. Assay of PI-PLC using the phosphorothiolate analog, DPsPI.

the bridging oxygen atom of PI and the composite general acid, His82•••Asp33 (*16,18*), except this was demonstrated through the replacement of an amino acid instead of a metal ion.

Composite Nature of General Base and General Acid

Our SDM study indicated (*14,18*) that the general base of PI-PLC is composite in nature, and is comprised of the His32•••Asp274 diad. The geometrical relationship between His32•••Asp274 should be critical to the function of the diad. Indeed, the crystallographic data of PI-PLC-inositol complex show that these residues are 2.8Å away from each other, and that they are coplanar (*11*). Such an arrangement is optimal for the proton shuttle function of these two residues and the 2-OH group of inositol. Consistent with strong interactions between His32 and Asp274, it has been recently shown by NMR that these residues form a short hydrogen bond (see the earlier chapter by Griffith et al.). As the carboxylate group of Asp274 is too far away from the 2-OH group of *myo*-inositol in the crystal structure of PI-PLC (*11*), it must play a supporting role, most likely maintaining orientation of His32 and altering its pK_a.

The observation of the reverse thio-effect in reactions of the phosphorothiolate analog with His82 and Asp33 mutants clearly points out a "composite" nature of the general acid. We have originally proposed that this composite acid is comprised of the His82•••Asp33 diad (*15,18*). The observation of the greatly reduced stereoselectivity of Asp33 mutants, and 10-fold reduction in stereoselectivity of H82A mutant, attest that both residues are additionally involved in interaction with Arg69. Since only one of the three amino acids can directly protonate the leaving group, the remaining two amino acids must play a supporting role. The larger rate effect upon mutation of His82 than Asp33, suggests that the former plays a primary role. A logical explanation is that His82 acts as a general acid, whereas the Asp33 residue stabilizes the positive charge on the imidazolium ion. In accord with this proposal we have found that the removal of Asp33 causes a decrease of the pK_a of His82 (*18*). These two residues further interact with Arg69 to form a triad which perform a dual function as discussed later.

Thio-Effects of PI-PLC and Ribonuclease A

The high stereoselectivity exhibited by PI-PLC contrasts with that of ribonuclease A. Ribonuclease A shows a thio-effect of only 5 for *both* isomers of cUMP (*33*). Additionally, the thio-effect for the R_p-isomer of U_pA is also small (2-fold in k_{cat}), but is somewhat larger for the S_p-isomer (70-fold in k_{cat}) (*34*). The large differences observed in P-stereoselectivities of the two enzymes may be explained in part by the differential interactions of lysine and arginine residues with the phosphate group. The interaction of Lys-41 of ribonuclease A is "monodentate" in nature, and results in formation of a single hydrogen bond between its ε-amino group and a nonbridging oxygen of the phosphate group. In addition, as shown by the diffused electron density maps (*35*), the lysine side chain is highly flexible since its orientation is not restricted by interactions with other amino acid residues. The high flexibility of the lysine side chain could provide an equally good access of its ammonium cation to both *pro-S* and *pro-R* oxygens.

In contrast, the guanidinium residue of PI-PLC is capable of forming unique patterns of hydrogen bonds with phosphate ligands (*36*), such as those shown in Figure 4 (*37*). Thus, the large stereoselectivity of PI-PLC can be partly explained in terms of "bidentate" hydrogen bonds between Arg69 and the *pro-S* oxygen such as Type III or Type V interactions. In addition, participation of Arg69 in the triad with Asp33 and His82, most likely imposes significant restrictions on the mobility and orientation of the guanidinium group relative to phosphate, allowing only its exclusive interactions with the *pro-S* oxygen.

The Overall Mechanism

The overall mechanism can be best described as involving participation of the "complex" general acid-general base pair because three elements are required for catalysis (Figure 5). The Asp274 abstracts the proton from His82 which in turn deprotonates the 2-OH group. The function of Asp274 is not only to enhance basicity of His32, but also to maintain its proper orientation for deprotonation of the 2-OH, and to keep it in the correct tautomeric state. The proton abstraction occurs with concomitant activation of the phosphodiester by a hydrogen bond between Arg69 and the *pro-S* oxygen of the phosphate group. As the attacking nucleophile approaches the phosphorus center, and as more negative charge develops on the *pro-S* oxygen, the hydrogen bond may strengthen and the proton transfer may occur.

The data presented here show that the function of Arg69 is augmented by *both* the Asp33 and His82 residues. One of the possibilities is that the three residues also function as a proton shuttle. During the early event, Arg69 protonates the pro-*S* oxygen atom thus facilitating the P-O bond formation. Once the new bond had been formed, the cleavage of the juxtaposed P-O bond could be facilitated by relaying the proton from the nonbridging to the bridging oxygen, using the Arg69-Asp33-H82A shuttle. Shifting the proton from the nonbridging to bridging oxygen should be beneficial to P-O bond breaking process, since it would increase the negative charge of the trigonal O-P-O fragment, and would increase electrostatic repulsion between this fragment and the leaving group. This explanation would require dissecting the cyclization process into two separate chemical events, involving formation and breakdown of the pentacoordinated trigonal bipyramidal (TBP) intermediate. While, such mechanism has been proposed for the chemical and enzymatic cyclization of ribonucleotides (*4*), it cannot be yet regarded as well established. Alternatively, the triad Arg69-Asp33-His82 could act cooperatively to donate two protons simultaneously to the nonbridging and bridging oxygens, assisting the formation of the single TBP transition state. It is now clear that *one of the functions of Asp-33 is to maintain communication between the Arg69 and His82*. In addition, the function of Asp33 is to modulate acid/base properties of Arg69 and His82. The removal of this residue enables greater flexibility of Arg69, so that it can reach both the *pro-S* and *pro-R* oxygens. The latter conclusion is corroborated by the fact that the lower stereoselectivity of the D33N mutant is due to both the reduced rate of R_p-DPPsI turnover and the increased rate of cleavage of S_p-DPPsI.

Conclusions

The combined use of protein engineering and thio-PI analogs as PI-PLC substrates has enabled in depth elucidation of the catalytic mechanism of this enzyme. In particular, our results allow unambiguous assignment of the roles of several active site residues in the overall mechanism. Our stereochemical analyses have provided not only the valuable insight into the mechanism of PI-PLC, but also have allowed a critical reappraisal of the well studied ribonuclease A mechanism.

Acknowledgments: This work was supported by NIH Grant GM 30327. This is paper 5 in the series "Mechanism of Phosphatidylinositol-Specific Phospholipase C". For paper 4, see ref. 18.

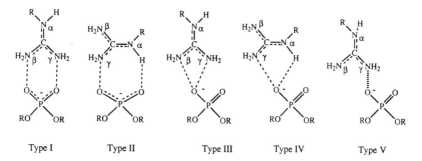

Type I
Type II
Type III
Type IV
Type V

Figure 4. Possible types of interactions between an arginine residue and a phosphate group (modified from ref. *37*).

Asp33
Arg69
His82
Asp274
His32

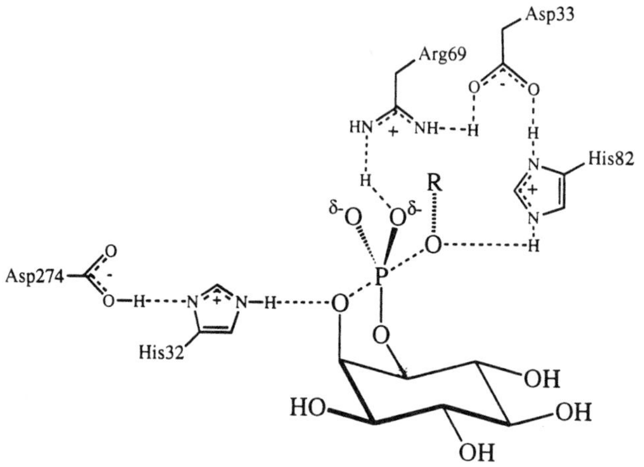

Figure 5. The proposed structure of the transition state showing the composite nature of the general base and general acid. This diagram is meant to show interactions between the phosphate, the leaving group, and the Arg69-Asp33-His82 triad, however, the exact structure of the triad remains unknown.

Literature Cited

1. Richards, F. M.; Wyckoff, H. W. *Enzymes* **1971**, *4*, 647
2. Eftink, M. R.; Biltonen, R. L. In *Hydrolytic Enzymes*; Neuberger, A., Brocklehurst, K., Eds.; Elsevier: New York. 1987; p 333.
3. Blackburn, P.; Moore, S. *Enzymes* **1982**, *15*, 317.
4. Breslow, R.; Xu, R. *Proc. Natl. Acad. Sci. U.S.A.* **1993**, *90*, 1201.
5. Messmore, J. M.; Fuchs, D. N.; Raines, R. T. *J. Am. Chem. Soc.* **1995**, *117*, 8057.
6. Gerlt, J. A.; Gassman, P. G. *Biochemistry* **1993**, *32*, 11943.
7. Volwerk, J. J.; Shashidhar, M. S.; Kuppe, A.; Griffith, O. H. *Biochemistry* **1990**, *29*, 8056.
8. Lin, G.; Bennet, F.; Tsai, M. D. *Biochemistry* **1990**, *29*, 2747.
9. Bruzik, K. S.; Morocho, A. M.; Jhon, D.-Y.; Rhee, S.-G.; Tsai, M.-D. *Biochemistry* **1992**, *31*, 5183.
10. Usher, D. A.; Erenich, E. S.; Eckstein, F. *Proc. Natl. Acad. Sci. U.S.A.* **1972**, *68*, 115.
11. Heinz, D.W.; Ryan, M.; Bullock, T. L.; Griffith, O. H. *EMBO J.* **1995**, *14*, 3855.
12. Thompson, J. E.; Venegas, F. D.; Raines, R. T. *Biochemistry* **1994**, *33*, 7408.
13. Cuchillo, C. M.; Pares, X.; Guasch, A.; Barman, T.; Travers, F.; Nogues, M. V. *FEBS Lett.* **1993**, *3*, 207.
14. Hondal, R. J.; Riddle, S. R.; Kravchuk, A. V.; Zhao, Z.; Bruzik, K. S.; Tsai, M.-D. *Biochemistry* **1997**, *36*, 6633.
15. Gässler, C. S.; Ryan, M.; Liu, T.; Griffith, O. H.; Heinz, D. W. *Biochemistry* **1997**, *36*, 12802.
16. Hondal, R. J.; Zhao, Z.; Bruzik, K. S.; Tsai, M.-D. *J. Am. Chem. Soc.* **1997**, *119*, 5477.
17. Hondal, R.J.; Zhao, Z; Riddle, S. R.; Kravchuk, A. V.; Liao, H; Bruzik, K. S.; Tsai, M.-D. *J. Am. Chem. Soc.* **1997**, *119*, 9933.
18. Hondal, R. J.; Zhao, Z.; Kravchuk, A. V.; Liao, H.; Riddle, S. R; Bruzik, K. S.; Tsai, M.-D. *Biochemistry* **1997**, *37*, 4568-4580.
19. Cheng, H. F.; Jiang, M. J.; Chen, C. L.; Liu, S. M.; Wong, L. P.; Lomasney, J. W.; King, K. *J. Biol. Chem.* **1995**, *270*, 5495.
20. Ellis, M. V.; Sally, U.; Katan, M. *Biochem J.* **1995**, *307*, 69.
21. Craik, C. S.; Roczniak, S.; Largman, C.; Rutter, W. J. *Science* **1987**, *237*, 909.
22. Thompson, J. E.; Raines, R. T. *J. Am. Chem. Soc.* **1994**, *116*, 5467.
23. Herschlag, D.; Piccirilli, J. A.; Cech, T. R. *Biochemistry* **1991**, *30*, 4844.
24. Hinrichs, W.; Steifa, M.; Saenger, W.; Eckstein, F. *Nucleic Acids Res.* **1987**, *15*, 4945.
25. Perreault, D. M.; Anslyn, E. V. *Angew. Chem. Int. Ed. Engl.* **1997**, *36*, 432.
26. Hendrickson, E. K.; Hendrickson, H. S.; Johnson, J. L.; Khan, T. H.; Chial, H. J. *Biochemistry* **1992**, *31*, 12169.
27. Hendrickson, E. K.; Johnson, J. L.; Hendrickson, H. S. *Bioorg. Med. Chem. Lett.* **1991**, *1*, 615.
28. Bruzik, K. S.; Guan, Z.; Riddle, S.; Tsai. M.-D. *J. Am. Chem. Soc.* **1996**, *118*, 7679.
29. Kuimelis, R. G.; McLaughlin, L. W. *Nucleic Acids Res.* **1995**, *23*, 4753.
30. Kuimelis, R. G.; McLaughlin, L. W. *J. Am. Chem. Soc.* **1995**, *117*, 11019.
31. Kuimelis, R. G.; McLaughlin, L. W. *Biochemistry* **1996**, *35*, 5308.
32. Piccirilli, J. A.; Vyle, J. S.; Caruthers, M. H.; Cech, T. R. *Nature* **1993**, *361*, 85.
33. Eckstein, F. *FEBS Lett.* **1968**, *2*, 86.

34. Herschlag, D. *J. Am. Chem. Soc.* **1994**, *116*, 11631.
35. Haydock, K.; Lim, C.; Brünger; A. T. Karlus, M. *J. Am. Chem. Soc.* **1990**, *112*, 3826.
36. Cotton, F. A.; Day, V. W.; Hazen, E. E. Jr.; Larsen, S. *J. Am. Chem. Soc.* **1973**, *95*, 4834.
37. Salunke, D. M.; Vijayan, M. *Int. J. Peptide Protein Res.* **1981**, *18*, 348.

Chapter 8

Structural Analysis of the Catalysis and Membrane Association of PLC-δ_1

L.-O. Essen[1,2], P. D. Brownlie[1], O. Perisic[1], M. Katan[3], and R. L. Williams[1]

[1]Laboratory of Molecular Biology, MRC Centre, Hills Road, Cambridge CB2 2QH, United Kingdom
[2] Department of Membrane Biochemistry, Max-Planck-Institute for Biochemistry, Am Klopferspitz 18a, D-82152 Martinsried, Germany
[3]CRC Centre for Cell and Molecular Biology, Chester Beatty Laboratories, Fulham Road, London SW3 6JB, United Kingdom

The structural analysis of the phosphoinositide-specific phospholipase C δ_1 showed that a catalytic TIM-barrel-like domain is accompanied by three accessory domains: a PH domain, a calmodulin-like EF-hand domain and a C-terminal C2 domain. Crystallographic studies on complexes with substrate analogues combined with mutagenesis data established a reaction pathway where calcium is intimately involved in substrate binding and catalysis. The new 2.4 Å structure of PLC-δ_1 complexed with glycero-phosphoinositol-4-phosphate, together with previously published complex structures, gives a structural basis for the substrate preference $PIP_2 > PIP \gg PI$: The 4-phosphoryl accomplishes a defined binding mode of the inositol head group, whereas the 5-phosphoryl determines its precise orientation in the active site. The refined 2.5 Å structure of the PLC-δ_1/CHAPSO complex identifies a hydrophobic ridge on the catalytic domain as a putative membrane penetration site. These structural features suggest a framework for the future design of PI-PLC specific drugs.

Mammalian phosphoinositide-specific phospholipases C (PI-PLC) catalyze the hydrolysis of phosphatidylinositol 4,5-bisphosphate (PIP_2), a minor phospholipid component of biological membranes, to the second messengers diacylglycerol (DAG) and D-myo-inositol-1,4,5-trisphosphate (IP_3). This reaction is a highly regulated step of numerous signal transduction cascades (*1-3*); the soluble product, IP_3, triggers the influx of calcium from intracellular stores into the cytosol, whereas the membrane-resident diacylglycerol controls cellular protein phosphorylation states by stimulating protein kinase C isozymes. Another important aspect of eukaryotic PI-PLC activity is the regulation of the cellular PIP_2 content, since many cellular functions like the cytoskeletal organisation depend on proteins which are regulated by interaction with PIP_2 (*4,5*).

Ten different isozymes of mammalian PI-PLCs which cluster in three major classes are currently known: β_1-β_4, γ_1-γ_2 and δ_1-δ_4. The β- and γ-isozymes (MW 145-150 kDa) are regulated by G-protein-coupled and tyrosine-kinase-linked receptors, respectively. The δ-isozymes are smaller (85 kDa) than PLC-β and -γ and their widespread occurrence among other eukaryotes such as plants, yeasts or slime molds indicates that this archetypal enzyme class evolved rather early in eukaryotes. All mammalian isozymes are apparently strictly regulated *in vivo* (*2*). PLC-β and -γ contain regulatory domain modules in addition to those found in PLC-δ: PLC-β carries a C-terminal domain that mediates activation by $G_q\alpha$ (*6*); PLC-γ has a complex multidomain array as an insertion in its catalytic domain that consists of a split PH domain, one SH2 and two SH3 domains. Both, SH2 domain interaction with tyrosine kinase receptors and phosphorylation, are required for activation of PLC-γ. The regulation of PLC-δ_1 is still a matter of debate. Proteins which are described to modulate PLC-δ_1 activity are p122-RhoGAP (*7*), or the α subunit of G_h (*8*), a novel G-protein class that combines the GTPase with transglutaminase activity. Calcium is indispensable for mammalian PI-PLC activity and probably represents the key activator of PLC-δ_1 *in vivo* (*9*). The lipids sphingosine and sphingomyelin, or polyamines like spermine were proposed as low molecular-weight modulators of PLC-δ_1 that might have some physiological relevance (*10*). The display of phospholipids in a membrane may also critically regulate PI-PLC activity (*11-14*).

We pursued an X-ray crystallographic approach to address questions concerning the architecture of PI-PLCs, the existence of membrane-binding regions and the mechanisms for catalysis and regulation. Structures of PLC-δ_1 and some of its complexes with cofactors and substrate analogues have been published (*15-18*). We present here an overview of PLC-δ_1 catalysis based on new structures that we have recently determined, and kinetic data.

Architecture of PLC-δ_1

PLC-δ1 is organised in four distinct domains (Figure 1): an N-terminal PH domain (residues 1-132) followed by a calmodulin-like EF-hand domain (residues 133-281), a catalytic TIM-barrel (residues 299-606), and finally a C2 domain (residues 626-756). The PH domain carries a non-catalytic substrate binding site for PIP_2, and was solved as a complex with 1,4,5-IP_3 (*19*, PDB code 1MAI). We determined the structure of the remaining, catalytically active portion, Δ(1-132) PLC-δ_1 (*15*, PDB code 2ISD). The C2 domain was found to be tightly sandwiched between the catalytic TIM-barrel and the second lobe of the calmodulin-like EF-hand domain (Figure 1). The latter two domains form no contacts with each other. The catalytic domain interfaces the C2 domain only with its C-terminal half, also referred to as the 'Y-region' (*20*). This region contributes only a few residues to substrate binding, but not to catalysis. The N-terminal half, 'X-region', carries most of the catalytically important residues and has no interdomain contacts. The domain arrangement of a minimal PLC core consisting of the second EF-hand lobe, the catalytic domain and the C2 domain is apparently invariant: (1) sequence analysis of eukaryotic PI-PLCs shows that residues involved in domain-domain contacts are highly conserved among residues on the domain surfaces, and (2) a comparison between four crystallographically independent PLC-δ_1 molecules in our cubic and a subsequently published triclinic crystal form (PDB code 1QAS) showed only minor structural differences (*16*). The minimal core is found in all eukaryotic PI-PLC enzymes (*21*). Deletion mutants of PLC-δ_1 (*22*) and the observation that plant PI-PLCs consist only of the minimal core (*23*, *24*) indicate that this region is sufficient for catalysis.

Two regions of PLC-δ_1 which might be relevant for regulation are currently not well resolved in structural terms. One is a long disordered peptide linker between the N-terminal and C-terminal half of the catalytic TIM-barrel ('XY-linker', residues 446-488). It is this region that is replaced in γ-isozymes by the regulatory multidomain array controlling PLC-γ acitivity. In PLC-δ_1, the linker is rich in acidic residues. Although this linker region is dispensable for catalytic activity (*25*), it might be a

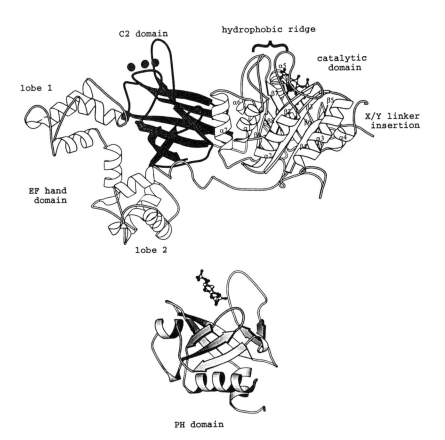

Figure 1. Ribbon drawing showing the domain organisation of PLC-δ_1. The upper panel illustrates the catalytic core domains. The C-terminal C2 domain is shaded to clarify the domain boundaries. The lower panel shows the N-terminal PH domain. Both panels illustrate complexes with 1,4,5-IP$_3$ shown in grey-shaded ball-and-stick representation. The position of the calcium co-factor required for catalysis is shown as a large grey sphere in the catalytic domain. The three sites of calcium/calcium analogue binding in the C2 domain are illustrated as grey spheres. The diagram was generated using MOLSCRIPT (59).

regulatory interaction site for spermine and sphingosine (*10*), two basic compounds known for their stimulatory effects on PLC-δ_1 (*26,27*). The other region showing structural disorder is the first EF-hand lobe consisting of EF-hands 1 and 2 that connects the N-terminal PH domain with the catalytic unit. The first EF-hand lobe is only visible for molecule B of the cubic crystal form, where it is restrained by crystal contacts, but even for this molecule, the "E-helix" of EF-hand 1 is not visible. No parts of the first EF-hand lobe could be observed for molecule A or the two independent molecules of the triclinic crystal form (*16*). Interestingly, the EF-hands 1 and 2 of PLC-δ_1 carry all ligands required for calcium binding. Despite mutational data indicating that this region is not involved in calcium-dependent PI-PLC activity (*22*), and our failure to observe unambiguous calcium binding in this region, it is possible that these sites might have some regulatory relevance. An intrinsic flexibility of the first EF-hand lobe is substantiated by the structural analysis of intact PLC-δ_1, which was crystallised in the same cubic form as $\Delta(1\text{-}132)$ PLC-δ_1. In these crystals, the PH domain and part of the first EF-hand lobe were not visible, although the presence of intact PLC-δ_1 in these crystals was confirmed by SDS-PAGE.

Catalysis in the Active Site of PLC-δ_1

Like in other TIM-barrel enzymes, the active site is located on the C-terminal end of the 8-stranded β-barrel. Its depression-like character (dimensions: $\approx$ 18 Å x 13 Å at its outer rim, maximal depth $\approx$ 10 Å) makes the active site highly accessible. The structural basis for the recognition of the PIP_2 head group and the mechanism of its hydrolysis was revealed by a series of complexes between the enzyme and the inositol phosphates: D-*myo*-inositol-1,4,5-trisphosphate (IP_3), D-*myo*-inositol-2,4,5-tris-phosphate and *myo*-inositol-2-methylene-1,2-cyclic-monophosphonate (*15*, *18*, PDB codes 1DJX, 1DJY and 1DJW; see also Table I). These structures support a sequential two-step mechanism of substrate hydrolysis analogous to the RNAses A or T1 (Figure 2). In the first step, a 1,2-cyclic phosphodiester intermediate is formed by nucleophilic attack of the axial 2-hydroxyl group of PIP_2 on the 1-phosphoryl diester. In the second step, the cyclic reaction intermediate is hydrolysed to yield the acyclic IP_3. His311 and His356 were identified by mutagenesis to be essential for catalysis (*28-30*). His356 is located at the beginning of a β-hairpin loop (residues 356-361) and probably acts as the general acid for protonating the diacylglycerol leaving group. This β-hairpin, along with the C-terminal ends of strands $\beta2$ and $\beta3$ (residues 343-345, 390-392), forms a small mixed β-sheet which is a structural feature unique to eukaryotic and prokaryotic PI-PLCs as compared with other TIM-barrel enzymes. His311 is part of a β-bulge which positions the protonated imidazole of His311 close to the highly negatively charged pentavalent 1-phosphoryl of the transition state.

A striking aspect of the active site is the presence of a calcium ion as a catalytic cofactor. Its direct coordination to the attacking 2-hydroxyl group of the inositol moiety not only contributes to substrate binding, but probably also enhances the nucleophilicity of this group by lowering its pK_a. The calcium ion is coordinated in the active site by the carboxyl groups of Asn312, Glu341, Asp343 and Glu390. Consequently, mutations of the calcium ligands to alanine or glycine lead to a complete loss of catalytic activity (*29, 31*). A second role of the calcium is the stabilisation of the pentavalent transition state by additional coordination to the prochiral O_S atom of the phosphodiester group (*18*). In the calcium-independent PI-PLC from the prokaryote *Bacillus cereus*, this function was attributed to Arg69 whose guanidinium group occupies an analogous position as the calcium cofactor in eukaryotic enzymes (*32, 33*). Attempts to engineer a calcium-independent PLC-δ_1 variant by replacing the calcium site with the arginine or lysine residues of the bacterial enzyme (e.g., D343R or D343R/E390K) failed to give active protein (*31*).

Table 1. Crystallographically analysed complexes of PLC-$\delta1$ with substrate analogues

Compound	Abbreviation	Source	Cofactor	Resolution (Å)	Ordered	Reference	Comments
D-myo-inositol-1,4,5-trisphosphate	1,4,5-IP$_3$	Alexis	Ca^{2+}	2.30	yes	1996,1997b	
D-myo-inositol-2,4,5-trisphosphate	2,4,5-IP$_3$	Robin Irvine	Ca^{2+}	2.80	yes	1997b	
D-myo-inositol-4,5-bisphosphate	4,5-IP$_2$	SIGMA	Ca^{2+}	2.95	yes	1997b	
myo-inositol-2-methylene-1,2-cyclic-monophosphonate	cICH$_2$P	Mary Roberts	Ca^{2+}	2.45	yes	1997b	
glycero-phospho-D-myo-inositol	GPI	Calbiochem	Ba^{2+}	2.5	no	unpublished	
glycero-thio-phospho-D-myo-inositol	GTPI	Karol Bruzik	Ca^{2+}		no	unpublished	
glycero-phospho-D-myo-inositol-4-phosphate	GPIP	SIGMA	Ca^{2+}	2.40	yes	unpublished	only IP$_2$ moiety
glycero-phospho-D-myo-inositol-4,5-bisphosphate	GPIP$_2$	SIGMA	La^{3+}	2.30	yes	unpublished	only IP$_3$ moiety
1,2-dihexanoyl-glycero-3-phospho-D-myo-inositol	diC$_6$-PI	Mary Roberts	La^{3+}	2.60	no	unpublished	
1,2-dihexanoyl-glycero-dithio-phospho-D-myo-inositol	diC$_6$-dithio-PI	Stephen Martin	Ca^{2+}	2.50	no	unpublished	
1,2-dihexanoyl-glycero-3-phospho-D-scyllo-inositol	diC$_6$-scyllo-PI	Stephen Martin	Ca^{2+}	2.70	no	unpublished	
D-myo-inositol-4-(hexadecyloxy)-3-metoxybutane-phosphonate	C$_4$-PI	Robert Bittman	Ca^{2+}	3.50	no	unpublished	
1,2-dibutyryl-glycero-thio-phospho-D-myo-inositol-4,5-bisphosphate	diC$_4$-PIP$_2$	Glenn Prestwich	Ca^{2+}	2.70	yes	unpublished	primarily IP$_3$ moiety
3-((3-cholamidopropyl) dimethylammonio)-2-hydroxy-1-propane-sulphonate	CHAPSO	Fluka	none	2.50	yes	1996	
Manoalide		Alexis	Ca^{2+}	2.60	no	unpublished	

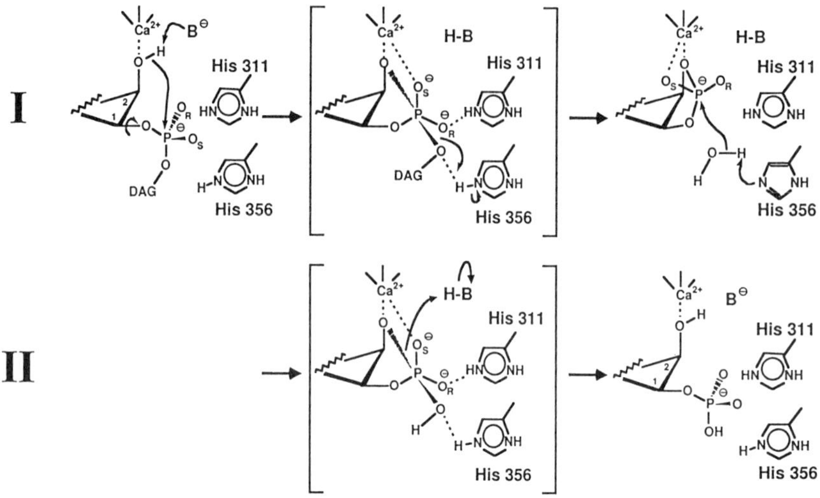

Figure 2. The two-step catalytic mechanism of eukaryotic PI-PLCs. The first step (I) is a phosphotransferase reaction where the 1-phosphoryl group is transferred from DAG onto the 2-hydroxy group of the inositol moiety. The subsequent phosphohydrolase reaction (II) converts the 1,2-cyclic intermediate to the acyclic inositol phosphate.

The identification of the general base responsible for the abstraction of the proton from the 2-hydroxyl group is still under investigation. In the bacterial enzyme, the residue analogous to His311 was shown to act as the general base, and to form a hydrogen bond with the 2-hydroxyl of the inositol moiety (*34*). In PLC-δ_1, such a hydrogen bond with His311 is not observed for any of the complexes. Only the calcium ligands Glu341 and Glu390 form hydrogen bonds with the 2-OH of the inositol. Mutagenesis of these residues to glutamine resulted in significant residual activity of E390G mutant, suggesting that the Glu341 residue would be more likely to act as a general base for the first step of the reaction (*31*).

Structural Aspects of Substrate Preference

At physiological calcium concentrations (10 nM - 10 μM) the substrate preference of eukaryotic PI-PLCs for phosphoinositides is $PIP_2 > PIP \gg PI$ (*35*). The same preference for substrates phosphorylated at the 4- and 5-position of the inositol moiety is exhibited against the deacylated and soluble substrates glycero-phosphoinositol (GPI), glycero-phosphoinositol-4-phosphate (GPIP) and glycero-phosphoinositol-4,5-bisphosphate ($GPIP_2$). The V_{max} of PLC-δ_1 is about six times higher for $GPIP_2$ (4.8 μmol/min/mg) than for GPIP (0.8 μmol/min/mg, see ref. *36*). In contrast, GPI lacking a 4-phosphoryl is such a poor substrate that V_{max} could not be determined. The importance of the 4-phosphoryl on substrate binding and hydrolysis can be rationalized in structural terms. In complexes with inositol phosphates containing both 4- and 5-phosphoryls, the 4-phosphoryl is tightly involved in hydrogen bonds and salt bridges with residues Lys438, Ser522 and Arg549. Several additional indirect interactions between the 4-phosphoryl and the active site are mediated by intervening water molecules. In contrast, the 5-phosphoryl group forms only a single salt bridge with Lys440. Consequently, mutations of the residues involved in the recognition of the 4-phosphoryl group led to reduced enzymatic activities towards PIP_2 and PIP, whereas the activity towards PI is mostly retained (*29, 31, 37*).

Unlike the 4-phosphoryl that is buried at the bottom of the active site, the 5-phosphoryl is solvent exposed in these complexes. The salt bridge of the 5-phosphoryl to Lys440 and two further water-mediated interactions with the active site seem not to contribute significantly to the stabilisation of an enzyme-substrate complex as is indicated by the similar K_m values for GPIP and $GPIP_2$ (1.8 mM and 4.4 mM, respectively). However, the removal of the 5-phosphoryl group affects the alignment of the substrate head group in the active site as shown by the recent 2.4 Å structure of the complex between PLC-δ_1, GPIP and calcium (Figure 3A, 3B, 3C). Only the 1,4-IP_2 moiety shows electron density, whereas the glycerol moiety is apparently lost by hydrolysis. In comparison to other inositol phosphates with 4- and 5-phosphoryls, the 1,4-IP_2 moiety rotates around the 1-phosphoryl by approximately 30° out of the active site, thus displacing the 4-phosphoryl by 4.4 Å (Figure 3B, 3D). The interactions between the 4-phosphoryl and the active site are now mediated by two bridging water molecules that invade the 4-phosphoryl binding pocket around Lys438, Ser522 and Arg549. Most of the interactions around the 1-phosphoryl and the catalytic calcium site whose importance for catalysis was recognized in the PLC-δ_1/IP_3 complex, can also be found in the PLC-δ_1/GPIP complex (Figure 3C, 3D). Significant structural differences among residues of the active site are observed solely for Glu341 and Arg549: Arg549 moves its guanidinium group by 2.0 Å towards the 4-phosphoryl; Glu341 preserves its salt bridge with Arg549 by shifting the carboxyl group by 1.8 Å. These small structural changes and the different alignment of the inositol head group might correlate with the sixfold decrease of catalytic turnover for GPIP.

128

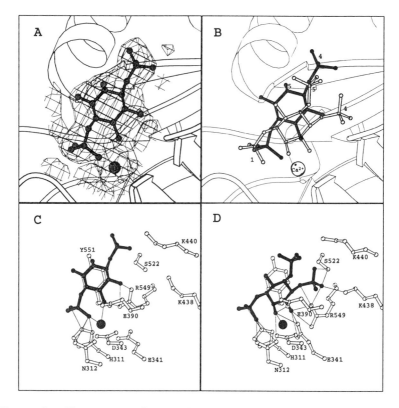

Figure 3. The structure of a complex between Δ(1-132) PLC-δ₁ and GPIP. Cubic crystals were soaked for 32 hours with 10 mM GPIP then for 14 min with 10mM GPIP and 1 mM calcium chloride. The model was refined to 2.4 Å resolution with the program TNT (*60*) using 98520 unique reflections (final R-factor 0.22, R-free 0.28, 667 water molecules). (A) The SIGMAA-weighted (*61*) m|Fo|-D|Fc| difference electron density contoured at 1.0 σ. The density within 3.0 Å of the GPIP is illustrated superimposed on the model of the 1,4-IP₂ head group of the GPIP. No density is visible for the glycerol moiety that was presumably hydrolysed by the enzyme. (B) Comparison of the observed 1,4-IP₂ moiety (grey-shaded ball-and-stick) with the inositide of the 1,4,5-IP₃/PLC-δ₁ complex (white ball-and-stick, see ref. *15*). The position of the calcium cofactor is identical in both complexes. (C) Interactions with the protein and co-factor closer than 3.5 Å for the GPIP/enzyme complex. (D) Interactions with the protein and co-factor closer than 3.5 Å for the 1,4,5-IP₂/enzyme complex. The diagram was generated using BOBSCRIPT (*62*).

In contrast, none of the PI analogues analysed so far (Table I) showed well resolved electron density for the inositol residue in difference Fourier syntheses for crystals soaked or cocrystallised with these compounds. Only the catalytic calcium and some globular density at the presumed position of the 1-phosphate were visible in these complexes. The absence of a 4-phosphoryl serving as an anchor for the inositol moiety in the active site is probably responsible for this disorder. Multiple modes for aligning the headgroup of such PI-like substrates in the active site might explain why these unphosphorylated phosphoinositides are so ineffective as substrates.

PI-PLCs release cyclic inositol phosphates as side products during phosphoinositide hydrolysis (38). A premature release of the cyclic reaction intermediates appears to be a consequence of the open nature of the active site. This would be consistent with the reported K_m of 28 mM for the hydrolysis of cIP by PLC-δ_1 that largely precludes the hydrolysis of already released cIP (36).

Structural Features Important for Interfacial Catalysis

As an interfacial catalyst, the full kinetic description of PLC-δ_1 activity is a demanding task. Three regions have been identified that might be important for interfacial catalysis and reversible membrane binding. The first region is the N-terminal PH domain which has an affinity towards PIP_2 by two orders of magnitude higher than the catalytic domain (K_D of 1.7 μM vs. greater than 100 μM). Its presence is clearly required for processive catalysis (39) and the tethering of PLC-δ_1 to the substrate-containing plasma membrane (40). A mutant with tighter binding of the PH domain to PIP_2, E54K, was recently reported to exhibit an increased processivity (41). The product IP_3 is capable of inhibiting PLC-δ_1 at physiological concentrations ($\approx$ 1 μM) in permeabilized cells (9) and residues forming the IP_3-binding surface were shown by mutagenesis to be critical for hydrolysis of PIP_2 and membrane attachment (42). The IP_3 inhibition is most likely caused by competition with PIP_2 for binding to the PH domain.

The second region which could affect binding to the lipid bilayer is the C-terminal C2 domain. Its β-sandwich topology is the same as found in cytosolic phospholipase A$_2$ (cPLA$_2$, see ref. 43) and circularly permuted relative to the C2 domain from synaptotagmin I (P-type vs. S-type: β-strands 8,1,2,3,4,5,6,7 permute to 1,2,3,4,5,6,7,8, see ref. 44). A novel feature is the calcium binding region (CBR) of the PLC-δ_1 C2 domain. Up to three metal ions can be bound in a crevice-like invagination as was demonstrated by analysis of several calcium and calcium analogue complexes (15-17, PDB codes 1DJG, 1DJH, 1DJI, 1QAT). This region resides on the same side of the enzyme as the active site and consists of three loops called CBR-1, CBR-2 and CBR-3 which contribute the calcium ligands. The C2A domain of synaptotagmin I and the C2 domain of cPLA2 mediate calcium-dependent binding to phospholipids (45, 46), and show similar features of calcium ligation (43,44,47). How calcium binding to the C2 domain of PLC-δ_1 could affect catalytic activity is not yet clear; binding of calcium ions might directly steer the docking of the minimal PLC core onto the lipid membrane.

The third structural feature that might affect interfacial catalysis is the ridge around the active site depression that has a distinct hydrophobic character. Kinetic measurements suggested that the enzyme might partially penetrate into the membrane during catalysis (48). The putative role of the hydrophobic ridge in the partial penetration of the enzyme into the hydrophobic zone of the lipid bilayer was suggested by early structural data of the complex between PLC-δ_1 and the detergent CHAPSO (15). In the refined 2.5 Å structure, the hydrophilic moiety of CHAPSO extends into the active site so that its terminal sulfonyl group occupies a position close to the 1-phosphoryl group of 1,4-IP$_2$ in the PLC-δ_1/GPIP complex (Figure 4). The binding site for the calcium cofactor is preserved and the observed metal ion coordinates to the sulphonyl group in a bidentate manner. The 2-hydroxyl group of CHAPSO resides close to the site where the 2- and 3-hydroxyl groups of inositol phosphates form hydrogen bonds with Glu341 and Glu390. The steroid portion binds to the hydrophobic ridge around the active site depression and makes van-der-Waals contacts with the sidechains of Leu320 and Thr557. Other residues nearby that contribute to the hydrophobic character of this surface area are Tyr358, Phe360 and Trp555. All of these residues belong to loops connecting Tβ1 with Tα1, Tβ2 with Tα2 and Tβ7 with Tα6. CHAPSO binding to PLC-δ_1 is probably very weak and partially dictated by the crystal packing of the cubic crystal form: (1) CHAPSO does not act as an inhibitor and is actually used in mixed-micelle assays for analysing PLC activity, and (2) binding is restricted only to molecule B in the asymmetric unit of the cubic crystals. Parts of the steroid moiety make additional contacts with the hydrophobic ridge of molecule A and thus stabilise an important crystal contact (Figure 4C). The hydrophobic character of the ridge is preserved in the whole eukaryotic PI-PLC family despite some sequence variation (Figure 5). The activity of PI-PLCs depends strongly on the membrane surface pressure (48-50). From these analyses, a protein surface area of about 1 nm^2 was estimated to penetrate into the membrane (49) which corresponds roughly to the extent of the hydrophobic ridge. Consistent with this view, mutations of Leu320, Phe360 and Trp555 to alanine decrease significantly the pressure dependence of PLC-δ_1 (31).

Insights for the Regulation of PLC-δ_1

Although the regulation of PLC-δ_1 is still not entirely clear, there is some evidence that calcium might be the key regulator for PLC-δ_1 activity (9, 51) whose potency in activation might be modulated by additional factors (7, 52). PLC-δ_1 shows multiple calcium binding sites: in the active site, in the C2 domain and probably also in the first and second EF-hands of the EF-hand domain (Figure 6). The calcium sites in the EF-hand and C2 domains of PLC-δ_1 are not highly conserved among other isozymes, e.g. β-isozymes lack calcium ligands in the C2 domain and most of the putative calcium ligands in EF-hand 1. A possible role of the C2 domain is the tethering of the minimal PLC core on the lipid membrane. While calcium represents a positive regulatory signal for PLC-δ_1 activity, IP$_3$ might provide a negative signal by competing with the anchoring of PLC-δ_1 to the lipid membrane via the PH domain/PIP$_2$ interaction (53).

Remarks for the Design of PI-PLC Specific Inhibitors

Mammalian PI-PLCs have been proposed as interesting targets in controlling certain human cancers due to their involvement in growth factor and oncogene signaling (54). Two general classes of inhibitors have been developed, phosphoinositide-like and non-phosphoinositide-like inhibitors (55). The phosphoinositide-like inhibitors are mostly derived from PI, only some of them have reported IC$_{50}$ values in the submicromolar range (56). Structural analysis of several complexes with PI analogues failed to demonstrate unambiguous electron density for the inositol ring in the active site, as outlined above. Therefore, addition of further polar functions to the 4-hydroxyl position might be beneficial for tighter binding of future inhibitors, although some price may have to be paid in terms of permeability.

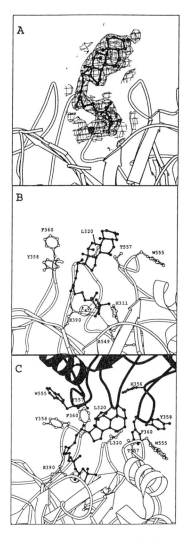

Figure 4. The structure of a complex of Δ(1-132) PLC-δ$_1$ with CHAPSO. The model of the complex was refined to 2.5 Å resolution (91451 unique reflections, R-factor 0.23, R-free 0.28, 513 water molecules). (A) The SIGMAA-weighted m|Fo|-D|Fc| difference electron density contoured at 1.0 σ. The density within 3.0 Å of the CHAPSO is illustrated superimposed on the model of the CHAPSO molecule. (B) A model of the CHAPSO (grey-shaded ball-and-stick) bound to molecule B of the crystal's asymmetric unit. The residues of the hydrophobic ridge and the residues with which the hydrophilic tail of the detergent makes putative hydrogen bonding interactions are shown in ball-and-stick representation (white). (C) The CHAPSO/enzyme complex viewed down the non-crystallographic two-fold axis relating the two molecules in the asymmetric unit. Molecule B (white) makes nearly all of the interactions with the CHAPSO molecule (black). Parts of the hydrophobic ridge of molecule A (shaded grey) contact the hydrophobic ridge of molecule B and the edge of the CHAPSO steroid moiety.

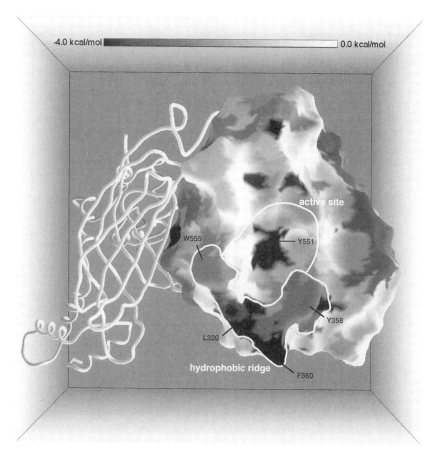

Figure 5. Surface representation of the eukaryotic PI-PLC catalytic domain viewed towards the active site. The surface is color-coded according to the underlying average hydrophobicity found for surface-exposed residues (black: hydrophobic, white: hydrophilic). For this purpose a multiple sequence alignment of 25 eukaryotic PI-PLCs was generated using CLUSTAL W (*63*) and adjusted manually to be consistent with the structure of PLC-δ_1. For each sequence position an average hydrophobicity was calculated using the relative occurence of each amino acid type and its hydrophobicity according to the corrected hydrophobicity scale of Sharp et al. (*64*). The average hydrophobicities thus obtained were mapped on the molecular surface of the catalytic domain (4 Å probe radius) using GRASP (*65*). The Figure shows a strong conservation for solvent-exposed hydrophobic residues corresponding to Leu 320 and Phe 360 of PLC-δ_1. These residues are located at the very tip of the hydrophobic ridge which comprises additionally Tyr 358, Trp 555 and Thr 557. For comparison, Tyr 551 which is located in the active site depression is also strictly conserved. This residue is responsible for substrate binding by packing against the inositol moiety of the lipid head group.

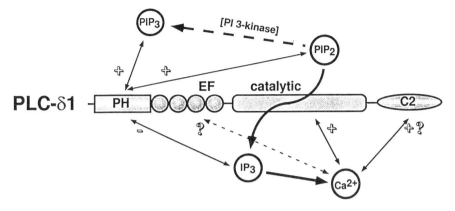

Figure 6. The PIP_2 signalling pathway (thick arrows) and its putative involvements in the regulation of PLC-δ_1 (thin arrows indicating activation, +, or inactivation, -, of PLC-δ_1 activity). A role of the calcium binding sites found in the C2 domain and postulated for EF-hands 1 and 2 for activity of PLC-δ_1 is still hypothetical.

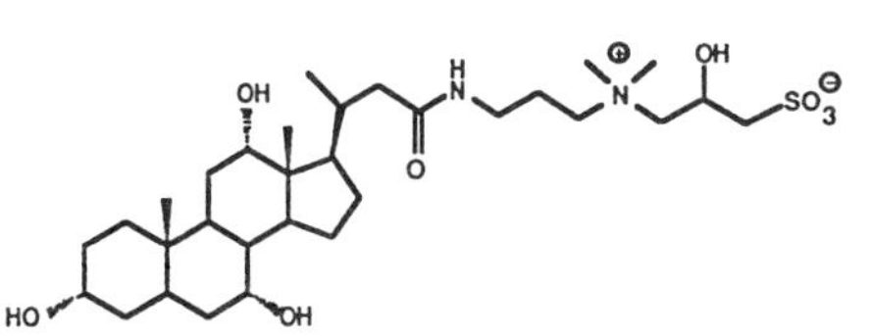

CHAPSO

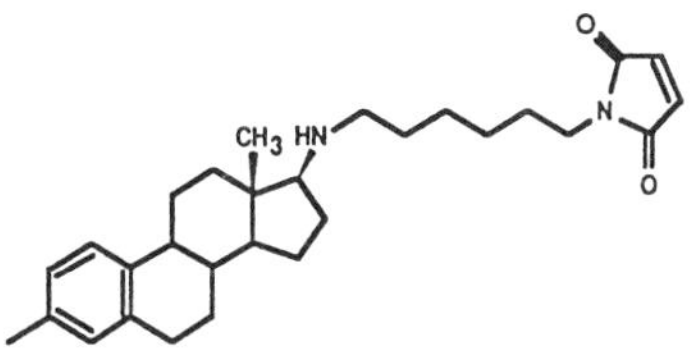

U73122

$IC_{50} \approx 1$ -10 µM, irreversible

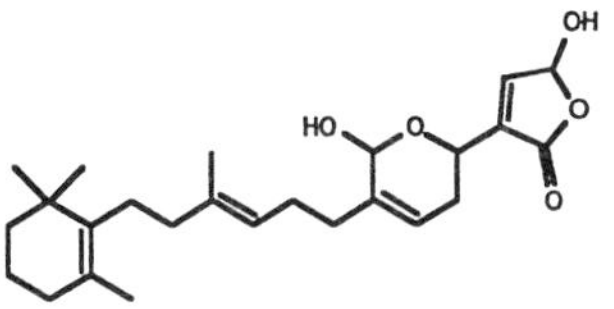

Manoalide

$IC_{50} \approx 3$ -6 µM, irreversible

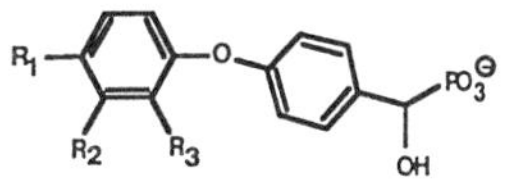

Merck, Patent No. WO 95/10286

$IC_{50} \approx 10$ - 1000 µM

Figure 7. Structural relationships between CHAPSO and non-phosphoinositide PI-PLC inhibitors. The IC_{50} values for manoalide, U73122 and nitrocoumarins are taken from references 57, 58 and 66.

The non-phosphoinositide-inihibitors have in common a large hydrophobic group connected to a polar or reactive group (Figure 7). The one mostly used in cell biology, U73122, is a steroid derivative and bears some structural similarity to the detergent CHAPSO. The binding mode of CHAPSO to PLC-δ_1 might serve as a model for a class of inhibitors where a hydrophobic group would cover parts of the hydrophobic rim and a reactive group, as found in U73122, manoalide (57) or 3-nitrocoumarins (57), would form covalent adducts with nucleophilic residues of the active site.

An unsolved problem is the lack of isozyme-specificity that is required to affect specifically oncogene-related pathways. Because the residues contributing to the active site are strictly conserved among PI-PLCs from simple eukaryotes to humans, an isozyme-specific approach would probably have to target the protein-protein interactions between the regulatory domains of PLC-β and PLC-γ and their respective binding partners.

Acknowledgements

L. O. E gratefully acknowledges the support from the European Commision, M. K. from the Cancer Research Campaign, and R. L. W. from the British Heart Foundation and from the ZENECA/DTI/MRC LINK Programme. We thank the staff of beamlines BL4/ID2 at ESRF, Grenoble, BW7B at EMBL Hamburg and Station 9.6 at Daresbury SRS for valuable technical support.

Literature Cited

1. Katan, M.; Williams, R. L. *Seminars in Cell Develop. Biol.* **1997**, *8*, 287-296.
2. Rhee, S. G.; Bae, Y. S. *J. Biol. Chem.* **1997**, *272*, 15045-15048.
3. Singer, W. D.; Brown, H. A.; Sternweis, P. C. *Annu. Rev. Biochem.* **1997**, *66*, 475-509.
4. Divecha, N.; Irvine, R. F. *Cell* **1995**, *80*, 269-278.
5. Lee, S. B.; Rhee, S. G. *Curr. Opin. Cell Biol.* **1995**, *7*, 183-189.
6. Morris, A. J.; Scarlata, S. *Biochem. Pharmacol.* **1997**, *54*, 429-435.
7. Homma, Y.; Emori, Y. *EMBO J.* **1995**, *14*, 286-291.
8. Hwang, K. C.; Gray, C. D.; Sivasubramanian, N.; Im, M. J. *J. Biol. Chem.* **1995**, *270*, 27058-27062.
9. Allen, V.; Swigart, P.; Cheung, R.; Cockcroft, S.; Katan, M. *Biochem. J.* **1997**, *327*, 545-552.
10. Pawelczyk, T.; Matecki, A. *Eur. J. Biochem.* **1997**, *248*, 459-465.
11. Glaser, M.; Wanaski, S.; Buser, C. A.; Boguslavsky, V.; Rashidzada, W.; Morris, A.; Rebecchi, M.; Scarlata, S. F.; Runnels, L. W.; Prestwich, G. D.; Chen, J.; Aderm, A.; Ahn, J.; McLaughlin, S. *J. Biol. Chem.* **1996**, *18*, 26187-26193.
12. Scarlata, S.; Gupta, R.; Garcia, P.; Keach, H.; Shah, S.; Kasireddy, C. R.; Bittman, R.; Rebecchi, M. J. *Biochemistry* **1996**, *35*, 14882-14888.
13. Hurley, J. H.; Grobler, J. A. *Curr. Opin. Struct. Biol.* **1997**, *7*, 557-565.
14. James, S. R.; Downes, C. P. *Cellular Signalling* **1997**, *9*, 329-336.
15. Essen, L. O.; Perisic, O.; Cheung, R.; Katan, M.; Williams, R. L. *Nature* **1996**, *380*, 595-602.
16. Grobler, J. A.; Essen, L. O.; Williams, R. L.; Hurley, J. H. *Nature Structural Biology* **1996**, *3*, 788-95.
17. Essen, L.-O.; Perisic, O.; Lynch, D.; Katan, M.; Williams, R. L. *Biochemistry* **1997**, *36*, 2753-2762.
18. Essen, L.-O.; Perisic, O.; Roberts, M.; Katan, M.; Williams, R. L. *Biochemistry* **1997**, *36*, 1704-1718.

19. Ferguson, K. M.; Lemmon, M. A.; Schlessinger, J.; Sigler, P. B. *Cell* **1995**, *83*, 1037-1046.
20. Rhee, S. G.; Choi, K. D. *J. Biol. Chem.* **1992**, *267*, 12393-12396.
21. Williams, R. L.; Katan, M. *Structure* **1996**, *4*, 1387-139422.
22. Nakashima, S.; Banno, Y.; Watanabe, T.; Nakamura, Y.; Mizutani, T.; Sakai, H.; Zhao, Y.; Sugimoto, Y.; Nozawa, Y. *Biochem. Biophys. Res. Comm.* **1995**, *211*, 364-369.
23. Hirayama, T.; Ohto, C.; Mizoguchi, T.; Shinozaki, K. *Proc. Natl. Acad. Sci. USA* **1995**, *92*, 3903-3907.
24. Shi, J.; Gonzales, R. A.; Bhattacharyya, M. K. *Plant J.* **1995**, *8*, 381-390.
25. Ellis, M. V.; Carne, A.; Katan, M. *Eur. J. Biochem.* **1993**, *213*, 339-347.
26. Haber, M. T.; Fukui, T.; Lebowitz, M. S.; Lowenstein, J. M. *Arch. Biochem. Biophys.* **1991**, *288*, 243-249.
27. Pawelczyk, T.; Lowenstein, J. M. *Arch. Biochem. Biophys.* **1992**, *297*, 328-333.
28. Smith, M. R.; Liu, Y.-L.; Matthews, N. T.; Rhee, S. G.; Sung, W. K.; Kung, H.-F. *Proc. Natl. Acad. Sci. USA* **1994**, *91*, 6554-6558.
29. Chen, H.-F.; Jiang, M.-J.; Chen, C.-L.; Liu, S.-M.; Wong, L.-P.; Lomasney, J. W.; King, K. *J. Biol. Chem.* **1995**, *270*, 5495-5505.
30. Ellis, M. V.; U.; S.; Katan, M. *Biochem. J.* **1995**, *307*, 69-75.
31. Ellis, M. V.; James, S. R.; Perisic, O.; Downes, C. P.; Williams, R. L.; Katan, M. *J. Biol. Chem.* **1998**, in press.
32. Hondal, R. J.; Riddle, S. R.; Kravchuk, A. V.; Zhao, Z.; Liao, H.; Bruzik, K. S.; Tsai, M.-D. *Biochemistry* **1997**, *36*, 6633-6642.
33. Heinz, D.; Essen, L.-O.; Williams, R. L. *J. Mol. Biol.* **1998**, *275*, 635-650.
34. Heinz, D. W.; Ryan, M.; Bullock, T. L.; Griffith, O. H. *EMBO J.* **1995**, *14*, 3855-3863.
35. Ryu, S. H.; Suh, P.-G.; Cho, K. S.; Lee, K.-Y.; Rhee, S. G. *Proc. Natl. Acad. Sci. USA* **1987**, *84*, 6649-6653.
36. Wu, Y.; Perisic, O.; Williams, R. L.; Katan, M.; Roberts, M. F. *Biochemistry* **1997**, *36*, 11223-1123312.
37. Wang, L.-P.; Lim, C.; Kuan, Y.-S.; Chen, C.-L.; Chen, H.-F.; King, K. *J. Biol. Chem.* **1996**, *271*, 24505-24516.
38. Kim, J. W.; Ryu, S. H.; Rhee, S. G. *Biochem. Biophys. Res. Commun.* **1989**, *163*, 177-182.
39. Cifuentes, M. E.; Honkanen, L.; Rebecchi, M. J. *J. Biol. Chem.* **1993**, *268*, 11586-11593.
40. Paterson, H. F.; Savopoulos, J. W.; Perisic, O.; Cheung, R.; Ellis, M. V.; Williams, R. L.; Katan, M. *Biochem. J.* **1995**, *312*, 661-666.
41. Bromann, P. A.; Boetticher, E. E.; Lomasney, J. W. *J. Biol. Chem.* **1997**, *272*, 16240-16246.
42. Yagisawa, H.; Sakuma, K.; Paterson, H. F.; Cheung, R.; Allen, V.; Hirata, H.; Watanabe, Y.; Hirata, M.; Williams, R. L.; Katan, M. *J. Biol. Chem.* **1998**, *273*, 417-424.
43. Perisic, O.; Fong, S.; Lynch, D. E.; Bycroft, M.; Williams, R. L. *J. Biol. Chem.* **1998**, *273*, 1596-1604.
44. Sutton, R. B.; Davletov, B. A.; Berghuis, A. M.; Sudhof, T. C.; Sprang, S. R. *Cell* **199**, *80*, 929-938.
45. Sugita, S.; Hata, Y.; Sudhof, T. C. *J. Biol. Chem.* **1996**, *271*, 1262-1265.
46. Nalefski, E. A.; McDonagh, T.; Somers, W.; Seehra, J.; Falke, J. J.; Clark, J. D. *J. Biol. Chem.* **1998**, *273*, 1365-1372.
47. Shao, X.; Davletov, B. A.; Sutton, B.; Sudhof, T. C.; Rizo, J. *Science* **1996**, *273*, 248-251.

48. James, S. R.; Demel, R. A.; Downes, C. P. *Biochem. J.* **1994**, *298*, 499-506.
49. Rebecchi, M.; Boguslavsky, V.; Boguslavsky, L.; McLaughlin, S. *Biochemistry* **1992**, *31*, 12748-12753.
50. Boguslavsky, V.; Rebecchi, M.; Morris, A. J.; Jhon, D.-Y.; Rhee, S. G.; McLaughlin, S. *Biochemistry* **1994**, *33*, 3032-3037.
51. Banno, Y.; Okano, Y.; Nozawa, Y. *J. Biol. Chem.* **1994**, *269*, 15846-15852.
52. Feng, J.-F.; Rhee, S. G.; Im, M.-J. *J. Biol. Chem.* **1996**, *271*, 16451-16454.
53. Lemmon, M. A.; Ferguson, K. M.; Schlessinger, J. *Cell* **1996**, *85*, 621-624.
54. Powis, G.; Alberts, D. S. *Eur. J. Cancer* **1994**, *8*, 1138-1144.
55. Bruzik, K. S. In *Carbohydrates in Drug Design*, Nieforth, C. and Witczak, Z., Eds.; Marcell Dekker, New York, 1997, p. 385-431.
56. Powis, G.; Seewald, M. J.; Gratas, C.; Melder, D.; Riebow, J.; Modest, E. J. *Cancer Research* **1992**, *52*, 2835-2840.
57. Bennett, C. F.; Mong, S.; Wu, H.-L. W.; Clark, M. A.; Wheeler, L.; Crooke, S. T. *Molec. Pharmacol.* **1988**, *32*, 587-593.
58. Perrella, F. W.; Chen, S. F.; Behrens, D. L.; Kaltenbach III, R. F.; Seitz, S. P. *J. Med. Chem.* **1994**, *37*, 2232-2237.
59. Kraulis, P. J. *J. Appl. Crystallogr.,* **1991**, *24*, 946-950.
60. Tronrud, D. E.; Ten Eyck, L. F.; Matthews, B. W. *Acta Crystallogr.* **1987**, *A43*, 489-501.
61. Read, R. J. *Acta Crystallogr.* **1986**, *A42*, 140-149.
62. Esnouf, R. M. *J. Mol. Graphics* **1997**, *15*, 133-138.
63. Thompson, J. D.; Higgins, D. G.; Gibson, T. J. *Nucl. Acid. Res.* **1994**, *22*, 4673-4680.
64. Sharp, K. A.; Nicholls, A.; Friedman, R.; Honig, B. *Science* **1991**, *252*, 106-109.
65. Nicholls, A.; Sharp, K. A.; Honig, B. *Proteins Structure Function Genetics* **1991**, *11*, 281-296.
66. Zheng, L.; Paik, W.-Y.; Cesnjaj, M.; Balla, T.; Tomic, M.; Catt, K. J.; Stojilkovic, S. S. *Endocrinology* **1995**, *136*, 1079-1088.

Chapter 9

Sequential Mechanism and Allosteric Activation of Phosphoinositide-Specific Phospholipase C-δ₁

Mary F. Roberts[1], Yiqin Wu[1], Olga Perisic[2], Roger L. Williams[2], and Matilda Katan[3]

[1]Merkert Chemistry Center, Boston College, Chestnut Hill, MA 02167
[2]MRC Laboratory of Molecular Biology, Hills Road, Cambridge CB2 2QH, United Kingdom
[3]CRC Centre for Cell and Molecular Biology, Chester Beatty Laboratories, Fulham Road, London SW3 6JB, United Kingdom

Kinetic analysis (using ^{31}P NMR spectroscopy) of full length and PH domain truncated phosphoinositide-specific phospholipase C-δ₁ (PI-PLC-δ₁) toward the soluble substrates inositol 1,2-cyclic phosphate (cIP) and glycerophosphoinositol phosphates (GPIP$_x$), as well as PI, in detergent micelles provides the following mechanistic results: (i) PI cleavage follows a two-step mechanism with slow release of cIP produced by the phosphotransferase step. This slow release of cIP from the E•cIP complex explains why both cIP and I-1-P are produced in parallel, and why cIP is not the dominant product (which would otherwise be expected based on the ~25 mM K$_m$ for cIP hydrolysis). (ii) In solution the enzyme exists in a less active and more active form; the equilibrium between these forms is affected by substrate (concentration as well as phosphorylation of the inositol ring), interfaces (notably PC and PIP₂), organic solvents, and temperature. (iii) The allosteric activation provided by interfaces requires an intact PH domain. These kinetic constraints are discussed in terms of the crystal structure of the enzyme with substrate / product analogues bound.

Mammalian phosphoinositide-specific phospholipase C (PI-PLC) enzymes catalyze the hydrolysis of phosphatidylinositol-4,5-bisphosphate (PIP₂) to produce two second messengers, water soluble D-*myo*-inositol 1,4,5-trisphosphate (IP₃), which elevates the intracellular calcium level, and membrane associated diacylglycerol (DAG), which activates many protein kinase C isozymes (*1, 2*). Mammalian PI-PLCs (ranging in size from 85 to 210 kDa) are Ca^{2+}-dependent and prefer phosphorylated inositols (*3, 4*), although this is difficult to evaluate because of changes in physical states of the substrates with phosphorylation (*5*). With PI as the substrate they generate both inositol cyclic-1,2-monophosphate (cIP) and inositol 1-phosphate (I-1-P) as products (at a constant ratio that depends on the isozyme class, substrate, pH and calcium (*6*)). Mammalian enzymes also have an N-terminal PH domain that has been reported to have a role in anchoring the enzyme to interfaces, and possibly an allosteric role, as well.

In contrast, bacterial PI-PLC enzymes are much smaller (~33 kDa), secreted, Ca^{2+}-independent, can not cleave phosphorylated PIs, generate cIP as the initial (and under most circumstances the only) product of PI cleavage, and have only limited homology to the catalytic portion of the mammalian sequences (3). The phosphodiesterase reaction (hydrolysis of cIP to I-1-P) is considerably less efficient than the phosphotransferase reaction since the enzyme has a much higher K_m for cIP and a much lower V_{max} (7). PI-PLC from *Bacillus thuringiensis* is an allosteric enzyme with both a phospholipid activator site and an active site. The activation of the bacterial PI-PLC enzyme is optimal with phosphatidylcholine (PC), which binds to the activator site and anchors the enzyme to the interface (7-9). Water miscible organic solvents such as isopropanol, dimethylformamide or dimethylsulfoxide also activate PI-PLC hydrolysis of cIP by mimicking the polarity of the interface, thereby stabilizing the active form of the enzyme (10).

A comparison of the structure of the bacterial enzyme (11) to PI-PLC-δ_1 from rat (12) shows them both to have imperfect α/β-barrels with similar placement of catalytic residues. There are a number of important mechanistic questions to address for the mammalian enzymes. For example, if mammalian enzymes use a similar mechanism to the bacterial enzymes, then cIP must be a substrate, however both cIP and I-1-P are generated at a fixed ratio. How can this be rationalized with a sequential mechanism which is supported by the retention of the configuration at the 1-phosphorus of acyclic inositol phosphates formed by PI-PLC b1 (13), and the lack of activity of the enzyme towards substrate analogs lacking the 2-hydroxyl group (14)?

Materials and Methods

^{31}P NMR assays of PI-PLC-δ_1 activity. Since PI, cIP and I-1-P have distinct ^{31}P chemical shifts, the action of PI-PLC can easily be followed with ^{31}P NMR spectroscopy. ^{31}P NMR (202.3 MHz) spectra were acquired as described previously (7). The buffer used in all cIP (enzymatically generated from crude PI (7)) assays was 50 mM Hepes, pH 7.5. The calcium concentration was optimized for different assay systems. The amount of full length and PH-domain truncated PI-PLC (Δ(1-132)PI-PLC-δ_1) (purified as described previously (15)) added to initiate hydrolysis of water-soluble substrates varied between 18 μg and 160 μg, depending on the substrates used and assay conditions. Hydrolysis rates were measured from the integrated intensity of the resonance corresponding to the phosphorylated product as a function of incubation time, typically 2-3 h, at 30°C, unless otherwise indicated.

In the cIP assays, 8 mM diC$_7$PC (predominantly micelles, given the CMC of 1.5 mM), and different amounts of DMSO or isopropanol were used to study the interfacial activation and organic solvent activation of the cyclic phosphodiesterase reaction. Similar assay conditions were used for and glycerophosphoinositol phosphates as substrates for PI-PLC-δ_1 (16). PI and PIP solubilized in mixed micelles with Triton X-100 (1:2 PIP$_x$/TX-100) were used as micellar substrates for PI-PLC-δ_1. TX-100 was used as the matrix to solubilize PI, since it has been shown that: (i) the extremely fast micelle exchange kinetics of TX-100 ensure that substrate depletion is not a problem (17); (ii) TX-100 is relatively noninteractive with Ca^{2+}; and (iii) TX-100 micelles did not affect PI-PLC-δ_1 hydrolysis of cIP (16). A variant of the ^{31}P NMR assays of cIP hydrolysis involved incubating PI-PLC (60 μg) with cIP in 50 mM Hepes, pH 7.5, 0.5 mM Ca^{2+} (total volume 0.35 mL) for a fixed time (typically one hour) followed by quenching with the addition of 10 μl 200 mM EDTA, then analysis by ^{31}P NMR.

Determination of the energetic parameters of PI-PLC kinetics. The temperature dependence of k_{cat} was analyzed according to transition state theory (18,

19). The activation free energy, ΔG, was calculated by $\Delta G^{\ddagger} = -RT\ln(k_{cat}h/k_BT)$, where h, k_B and R are Planck, Boltzmann and the gas constants, respectively, and k_{cat} (s^{-1}) is the experimentally determined turnover number. The transmission coefficient (*18*) is assumed to be unity (*19*) and can be ignored. The activation enthalpy, $\Delta H^{\ddagger}$, was calculated from the slope of plotting $\ln(k_{cat}/T)$ versus $1/T$, based on the Eyring equation: $\ln(k_{cat}/T) = \ln(k_B/h) + DS^{\ddagger}/R - \Delta H^{\ddagger}/RT$. The activation entropy, $\Delta S^{\ddagger}$, was estimated from $\Delta S^{\ddagger} = (\Delta H^{\ddagger} - \Delta G^{\ddagger})/T$.

Results

PIP$_x$ / Triton X-100 mixed micelles as substrates for PI-PLC-δ_1. Amphiphilic PIP$_x$ substrates solubilized in TX-100 micelles were used as substrates for PI-PLC-δ_1. For the enzyme without the PH domain, no product was detected if Ca^{2+} was omitted from reaction mixtures. The dependence of activity on Ca^{2+} at fixed PI is shown in Figure 1. From these data a Ca^{2+} concentration of 0.5 mM was chosen for all assays of PIP$_x$ hydrolysis, as it is sufficiently above the threshold needed for maximum activation of PI-PLC-δ_1 toward PI so that competition of phosphoinositide substrates for Ca^{2+} should not inhibit the enzyme. For 8 mM PI solubilized in PI/TX-100 mixed micelles (the apparent K$_m$ appears to be in the 1-2 mM range), the rates of PI hydrolysis at 30°C were 16 and 36 μmol min^{-1} mg^{-1} for Δ(1-132)PI-PLC-δ_1 and full-length PI-PLC, respectively. The ratio of cIP/I-1-P appeared to be constant throughout each reaction. This phosphotransferase rate was considerably lower than that for bacterial PI-PLC hydrolysis of PI in TX-100, which has a V$_{max}$ of 1000 μmol min^{-1} mg^{-1} with an apparent K$_m$ of 1-2 mM (*7*). A key observation is that full-length PI-PLC-δ_1 was about 2-fold more active than the truncated one for PI hydrolysis.

Addition of phosphate groups to PI made a much more effective substrate for the enzyme (*16*). The truncated enzyme hydrolyzed PIP almost a 100 times faster than PI (1000-1200 μmol min^{-1} mg^{-1}), suggesting that the 4'-phosphate group plays a key role in binding and catalysis as suggested by the crystal structure (*15*). Unlike PI, which when hydrolyzed by PI-PLC is converted to comparable amounts of both cyclic and acyclic phosphate products, PIP hydrolysis by PI-PLC-δ_1 produced very little cyclic inositol phosphate (cIP$_2$) and mostly IP$_2$ as the product.

cIP as a substrate for PI-PLC-δ_1. If a sequential mechanism of PI hydrolysis for mammalian PI-PLCs is operational, cIP must be a substrate for the enzyme. The dependence of PI-PLC-δ_1 cyclophosphodiesterase activity on Ca^{2+} was similar to that for PI (Figure 1) with an apparent K$_D$ for metal ion binding at this cIP concentration of 50 mM (Ca^{2+} interacts with the negatively charged PI and cIP species, as well as the PI-PLC making estimate of the real K$_D$ difficult). Again 0.5 mM Ca^{2+} appeared to be an optimal level of ion for assays of soluble substrates. The dependence of PI-PLC activity on cIP concentration was sigmoidal rather than hyperbolic, with half-saturation concentrations, [S]$_{0.5}$, for cIP of 25 and 28 mM, and Hill coefficients of was 1.7±0.2 and 1.5±0.1 for Δ(1-132)PI-PLC-δ_1 and full length enzyme, respectively. The V$_{max}$ values for truncated and full length enzyme action on cIP were essentially identical, 0.40±0.03 and 0.39±0.03 μmol min^{-1} mg^{-1} (Table I). Thus, the presence of the PH domain has no significant effect of the hydrolysis of soluble cIP under these conditions. Also of note are the high values for [S]$_{0.5}$; these are problematic in light of the observation that PI hydrolysis yields both cIP and I-1-P products.

Glycerophosphoinositol phosphates are substrates of PI-PLC-δ_1. In a micellar assay system, PI-PLC-δ_1 prefers phosphorylated PI molecules as substrates (*4, 20*). This may also be the case for water-soluble substrates. Efficient preparation of cyclic-IP$_x$ is difficult, so a series of glycerophosphoinositol phosphates (GPIP$_x$)

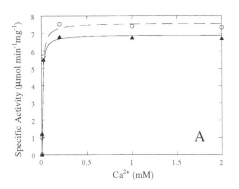

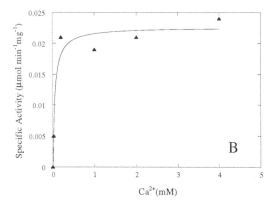

Figure 1. Ca^{2+} dependence of $\Delta(1\text{-}132)$ PI-PLC-δ_1 (A) cyclic phosphodiesterase activity toward cIP and (B) hydrolysis of PI to generate cIP (Δ) and I-1-P (O). For cIP hydrolysis, assay conditions included 10 mM cIP, 50 mM Hepes, pH 7.5, 30°C, and 60 µg enzyme; the curve is well fit by an apparent K_D of 0.05 ± 0.02 mM. For PI hydrolysis, assay conditions included 8 mM PI dispersed in 16 mM TX-100, 50 mM Hepes buffer, pH 7.5, 30°C, and 6 µg PI-PLC; the curve is well fit by an apparent K_D of 0.008 ± 0.003 mM. (Adapted from ref. *16*).

were examined as substrates for $\Delta(1\text{-}132)$PI-PLC-δ_1 to screen for the effect of adding phosphates to the inositol ring without the complication of different aggregate states of the substrate (Table II). These water-soluble molecules contain the same head group and the glycerol backbone as PIP_x without the hydrophobic fatty acyl chains. PI-PLC had a much higher activity toward cIP than toward GPI, likely due to release of the strain energy of cIP upon hydrolysis to I-1-P (21). However, phosphorylation of the inositol ring of GPI generated a substrate that was more efficiently hydrolyzed than cIP (higher V_{max} and lower K_m); only the acyclic phosphoinositol phosphate (and not cIP_x) was detected as product. The substrate saturation curve also became hyperbolic. From the kinetic parameters, it is obvious that the 4'-phosphate of the inositol ring is critical for both substrate binding and catalysis, while the 5'-phosphate enhances only catalysis.

Effect of PC and PIP_2 on PI-PLC hydrolysis of cIP. PI-PLC from *B. thuringiensis* was activated 21-fold toward cIP (as measured by the enzyme efficiency, V_{max}/K_m) by the presence of micellar diC_7PC (7); PC vesicles also enhanced PI-PLC activity towards cIP. The diC_7PC micellar interface (data indicated by the circles in Figure 2) activated the full-length enzyme, but only produced a modest two-fold increase in V_{max} with little change of K_m, and unaltered cooperativity in substrate binding (Figure 2A, Table I). In contrast to the results with the full-length enzyme, diC_7PC micelles had no effect on kinetic parameters for $\Delta(1\text{-}132)$PI-PLC-δ_1 (Figure 2B). A variety of interfaces were screened for their effect on PI-PLC specific activity at low cIP (Table II). The activation by micellar PC was moderately specific since TX-100 had no effect on the kinetic parameters of the enzyme acting on cIP. Interface physical state was also important since small unilamellar vesicles of POPC were not as effective as micellar PC in increasing PI-PLC activity toward cIP (Table II). The PH domain binds PIP_2 tightly and the addition of this ligand along with the PC interface enhanced PI-PLC activity. As shown in Table II, incubation of the enzyme with 4 mM PIP_2 / 4 mM POPC or 8 mM PIP_2 / 8 mM POPC / 8 mM diC_7PC increased cIP hydrolysis about two-fold over hydrolysis of 10 mM cIP alone. Increasing the Ca^{2+} to 2 mM did not increase the activity with these interfaces. The PI-PLC-δ_1 PH domain can interact with a PC interface and in turn alter the catalytic domain conformation in some way that enhances its activity toward cIP. However, PIP_2 is more effective in inducing this conformational change and has a synergistic effect with PC.

Organic solvent activation of PI-PLC-δ_1. It has been shown that organic solvents miscible in water activate bacterial PI-PLC hydrolysis of cIP (10). The solvent changes the polarity around the active site to a value achieved when a long-chain PI or when an activator surface is present. In this altered polarity environment, the enzyme is converted to a more active form. Given the structural similarity in the catalytic domain of the PI-PLC-δ_1 to *Bacillus cereus* PI-PLC, the mammalian-δ_1 enzyme may also be activated by an organic solvent such as DMSO. As shown in Figure 3A, increasing DMSO leads to a sigmoidal increase in hydrolysis of 8 mM cIP by both $\Delta(1\text{-}132)$PI-PLC-δ_1 and full length PI-PLC-δ_1. Explanations for this behavior include a conformational change associated with the addition of organic solvent or organic solvent dehydration of PI-PLC active site (10). The active site of PI-PLC-δ_1 is solvent-accessible (12, 15), and removal of water may be a precursor to soluble substrate binding. If many water molecules are involved, the expulsion of active site water could occur in a cooperative fashion, hence the sigmoidal appearance of PI-PLC activity versus organic solvent concentration. At a fixed concentration of solvent (45% or 0.xx mole fraction), the V_{max} for cIP was increased 4-5 fold (an enhancement similar to that observed with bacterial PI-PLC) with little change in the $[S]_{0.5}$ or the K_m determined by the fit with the Hill equation (Figure 3B, Table I). The lack of effect

Table I. Kinetic parameters for $\Delta(1\text{-}132)$PI-PLC-δ_1 and full length PI-PLCç catalyzed hydrolysis of cIP and $GPIP_x$.[a]

Substrate	$\Delta(1\text{-}132)$PI-PLC-δ_1			full length PI-PLC-δ_1		
	V_{max} μmol/min mg	K_m (mM) $(n)^b$	k_{cat}/K_m $(M^{-1} s^{-1})$	V_{max} μmol/min mg	K_m (mM) $(n)^b$	k_{cat}/K_m $(M^{-1} s^{-1})$
cIP alone[c]	0.40±0.03	25.2±2.9 (n=1.7±0.2)	22	0.39±0.03	28.0±3.0 (n=1.5±0.1)	20
cIP + 8 mM diC$_7$PC	0.40	25 (n=1.7)	22	0.85±0.06	24.7±2.9 (n=1.7±0.2)	49
cIP + 45% DMSO	-[d]			0.85±0.06	24.7±2.9 (n=1.7±0.2)	94
GPI	0.001[e]	>10	<0.1			
GPIP	0.79±0.08	1.8±0.4 (n=1)	620			
GPIP$_2$	4.8±0.6	4.4±1.1 (n=1)	1550			

[a] Adapted from ref. *16*.
[b] Hill coefficient, n, and K_m calculated from $V = V_{max}[S]^n / (K_m^n + [S]^n)$.
[c] Assay conditions include 50 mM Hepes pH 7.5, 0.5 mM CaCl$_2$, 30°C. Enzyme (18-120 μg depending on the substrate) was added to a total volume of 350 μL cIP solution to initiate the reaction.
[d] Activation identical to what is observed with the full length clone.
[e] This is the specific activity for 2 mM GPI.

Table II. Effect of interfaces on the specific activity of full length PI-PLC-δ_1 towards cIP.[a]

Interface	Specific Activity (μmol min^{-1} mg^{-1})
	0.100±0.015
diC$_7$PC (8 mM)	0.16±0.02
diC$_7$PC (8 mM) + 8 mM PIP$_2$ / 8 mM POPC[b]	0.19±0.03
POPC (4 mM)	0.12
POPC (4 mM) + 4 mM PIP$_2$	0.20±0.02

[a] Assay conditions included 10 mM cIP, 50 mM Hepes pH 7.5, 0.5 mM CaCl$_2$, 30°C.
[b] The PIP$_2$ was solubilized in small unilamellar vesicles of POPC and added to assays mixtures; the addition of mM POPC vesicles to a cIP assay mixture with diC$_7$PC micelles has no effect on cIP hydrolysis.

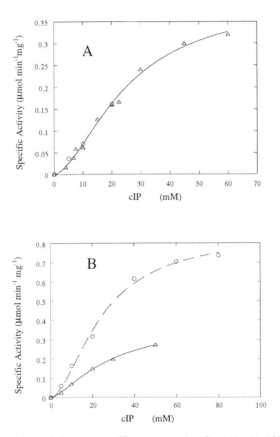

Figure 2. Specific activity versus cIP concentration for (A) $\Delta(1\text{-}132)$PI-PLC-δ_1, and (B) full-length PI-PLC-δ_1 in the absence (Δ) and in the presence (O) of 8 mM diC$_7$PC. Assay conditions include 50 mM Hepes pH 7.5, 0.5 mM CaCl$_2$, 30°C, and 60 μg enzyme. The curves were drawn by fitting the data with the Hill equation, $V = V_{max}[S]^n/(K_m^n + [S]^n)$, and the parameters are summarized in Table I. (Reproduced with permission from ref. *16*, Copyright 1997 ACS).

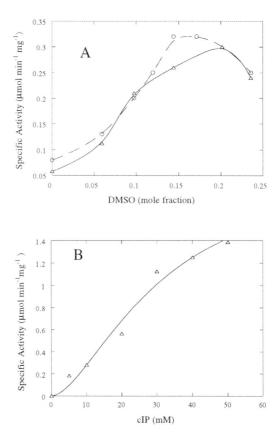

Figure 3. (A) Specific activity versus mole fraction DMSO for Δ(1-132) PI-PLC-δ_1 (Δ) and full-length PI-PLC-δ_1 (O). (B) Specific activity versus cIP concentration for full-length PI-PLC-δ_1 in the presence of 45% DMSO (0.17 mole fraction). Assay conditions include 50 mM Hepes pH 7.5, 10 mM cIP in (A) and variable cIP in (B), 0.5 mM $CaCl_2$, 30°C, and 60 μg enzyme. The curve in (B) was drawn by fitting the data with the Hill equation, $V = V_{max}[S]^n/(K_m^n + [S]^n)$, and the parameters are summarized in Table I. (Adapted from ref. *16*).

of DMSO on the K_m for cIP or the Hill coefficient is in sharp contrast to the bacterial enzyme.

Effect of modulating water activity on PI hydrolysis. The generation of I-1-P from cIP involves attack of a water molecule on the enzyme-bound cIP. A decrease in water activity (α_w) should increase production of cIP. Two types of solvent perturbations were used to reduce α_w: addition of (i) NaCl or (ii) water-soluble organic solvents. The effect of moderately high concentrations of NaCl on $\Delta(1-132)$PI-PLC-δ_1 activity toward PI is shown in Table III. The overall rate of PI cleavage (I-1-P and cIP production) was almost unaffected from 0 to 800 mM NaCl. However, NaCl slightly increased the percent of cIP production and decreased the amount of IP generation, presumably due to the decreased water activity.

Water activity can also be reduced by the addition of organic solvents (*22, 23*). For PI-PLC, water-miscible solvents also activate the enzyme for cIP hydrolysis (*10*). This complicates what one would expect as the outcome of adding solvent to PI/TX-100 mixed micelle system (water-miscible solvents can affect micelle size, but in general don't abolish them). The presence of 45% DMSO leads to a 2-3 fold increase in the rate of PI-PLC-δ_1 cleavage of PI. The solvent also decreased the cIP/I-1-P ratio significantly (Table III). iPrOH (25%) also decreased the amount of cIP generated from PI. The increase in I-1-P formation caused by organic solvent is not due to the rebinding and hydrolysis of free cIP, because in the assay conditions used the concentration of cIP was well below its K_m. Thus, the solvent must affect the balance of the steps leading to cIP and I-1-P production. A possible explanation for the increased I-1-P production is that cIP generated *in situ* has a slow off rate compared to its attack by a water molecule and that solvent, while it reduces water activity and the actual attack of a water molecule, can stabilize the enzyme-bound cIP, so that attack of a water is more likely.

Effect of temperature on PI and cIP hydrolysis. The temperature dependence of PI-PLC activity can also shed light on the proposal of a slow off-rate for cIP generated *in situ*. Since both full-length and PH domain truncated PI-PLC-δ_1 were completely inactivated if the assay temperature was above 60°C, initial rates of activity toward PI were determined from 24°C to 55°C for $\Delta(1-132)$PI-PLC-δ_1 and 24-45°C for full length PI-PLC-δ_1 (Figure 4). The generation of I-1-P was virtually temperature independent, while cIP generation increased linearly with temperature (Figure 4B). Thus, increasing the assay temperature increased the amount of cIP product. The temperature profile showed very similar effects on the ratio of cIP/I-1-P for both full length and truncated enzymes (Figure 4C). The linear relationship between $\log(k_{cat}/T)$ and $1/T$ allows an estimate of apparent $\Delta H^{\ddagger}$, $\Delta G^{\ddagger}$ and $\Delta S^{\ddagger}$ for production of each product. The apparent activation enthalpy, $\Delta H^{\ddagger}$, for PI hydrolysis, cIP production and the generation of I-1-P, were determined to be 19.6, 34.0, and 0.0 kJ/mol respectively, for the truncated enzyme, and 14.6, 28.5, and 0.0 kJ/mol respectively, for the full length PI-PLC-δ_1. $\Delta S^{\ddagger}$ values for PI hydrolysis, release of cIP and the generation of I-1-P, were similar for both full length and truncated enzymes and estimated as -160, -120, and -220 J mol^{-1} K^{-1}. The higher activation entropy for I-1-P production versus cIP release presumably reflects the energetic cost for ordering a water molecule to attack the bound cIP. These values indicate that $\Delta G^{\ddagger}$ (in the range of 63 kJ/mol to 74 kJ/mol) is dominated by the negative apparent activation entropy $\Delta S^{\ddagger}$.

When cIP was used as the substrate for PI-PLC, the temperature dependence was quite complex and did not follow Arrhenius' law (Figure 5), although activity clearly increased with increasing temperature. As the assay temperature increased, the V_{max} for cIP hydrolysis approached the specific activity of PI hydrolysis (a good

Table III. Effect of reducing water activity on the amount of cIP produced in PI hydrolysis by PI-PLC-δ_1.[a]

Additive	$\Delta(1\text{-}132)$PI-PLC-δ_1 cIP (%)	full-length PI-PLC-δ_1 cIP (%)
NaCl:		
control	39	
0.2 M	42	
0.8 M	47	
Organic solvent: [b]		
control	46	43
25% iPrOH	26	33
45% DMSO	17	19

[a] Standard assay conditions included 8 mM PI in 16 mM TX-100 mixed micelles, 50 mM Hepes, pH 7.5, 0.5 mM CaCl$_2$.

[b] Volume % of each solvent that leads to a maximum activation of PI-PLC (*10*). SOURCE: Adapted from ref. *16*.

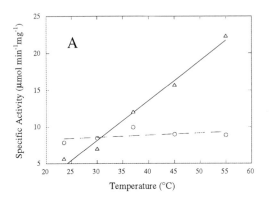

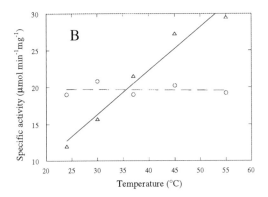

Figure 4. Temperature dependence of (A) Δ(1-132)PI-PLC-δ₁ and (B) full-length PI-PLC-δ₁ activities toward PI in TX-100 mixed micelles (1:2 PI/Triton X-100): (Δ), generation of cIP; and (O), generation of I-1-P. In (C) the ratio of cIP to I-1-P for the hydrolysis of PI in TX-100 micelles is shown as a function of assay temperature: (Δ) Δ(1-132)PI-PLC-δ₁; (O) full-length PI-PLC-δ₁. Assays used 3 μg of full-length PI-PLC-δ₁ and 6 μg of Δ(1-132)PI-PLC-δ₁ and 0.5 mM CaCl₂. (Reproduced with permission from ref. *16*, Copyright 1997 ACS).

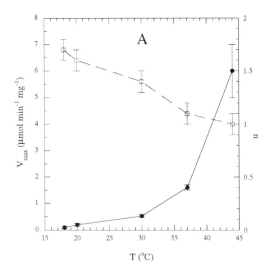

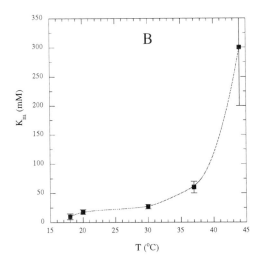

Figure 5. Temperature dependence of $\Delta(1\text{-}132)$PI-PLC-δ_1 kinetic parameters for cIP hydrolysis: (A) ($\bullet$) V_{max} ($\mu mol\ min^{-1}\ mg^{-1}$) and ($\square$) n, the Hill coefficient; ($\blacksquare$) K_m (mM).

approximation of V_{max} under these conditions), indicating at higher temperature, cIP is as efficient as PI in switching PI-PLC from a less active to a more active form. Furthermore, the cooperativity was decreased as the temperature increased while the K_m increased significantly. The latter would be expected if the E•cIP off-rate were accelerated at higher temperatures. Such behavior suggests that at higher temperature, the substrate induced enzyme conformational change to the more active form is no longer rate-limiting.

Discussion

Efficiency of Phosphotransferase and Cyclophosphodiesterase Reactions. What do the fatty acyl chains of the PI substrate contribute to enzyme activity? At 30°C, enzyme ($\Delta(1\text{-}132)$PI-PLC-δ_1) efficiency for the phosphotransferase activity for PI ($k_{cat}/K_m \sim 10^4$ to 10^6 M^{-1} s^{-1} for K_m = 0.02 and 1 mM, respectively) is much greater than for the cyclic phosphodiesterase activity toward cIP (k_{cat}/K_m = 22.5 M^{-1} s^{-1}). Even as the assay temperature is increased, the efficiency for cIP hydrolysis is much less than PI cleavage, in part due to weak binding of this water-soluble substrate to the enzyme. This strongly suggests that the lipophilic portion of the substrate greatly enhances the catalysis in a way that is not duplicated by either a PC interface or organic solvent activation of the enzyme (k_{cat}/K_m = 94 M^{-1} s^{-1}). However, ring strain contributes to cIP hydrolysis, and therefore IcP is not the best substrate for comparing the effects of the lipophilic chains. The acyclic water-soluble series GPIP$_x$ provide a better dissection of the effect of the fatty acyl chains as well as inositol ring phosphorylation. GPI is such a poor substrate for the enzyme that it is hard to measure kinetic parameters accurately: the k_{cat}/K_m value for GPI is less than 0.1 M^{-1} s^{-1}. Thus, the presence of the fatty acyl chains has a tremendous effect on PI-PLC activity.

Phosphorylation of the inositol ring partially makes up for the poor efficiency with GPI or cIP and much of this effect is through improved substrate binding. The addition of the 4'-phosphate enhances both substrate recognition and catalysis, and leads to a value for k_{cat}/K_m of 620, several thousand-fold greater than for GPI, although still well below the efficiency for PI cleavage. Perhaps most interestingly, the presence of this phosphate on the inositol ring abolishes the cooperativity observed in the cIP saturation curve and leads to only the acyclic inositol phosphate as product. GPIP binding appears very effective in switching the enzyme to the more active conformation. The addition of a phosphate group to the inositol C-5' position adds another factor of two to efficiency with the major contribution from increased k_{cat}. This was the maximal value observed for soluble substrates. All these observations are consistent with the crystallographic studies of $\Delta(1\text{-}132)$PI-PLC-δ_1 complex with IP$_3$ (*12, 15*) which showed that the 4'-phosphoryl group had the lowest B-factor (34 Å), and hence the lowest degree of structural disorder in the active site. This was followed by the 1'-phosphoryl group (40 Å) and the 5'-phosphoryl group (49 Å). The 4'-phosphoryl group interacts directly with Lys438, Ser522 and Arg549; indirect interactions with the enzyme mediated by water molecules are also observed. These interactions could be critical in both aligning substrate and in converting the enzyme to its activated form. The exposed 5'-phosphoryl group has far fewer interactions with enzyme - one salt bridge with Lys440 and two water-mediated interactions.

Interfacial activation and organic solvent activation. The enhancement of k_{cat} when an enzyme binds to an interface, interfacial activation, is a characteristic feature of water-soluble phospholipases. Dissecting whether this arises from substrate-based effects (such as changes in the conformation or surface concentration of substrates) or enzyme-based effects (changes in the conformation or enzyme dimerization or allosteric binding etc.) is difficult with an aggregated substrate. Using a water soluble substrate (e.g. cIP), that has no tendency to partition into surfaces, can provide evidence that interfaces serve an allosteric role in catalysis (7-9). For rat PI-PLC-δ_1, PC interfaces, and more effectively PIP$_2$ in that interface, allosterically enhance k_{cat} for cIP hydrolysis. This allosteric activation clearly involves the PH domain of the protein, a structural module with high affinity for PIP$_2$ (24-26). With a phospholipid substrate, PIP$_2$ binding to the PH domain affects the catalytic activity of PI-PLC-δ_1 enabling the processive hydrolysis of membrane-bound substrates (27, 28). With a soluble substrate, it shifts the distribution of protein to a more active form.

How does the allosteric binding of the interface enhance k_{cat}? Phospholipid interfaces provide a microenvironment that affects solvent polarity. This in turn can be critical for optimizing enzyme activity, particularly for a hydrolysis reaction. The combination of a water-soluble substrate and the addition of water miscible organic solvents (e.g. isopropanol and DMSO) provides a way of assessing the catalytic importance of these local environments (10). DMSO causes a 4-fold increase in V_{max} for both PI-PLC-δ_1 and the much smaller bacterial PI-PLC. This has been postulated to involve partial dehydration of substrate and active site in the case of the bacterial PI-PLC (10). Anchoring the enzyme to a phospholipid interface may likewise affect the hydration environment of the PI-PLC active site, since it is not buried but relatively well exposed to solvent (12, 15). A likely explanation for both this 'interfacial activation' is that in aqueous solution there is an equilibrium between less active and more active forms of PI-PLC. Both PC interfaces (by interacting with the PH domain) and organic solvent effectors modulate PI-PLC-δ_1 by changing the equilibrium between these two forms of the enzyme. Substrate phosphorylation also biases the equilibrium to the more active form.

A conformational switch in PI-PLC-δ_1? The cooperative behavior of the cIP saturation curve strongly suggests that the enzyme can be converted from a less active to a more active form by substrate binding. This cooperativity could be explained by multiple sites binding for cIP (one at the active site, another at a site where its binding enhances catalysis), enzyme dimerization, or other conformational changes that perturb the active site including active site hydration (10). The crystal structure is most consistent with PI-PLC-δ_1 acting as a monomeric enzyme with one active site (12). Hence, it is very unlikely that the cooperativity observed for both the full-length and the PH domain truncated enzyme hydrolysis of cIP is the result of enzyme site-site interactions, especially when taking into account that the cooperativity is persistent and independent of several assay conditions (i.e., addition of diC$_7$PC micelles or organic solvents). The most plausible explanation for the cooperativity observed in the cIP saturation curve at low temperature ($\leq 30°C$) is that a second molecule of cIP is needed to induce the transition to an activated form. As the assay temperature is increased, perhaps more of the enzyme is already in the activated form, thus activity increases. The possibility of a second cIP site is intriguing since the cooperativity of the enzyme is abolished when phosphorylated GPIP$_x$ is the substrate. Clearly, the enzyme has a site with a strong affinity for phosphates.

There are very few structural differences in the active site induced by substrate / product binding Glu341 and Arg549 are the only residues that move significantly. In the case of IP$_3$, these side chains move so that Arg549 interacts with the 4-phosphoryl group of the ligand. An alternative candidate for the conformational switch might be the X/Y linker sequence (residues 443-488). This region of the protein, disordered in the crystal structure, may influence substrate entrance or product exit from the active site. The much larger X/Y linker of PLC-γ has been shown to act as an inhibitor of the enzyme (29).

PI hydrolysis - a sequential mechanism with comparable release of cIP and hydrolysis to I-1-P. cIP is such a poor substrate for PI-PLC-δ_1, that it should be the major product if E•cIP and free cIP are in equilibrium after the phosphotransferase reaction. The observation that mammalian PI-PLC enzymes generate both cyclic and acyclic inositol phosphates simultaneously (and in a fixed ratio) adds a critical constraint to a simple sequential mechanism for PI hydrolysis: release of cIP produced *in situ* by the enzyme must be slow and comparable to the attack by a water molecule to produce I-1-P. Structural studies are consistent with this view (*15*). With this in mind, we can present a detailed scheme for the action of PI-PLC (Figure 6). PI-PLC-δ_1 is proposed to exist in two major states or conformations, a less active form (square) and a more active form (circle). In solution, in the absence of a ligand, the less active form is dominant. The presence of an effective interface or certain amount of water-miscible organic solvent can switch the enzyme to the partially active form (triangle), a metastable state between the most active and least active forms. Binding of substrates switches the distribution of enzyme toward the more active form. The switch from the enzyme less active form to more active form is slow at low temperature (≤ 30°C), and therefore could be rate limiting for cIP hydrolysis at low temperature and low cIP concentration. At higher temperature (≥ 37°C), this switch is faster, and therefore no longer rate-limiting for cIP hydrolysis.

A difference between the water-soluble (Fig. 6A) and micellar (Fig. 6B) substrates is that the enzyme is working in a processive mode for micellar substrates (for recent reviews of processive mode kinetics see refs. (*30-32.*)). When PI-PLC is anchored to the interface, another substrate molecule can readily diffuse to the enzyme active site as soon as product is released. Furthermore, the binding to an interface may enhance the release of product from the active site. However, for water-soluble monomeric substrates such as cIP, a second cIP may not be able to bind to the enzyme active site soon enough after the release of product molecule I-1-P, to maintain the more active form of PI-PLC, especially when the substrate concentration is low compared to the K_m (as is the case for cIP). The enzyme then relaxes back to its less active form (step 4). However, at high substrate concentration, a second cIP molecule may have a better opportunity to bind to the more active form of the enzyme as soon as the product molecule is released, or bind to an allosteric site. Under these conditions, the enzyme could remain activated for many substrate turnovers (step 5). The observation that the GPIP$_x$ saturation curves are not sigmoidal but hyperbolic, suggests that the strong interactions with the enzyme of the phosphate at inositol C-4' stabilize PI-PLC in its activated form even at low substrate concentrations and low temperature. An alternate explanation of the observed cooperativity for cIP hydrolysis at low temperature (≤ 30°C) is cIP induced slow transition between the enzyme less active and more active states. The slow transition from the enzyme less active form to more active form becomes more rapid at higher temperature (37°C and 44°C), and is no longer rate-limiting. These actually reflect the observed hyperbolic cIP saturation curve at higher temperature.

For PI hydrolysis, initial steps (through step 3) to produce E•cIP are rate-determining. Release of free cIP (step 4) or hydrolysis to I-1-P (steps 5 and 6) represent parallel pathways for decomposition of the intermediate E•cIP, leading to a constant ratio of the two products (equivalent to the ratio of the rate constants for the two steps) at a given temperature. The temperature effect on PI hydrolysis is relatively straightforward because as soon as PI-PLC is associated with the PI/TX-100 mixed micelles, it remains activated for many catalytic cycles. The temperature dependence for the cIP/I-1-P ratio reflects the difference in activation enthalpy difference for release of cIP versus hydrolysis to I-1-P (e.g., $\Delta\Delta H^{\ddagger}$). $\Delta\Delta H^{\ddagger} = 30$ kJ/mol higher for release of cIP versus hydrolysis of cIP; the corresponding $\Delta\Delta S^{\ddagger} = 0.1$ kJ/K-mol. Around room temperature, these two terms have comparable contributions to $\Delta\Delta G^{\ddagger}$,

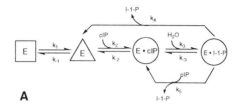

A

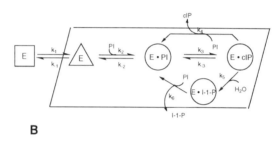

B

Figure 6. (A) Mechanism proposed for PI-PLC-δ_1 hydrolysis of cIP: the circle, the square and the triangle represent the more active, the less active, and partially active conformations of PI-PLC, respectively. The switch from less active to partially activated form (k_1 step) can be achieved by the addition of an effective interface (i.e., diC$_7$PC), a critical amount of water miscible organic solvents (i.e., DMSO or iPrOH), a ligand (i.e., cIP), or a combination of the above. cIP binding to the partially active enzyme (k_2 step) can further switch the enzyme to a more active state. At low temperature ($\leq 30°C$), the transition between the enzyme less active form and more active forms is slow. However, at higher temperature ($\geq 37°C$), this transition becomes faster, and is no longer rate-limiting. At low cIP concentration, the release of the product, I-1-P (the k_4 step) is accompanied by the partially return of enzyme to a less active state. At high cIP concentration, the release of the product, I-1-P, (the k_5 step) is accompanied by the rapid addition of a second cIP. Therefore, the enzyme has a higher probability of being maintained in its active form for many cycles of catalysis. (B) The sequential mechanism proposed for PI hydrolysis by PI-PLC-δ_1 is based on the formation of E•cIP as an intermediate with the release of cIP and attack of a water molecule parallel reactions from this intermediate. The relative rates of cIP release from the enzyme active site, which can be accelerated by a second PI binding, and an activated water molecule attack on the E•cIP complex control the ratio of cIP / I-1-P. (Reproduced with permission from ref. *16*, Copyright 1997 ACS).

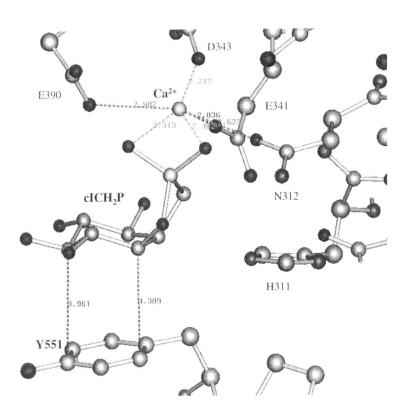

Figure 7. cICH$_2$P at the active site of PH domain truncated PI-PLC-δ_1 active site. This view of the active site was generated with Quanta using coordinates retrieved from the Brookhaven Protein Data Bank. (Accesssion Number 1DJW).

and equal amounts of cIP and I-1-P are produced). With increasing temperature, $\Delta\Delta G^{\ddagger}$ for cIP release compared to hydrolysis becomes increasing negative, favoring release of cIP. Other energetic terms provide insights into the PI-PLC mechanism. The attack of an activated water molecule on the E•cIP complex removes a water molecule from the liquid state and orders it in an enzyme active site. The entropy change for this has been estimated as -10 J mol^{-1} K^{-1} (*33*). The $\Delta\Delta S^{\ddagger}$ for the generation of I-1-P versus overall PI hydrolysis (-55 J mol^{-1} K^{-1}) is considerably greater than this value, and may reflect the cooperative behavior of water molecules in the enzyme active site. Activating one water molecule in the enzyme active site so that it can effectively attack the E•cIP complex could rearrange and alter several water molecules.

Differences in sequential mechanisms for bacterial and mammalian PI-PLC. While bacterial PI-PLC and PI-PLC-δ_1 both operate in a two step sequential mechanism, they have very different behavior with respect to the release of cIP from E•cIP. For the bacterial enzyme, the release of cIP from the enzyme active site is fast. Therefore, cIP accumulates during PI hydrolysis since the K_m for cIP rebinding and for hydrolysis is >25 mM. With PI-PLC-δ_1, the release of cIP from enzyme active site is relatively slow and attack of a water molecule to generate I-1-P is more likely. A rationale for the slower release of cIP is provided by the crystal structure of a nonhydrolyzable cIP analogue bound to PI-PLC-δ_1 (*15*): there is a bidentate interaction of the phosphate oxygens with the Ca^{2+}. Breaking this interaction should have a high activation energy (Figure 7). Since the bacterial enzyme works in a Ca^{2+}-independent fashion, such an interaction cannot occur.

Acknowledgments: This work has been supported by N.I.H. GM 26762 (MFR), British Heart Foundation (RLW), and the MRC/DTI/ZENECA/LINK Programme (RLW).

Literature Cited

1. Lee, S. B.; Rhee, S. G. *Curr. Opin. Cell Biol.* **1995**, *7*, 183-189.
2. Rhee, S. G.; Choi, K. D. *J. Biol. Chem.* **1992**, *267*, 12393-12396.
3. Roberts, M. F.; Zhou, C. In *Encyclopedia of Molecular Biology and Molecular Medicine*, VCH, Germany, **1996**, pp 415-432.
4. Ryu, S. H.; Suh, P.-G.; Cho, K. S.; Lee, K.-Y.; Rhee, S. G. *Proc. Natl. Acad. Sci. USA* **1987**, *84*, 6649-6653.
5. Flanagan, L. A.; Cunningham, C. C.; Prestwich, G. D.; Kosik, K. S.; Janmey, P. A. *Biophys. J.* **1997**, 1440-1447.
6. Kim, J. W.; Ryu, S. H.; Rhee, S. G. *Biochem. Biophys. Res. Comm.* **1989**, *163*, 177-182.
7. Zhou, C.; Wu, Y.; Roberts, M. F. *Biochemistry* **1997**, *36*, 347-355.
8. Wu, Y.; Zhou, C.; Roberts, M. F. *Biochemistry* **1997**, *36*, 356-363.
9. Zhou, C.; Qian, X.; Roberts, M. F. *Biochemistry* **1997**, *36*, 10089-10097.
10. Wu, Y.; Roberts, M. F. *Biochemistry* **1997**, *36*, 8514-8521.
11. Heinz, D. W.; Ryan, M.; Bullock, T. L.; Griffith, O. H. *EMBO J.* **1995**, *14*, 3855-3863.
12. Essen, L.-O.; Perisic, O.; Cheung, R.; Katan, M.; Williams, R. L. *Nature* **1996**, *380*, 595-602.
13. Bruzik, K .S.; Moorish, A. M.; John, D. Y.; Rhee, S. G.; Tsai, M. D. *Biochemistry* **1992**, *31*, 5183-5193.
14. Seitz, S. P.; Kaltenbach, R. F.; Vreekamp, R. H.; Calabrese, J. C.; Perrela, F. W. *Bioorg. Med. Chem. Lett.* **1992**, *2*, 171-174.
15. Essen, L.-O.; Perisic, O.; Katan, M.; Wu, Y.; Roberts, M. F.; Williams, R. L. *Biochemistry* **1997**, *36*, 1704-1718.

16. Wu, Y.; Perisic, O.; Williams, R. L.; Katan, M.; Roberts, M. F. *Biochemistry* **1997**, *36*, 11223-11233.
17. Soltys, C.; Roberts, M. F. *Biochem.* **1994**, *33*, 11608-11617.
18. Eyring, H. *Chem. Rev.* **1935,** *17*, 65-77.
19. Ferscht, A. *Enzyme Structure and Mechanism,* W. H. Freeman & Company, New York NY, 1985.
20. Ellis, M. V.; Carne, A.; Katan, M. *Eur. J. Biochem.* **1993,** *213*, 339-347.
21. Bruzik, K. S.; Guan, Z.; Riddle, S.; Tsai, M. D. *J. Am. Chem. Soc.* **1996,** *118*, 7679-7688.
22. Suzuki, H.; Kanazawa, T. *J. Biol. Chem.* **1996,** *271*, 5481-5486.
23. Colombo, M. F.; Bonilla-Rodriguez, G. O. *J. Biol. Chem.* **1996,** *271*, 4895-4899.
24. Yagisawa, H.; Hirata, M.; Kanematsu, T.; Watanabe, Y.; Ozaki, S.; Safuma, K.; Tanaka, H.; Yabuta, N.; Kamata, H.; Hirata, H.; Nojhima, H. *J. Biol. Chem.* **1994,** *269*, 20179-20188.
25. Lemmon, M. A.; Ferguson, K. M.; O'Brien, R.; Sigler, P. B.; Schlessinger, J. *Proc. Natl. Acad. Sci.* **1995,** *92*, 10472-10476.
26. Garcia, P.; Gupta, R.; Shah, S.; Morris, A. J.; Rudge, S. A.; Scarlata, S.; Petrova, V.; McLaughlin, S.; Rebecchi, M. J. *Biochemistry* **1995,** *34*, 16228-16234.
27. Cifuentes, M. E.; Honkanen, L.; Rebecchi, M. J. *J. Biol. Chem.* **1993,** *268*, 11586-11593.
28. Lomasney, J. W.; Cheng, H.-F.; Wang, L.-P.; Kuan, Y.-S.; Liu, S.-M.; Fesik, S. W.; King, K. *J. Biol. Chem.* **1996,** *271*, 25316-25326.
29. Horstman, D. A.; Destefano, K.; Carpenter, G. *Proc. Natl. Acad. Sci. USA* **1996,** *93*, 7518-7521.
30. Jain, M. K.; Gelb, M. H.; Rogers, J.; Berg, O. G. *Meth. Enzymol.* **1995,** *249*, 567-614.
31. Gelb, M. H.; Jain, M. K.; Hanel, A. M.; Berg, O. G. *Ann. Rev. Biochem.* **1995,** *64*, 653-688.
32. Carman, G. M.; Deems, R. A; Dennis, E. A. *J. Biol. Chem.* **1995,** *270*, 18711-18714.
33. Dunitz, J. D. *Science* **1994,** *264*, 670.

SYNTHESIS AND BIOLOGICAL PROPERTIES OF ANALOGS OF PHOSPHOINOSITIDES

Chapter 10

Structure–Activity Relationships of Adenophostin A and Related Molecules at the 1-D-*myo*-Inositol 1,4,5-Trisphosphate Receptor

Barry V. L. Potter

Wolfson Laboratory of Medicinal Chemistry, Department of Pharmacy and Pharmacology, University of Bath, Claverton Down, Bath BA2 7AY, United Kingdom

D-*myo*-Inositol 1,4,5-trisphosphate [Ins(1,4,5)P$_3$] is a well established, Ca^{2+}-releasing, biological second messenger. Our aim has been to radically redesign the Ins(1,4,5)P$_3$ molecule to explore whether ring-contracted and carbohydrate-based mimics might be feasible targets for development of novel ligands for pharmacological intervention in the polyphosphoinositide pathway of cellular signaling. We have primarily employed two approaches: first, the use of carbohydrates as chiral starting materials for synthesis of cyclopentane-based Ins(1,4,5)P$_3$ mimics using ring-contraction methodologies; second, the design of mono- and disaccharide based polyphosphates both *de novo* and based upon the novel hyperagonist adenophostin A, which is more potent than the natural messenger itself, as novel Ins(1,4,5)P$_3$ receptor ligands. Our studies demonstrate that both approaches possess great potential to produce highly active mimics of this fundamental second messenger.

Intracellular Ca^{2+} mobilization mediated by the second messenger D-*myo*-inositol 1,4,5-trisphosphate [Ins(1,4,5)P$_3$, **1**, Figure 1] is the prime response to phosphoinositidase C activation *via* stimulation of an extracellular G-protein-coupled receptor in a vast array of cell types (*1*). Understanding the subtleties of the polyphosphoinositide signaling pathway has been a fundamental biological aim since the discovery of the Ca^{2+} releasing activity of Ins(1,4,5)P$_3$ in 1983 (*2*). Since 1986 there has been an intensive chemical focus upon the synthesis of inositol polyphosphates, and on understanding the structure-recognition parameters at the Ins(1,4,5)P$_3$ receptor and other binding proteins (*3*). The synthesis of structurally-modified Ins(1,4,5)P$_3$ analogs offers the prospect of pharmacological intervention in such signaling pathways. We have been active in this area for several years (*3-7*), primarily with the synthesis of analogs and mimics of Ins(1,4,5)P$_3$. In this chapter we review several areas of current interest where the use of carbohydrates as starting materials can lead to the synthesis of chiral cyclitol and even carbohydrate based polyphosphates for potential pharmacological intervention in the polyphosphoinositide pathway of cellular signaling.

Ins(1,4,5)P$_3$ is metabolized by two pathways: hydrolysis of the phosphate at position 5 by a low affinity, high capacity Ins(1,4,5)P$_3$ 5-phosphatase giving Ins(1,4)P$_2$ **2**; or phosphorylation at position 3 by a high affinity, low capacity Ins(1,4,5)P$_3$ 3-kinase giving Ins(1,3,4,5)P$_4$ **3**. Ins(1,3,4,5)P$_4$ was discovered in 1985 (8), and the kinase was subsequently characterized (9,10). Its metabolism has been studied and the main route of production is from Ins(1,4,5)P$_3$ by 3-kinase; it has also been isolated *in vitro* as a product of Ins(1,3,4)P$_3$ 5/6-kinase action on D-*myo*-inositol 1,3,4-trisphosphate, Ins(1,3,4)P$_3$ **4** (11), but this route is not believed to be physiologically significant (11). The main route of Ins(1,3,4,5)P$_4$ catabolism is dephosphorylation by 5-phosphatase giving Ins(1,3,4)P$_3$ (12); it is also dephosphorylated by a 3-phosphatase, regenerating Ins(1,4,5)P$_3$ (13-15). It has been suggested that Ins(1,3,4,5)P$_4$ is involved in Ca^{2+} homeostasis at the plasma membrane, helping to control entry of extracellular Ca^{2+} into the cell (13). In support of this hypothesis, an Ins(1,3,4,5)P$_4$-sensitive Ca^{2+}-permeable channel has been characterized from endothelial cells (14); Ins(1,4,5)P$_3$ failed to induce an increase in this channel's activity. Intracellular sites that specifically bind Ins(1,3,4,5)P$_4$ with high affinity [and with high selectivity over Ins(1,4,5)P$_3$] have been noted in many cells. One example from porcine platelets has received particular recent interest: this protein has been tentatively proposed as an Ins(1,3,4,5)P$_4$ receptor (15-17), has shown a high specificity for Ins(1,3,4,5)P$_4$ over all other inositol tetrakis- (16) and various polyphosphates (15), and *in vitro* Ins(1,3,4,5)P$_4$-stimulated GAP activity against the oncogene *ras* has been demonstrated (17).

We have been attracted in recent years to the idea of engendering diverse molecules with Ins(1,4,5)P$_3$ and related activity, and for this reason we decided to explore the synthesis of radically redesigned Ins(1,4,5)P$_3$ mimics.

Synthesis of Ring-Contracted Analogs of Ins(1,4,5)P$_3$

All previously reported approaches to the structural modification of Ins(1,4,5)P$_3$ producing agonists have focused upon modifications at phosphorus or hydroxyl group deletion, reorientation, alkylation or replacement with isosteres and other groups in the six-membered ring. Despite numerous single and multiple modifications, the fundamental requirement of a six-membered ring has not been addressed. Even in the adenophostins (*vide infra*), which differ in many respects from Ins(1,4,5)P$_3$, the important 3,4-bisphosphate/2-hydroxyl triad, analogous to the 4,5-bisphosphate/6-hydroxyl arrangement of Ins(1,4,5)P$_3$ is contained within the six-membered pyranoside ring.

Since many studies have demonstrated that positions 2 and 3 of Ins(1,4,5)P$_3$ are tolerant to extensive modification, it was reasoned that a contracted structure such as **5** (Figure 2), obtained essentially by deletion of the 2-position carbon of Ins(1,4,5)P$_3$ with its associated hydroxyl group, should also fulfill the recognition requirements of the Ins(1,4,5)P$_3$ receptor. A carbohydrate based method to such a "pentagon IP$_3$" (**5**) was therefore devised (18). Many methods for converting carbohydrates into cyclopentane derivatives have been described (19), but the disclosure by Ito *et al.* (20) that treatment of methyl 2,3,4-tri-*O*-benzyl-6,7-dideoxy-α-D-*gluco*-hex-6-enopyranoside(1,5) **6** with zirconocene ("Cp$_2$Zr") followed by boron trifluoride etherate produced vinylcyclopentane **7** (see ref. 21 for a review) was of particular interest, as positions 1,2,3 and 4 of **7** possess the same relative stereochemistry as the equivalent positions in **5**, and therefore as positions 4,5,6 and 1, respectively in Ins(1,4,5)P$_3$. It was decided to apply this methodology in an attempt to prepare (1*R*, 2*R*, 3*S*, 4*R*, 5*S*)-3-hydroxy-1,2,4-trisphospho-5-vinylcyclopentane **8**, an analog of **5** which would allow the initial viability of cyclopentane-based Ins(1,4,5)P$_3$ mimics to be assessed.

Trisphosphate **8** would require a ring-contracted intermediate in which the protecting groups at positions 1 and 2 are different from that at position 3, to allow selective removal to provide the appropriate triol for phosphorylation. It was decided to employ *p*-methoxybenzyl ethers at positions 1 and 2, and a benzyl ether at position 3 (*i.e.* giving intermediate **9**), as this combination was as close as possible to the tribenzyl arrangement used by Ito *et al.* (*20*). The carbohydrate precursor to **9** is heptoside **10**. The 5-vinyl moiety of **10** would be introduced by successive Swern oxidation and Wittig methylenation (*22*) of methyl 2-*O*-benzyl-3,4-di-*O*-(*p*-methoxybenzyl)-α-D-glucopyranoside **11**. We first examined methods to prepare primary alcohol **11**.

Preparation of Methyl 2-*O*-Benzyl-3,4-di-*O*-(p-Methoxybenzyl)-α-D-Glucopyranoside 11.

p-Methoxybenzylidene acetals have the property that either of the acetal C-O bonds may be selectively cleaved to furnish a *p*-methoxybenzyl ether. The direction of cleavage depends upon steric and electronic factors as well as upon the choice of cleavage reagent. Various methods (*23,24*) have been used to cleave the acetal of 2,3-disubstituted derivatives of methyl 4,6-*O*-(*p*-methoxybenzylidene)-α-D-glucopyranoside **13** (Figure 3) resulting in selective formation of 4-*O*-(*p*-methoxybenzyl) ethers.

The known (*23*) **13** was prepared by reaction of methyl α-D-glucopyranoside with *p*-methoxybenzaldehyde dimethyl acetal. Benzylation of **13** with dibutyltin oxide and benzyl bromide gave major and minor products that were easily separated by chromatography, and were identified as the 2-benzyl ether **14** (48%) and the 3-benzyl isomer **15** (11%), respectively. Compound **14** was smoothly *p*-methoxybenzylated to furnish **16** and attention turned to cleavage of the acetal. When **16** was treated with sodium cyanoborohydride and trimethylsilyl chloride (*23*), the chromatographically separable methyl 2-*O*-benzyl-3,6-di-*O*-(*p*-methoxybenzyl)-α-D-glucopyranoside and the required 4-substituted isomer **11** were obtained in a ratio of *ca.* 1:1.7. As the selectivity of cleavage using the above method was disappointing, alternative conditions were explored. Joniak *et al.* (*24*) used LiAlH$_4$-AlCl$_3$ to convert methyl 4,6-*O*-(*p*-methoxybenzylidene)-2,3-di-*O*-methyl-α-D-glucopyranoside **17** exclusively to the 4-*O*-(*p*-methoxybenzyl) ether. Reaction of **16** with LiAlH$_4$-AlCl$_3$ gave exclusively the required **11**.

Preparation of (1*R*, 2*R*, 3*S*, 4*R*, 5*S*)-3-Hydroxy-1,2,4-Trisphospho-5-Vinylcyclopentane 8.

Swern oxidation of **11** gave aldehyde **12**, which was converted by Wittig methylenation to the vinyl carbohydrate **10**. Ring contraction was carried out as described (*20*) by treatment with Cp$_2$Zr(*n*-Bu)$_2$ followed by boron trifluoride etherate (Figure 4). One complication encountered was that gradual loss of *p*-methoxybenzyl groups after addition of boron trifluoride etherate became a competing reaction. Nevertheless, the desired vinylcyclopentane **9** was obtained in fair yield, together with a small amount of the kinetically disfavored product **9a**.

Removal of the *p*-methoxybenzyl protecting groups from **9** gave the triol **18** (Figure 5). Phosphitylation, followed by oxidation with MCPBA gave the fully protected trisphosphate **19**. Deprotection using sodium in liquid ammonia removed the benzyl groups, leaving the vinyl group intact. The vinyl group of **9** should provide a convenient starting-point for a range of modifications at this position, while maintaining the desired stereochemistry. Purification by ion-exchange chromatography gave the target trisphosphate **8**, which was then examined for Ca^{2+} mobilizing activity at the platelet Ins(1,4,5)P$_3$ receptor using fluorescence techniques, and also using saponin-permeabilized platelets loaded with ^{45}Ca^{2+}. It was found to be a full agonist, although with an EC$_{50}$ some 65-fold higher than Ins(1,4,5)P$_3$. The effect was inhibited by addition of heparin, and **8** was also active in Jurkat T-lymphocytes. These

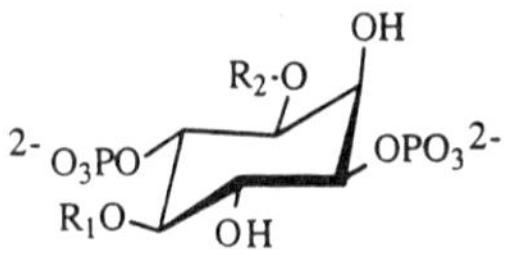

1 $R_1=PO_3^{2-}$; $R_2=OH$
2 $R_1=R_2=H$
3 $R_1=R_2=PO_3^{2-}$
4 $R_1=H$; $R_2=PO_3^{2-}$

Figure 1. Structures of inositol polyphosphate second messengers and some of their metabolites.

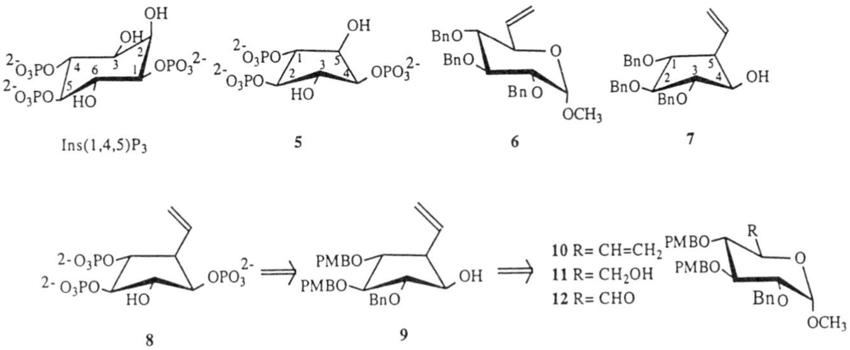

Figure 2. Structures of ring-contracted analogs of IP_3 (**5** and **8**) and their retrosynthetic analysis.

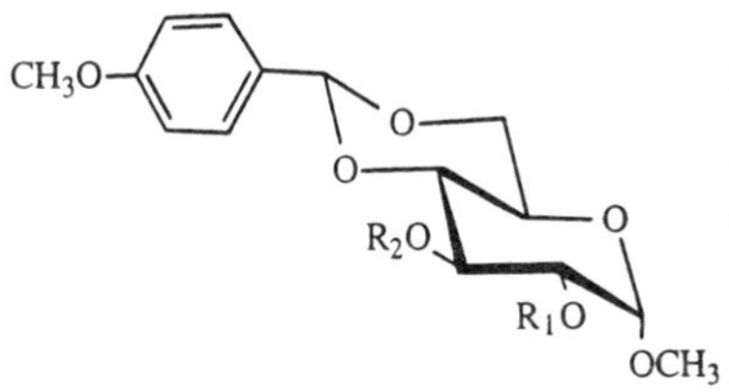

13 $R_1=R_2=H$
14 $R_1=Bn$; $R_2=H$
15 $R_1=H$; $R_2=Bn$
16 $R_1=Bn$; $R_2=PMB$
17 $R_1=R_2=CH_3$

Figure 3. Structure of intermediates obtained during synthesis of the vinycyclopentyl trisphosphate **8**.

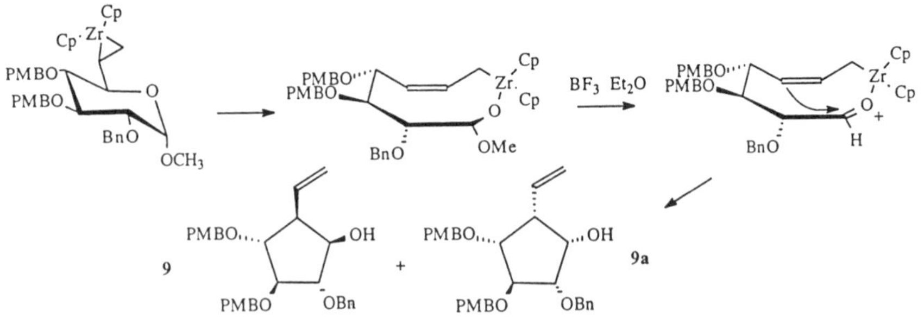

Figure 4. Ring contraction of the glucosyl derivative **6** to the cyclopentyl derivatives **9** and **9a**.

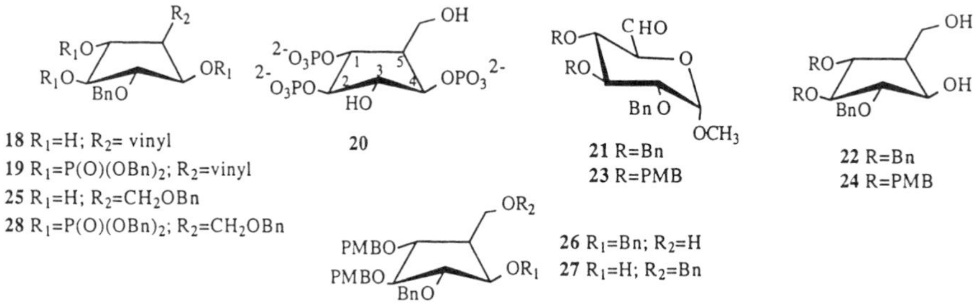

18 R_1=H; R_2= vinyl
19 R_1=P(O)(OBn)$_2$; R_2=vinyl
25 R_1=H; R_2=CH$_2$OBn
28 R_1=P(O)(OBn)$_2$; R_2=CH$_2$OBn

20

21 R=Bn
23 R=PMB

22 R=Bn
24 R=PMB

26 R_1=Bn; R_2=H
27 R_1=H; R_2=Bn

Figure 5. Structures of cyclopentyl intermediates in the synthesis of trisphosphate **20**.

results demonstrate for the first time that Ins(1,4,5)P$_3$ receptor mediated Ca^{2+} mobilization does not necessarily require a cyclohexyl (or equivalent) structural motif. A smaller ring phosphate which retains crucial recognition elements of Ins(1,4,5)P$_3$, *i.e.* three appropriately oriented phosphates and a pseudo-6-hydroxyl group, can still exhibit agonistic activity.

Preparation of (1R, 2R, 3S, 4R, 5S)-3-Hydroxy-5-hydroxymethyl-1,2,4-trisphosphocyclopentane 20.

Noting that DL-3-*O*-ethyl-Ins(1,4,5)P$_3$ and DL-3-*O*-propyl-Ins(1,4,5)P$_3$ both have EC$_{50}$ values of greater than 100 μM in SH-SY5Y neuroblastoma cells [*cf.* Ins(1,4,5)P$_3$ 0.18 μM] (*26*), it seemed reasonable that replacement of the vinyl group of **8** with a hydroxyl-containing side chain should increase its potency. An obvious target was **20**, in which the vinyl substituent is replaced by hydroxymethyl. A recent report by Chénedé *et al.* (*27*) offered the possibility of a simple route. These workers treated aldehyde **21** with samarium (II) iodide in the presence of *t*-butanol and HMPA, and obtained cyclopentane **22**. The structure of **22** was established by conversion to known compounds, and a mechanism for the ring contraction has been proposed (*27*). Applying this rearrangement to aldehyde **23** should give **24**. Benzylation of the primary hydroxyl group and acidic hydrolysis should furnish triol **25**, which could be elaborated to **20**.

Aldehyde **23** was prepared by Swern oxidation of **11**. Treatment with samarium (II) iodide in THF in the presence of *t*-butanol and HMPA with rigorous exclusion of air and moisture gave alcohol **11** (19%), arising from reduction of the aldehyde, and **24** (37%). Benzylation of **24** gave a mixture of **26** and **27** in a ratio of 3:2, respectively. Acidic hydrolysis of **27** provided triol **25**. Phosphitylation of **25** and oxidation with MCPBA provided **28** which was deprotected using sodium in liquid ammonia. The target trisphosphate **20** was purified by ion-exchange chromatography. Preliminary evaluation of the Ca^{2+} mobilizing activity of **20** in Jurkat T-lymphocytes has demonstrated an EC$_{50}$ value only *ca* 4-fold higher than that of Ins(1,4,5)P$_3$ itself (*25*).

Synthesis of Chiral, Ring-Contracted 1D-*myo*-Inositol 1,4,5-Trisphosphate and 1,3,4,5-Tetrakisphosphate Analogs with C-2 Excised.

Noting that the potency of vinylcyclopentane **5** was significantly increased when the vinyl substituent was replaced by hydroxymethyl, in **20** (*25*), we felt that, in principle, the most desirable ring contracted trisphosphate would be **5**, representing the Ins(1,4,5)P$_3$ derivative in which only the carbon atom at position 2 and its associated hydroxyl group have been deleted. We reasoned that chiral **28** ought to be available by a route involving a samarium (II) iodide-mediated pinacol coupling (*28*) of a suitably protected D-*xylo*-pentodialdose such as **29**. We synthesized **28**, together with the corresponding ring-contracted Ins(1,3,4,5)P$_3$ tetrakisphosphate **30**, from D-xylose (*30*). Fisher glycosidation of D-xylose with allyl alcohol gave a mixture of pyranosides from which the α-anomer **31** (Figure 6) could be crystallized. Stannylene-mediated benzylation (*31,32*) of **31** gave the 2-*O*-benzyl derivative **32** in poor yield, and consequently the use was made of the recently described (*33*) butane 2,3-bisacetal (BBA) protecting group with **31**. Acid-catalyzed reaction of **31** with 2,2,3,3-tetramethoxybutane gave a 1:1 mixture of **33** and **34**, which were separated by chromatography. The lack of selectivity in protection of **31** with the BBA protecting group is consistent with experiments on methyl α-D-glucopyranoside, which also gave a 1:1 mixture of products (*33*), but is in contrast with kinetic acetonation of methyl (*34*) and benzyl (*35*) β-D-xylopyranosides, which gave the 2,3-*O*-isopropylidene derivatives.

164

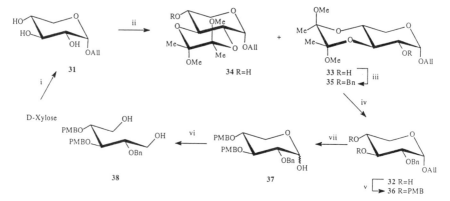

Figure 6. *Reagents and Conditions:* i, CH$_2$=CHCH$_2$OH, HCl, reflux, 16 h
(α-anomer by crystallization, 28%); ii, MeC(OMe)$_2$C(OMe)$_2$Me, CSA, MeOH,
(MeO)$_3$CH, reflux, 90 min (93%); iii, NaH, BnBr, DMF, 0°C, 2 h; iv, 95%
aq. TFA-CH$_2$Cl$_2$ (1:1), room temp., 15 min (84% from **10**); v, NaH, PMBCl,
DMF, 60°C, 2.5 h (74%); vi, (*a*) *t*-BuOK, DMSO, 50 °C, 3.5 h; (*b*) Me$_2$CO-
MHCl (10:1), 50°C, 30 min (87%); vii, NaBH$_4$, THF-H$_2$O (3:2), room temp.,
2 h (77%).

Benzylation of **33** with benzyl bromide gave the 2-*O*-benzyl derivative **35** and the BBA group was then removed with trifluoroacetic acid to give **32**. It was felt necessary to replace the BBA group with *p*-methoxybenzyl ethers for two reasons. First, the presence of acetals next to an aldehyde group has been reported to cause side reactions in the SmI$_2$-pinacol coupling, whereas ethers do not, and second, ethers adjacent to the aldehyde groups tend to direct *cis*-diol formation in the cyclitol products (*29*). Isomerization of the allyl group of **36** followed by acidic hydrolysis of the resulting enol ethers gave xylopyranose **37** as a *ca.* 7:3 α:β anomeric mixture. Reduction of **37** with sodium borohydride cleanly furnished xylitol **38**. Swern oxidation of xylitol **38** gave the required dialdose **29** (Figure 7), which was reacted with excess samarium (II) iodide (*28*) to give the products **39** and **40** in a ratio of *ca.* 1:3. The identity of the major product **40** was established by chemical correlation with known compounds. The minor product **39** was characterized as its isopropylidene acetal. Acidic hydrolysis of **40** gave the tetrol **41**. Hydrogenation of **41** gave the known (*36,37*) 1,2,4/3,5-cyclopentanepentol, which on benzoylation gave the known (*36*) pentabenzoate, thereby confirming the stereochemistry of **40** and derivatives, and since **39** contains a *cis*-diol (as deduced from the formation of an isopropylidene derivative), also indirectly confirming its stereochemistry. Tetrol **41** was phosphitylated and oxidized to give **42**, which was deprotected by hydrogenation to furnish the target tetrakisphosphate **30**, purified by ion-exchange chromatography.

For the target trisphosphate **28**, stannylene-mediated benzylation of *cis*-diol **40** gave a *ca.* 1:1 mixture of dibenzyl ethers **43** and **44**, which could not be separated, but upon acidic hydrolysis of the mixture the triols **45** and **46** were easily separated by chromatography. Phosphitylation and oxidation of **45** gave **47**, and the subsequent hydrogenation afforded the target trisphosphate **28** which was purified by ion-exchange chromatography. A preliminary examination showed that when **28** was microinjected into *Xenopus* oocytes it was able to induce Ins(1,4,5)P$_3$-like oscillations indicative of release of Ca^{2+} stores, but a higher threshold concentration was required relative to Ins(1,4,5)P$_3$ (at least 100-fold more). Biological and physicochemical evaluation of polyphosphates **28** and **30** is in progress, and full synthetic details have been published (*30*).

Synthesis of Carbohydrate-Based Inositol Polyphosphate Mimics Based on Adenophostin A

Although most of the active inositol polyphosphates and related compounds tested for Ca^{2+} release at the Ins(1,4,5)P$_3$ receptor during 1987-93 were full agonists, few exhibited a potency comparable to the natural ligand. However, in late 1993 a Japanese group reported the isolation of two potent trisphosphates from a culture broth of *Penicillium brevicompactum* (*38*). These were named adenophostins A and B, and identified as 3'-*O*-(α-D-glucopyranosyl)-adenosine-2',3'',4''-trisphosphate **48** (Figure 8) and its 6''-*O*-acetyl derivative **49** respectively (*39*). The structure of **48** has been confirmed by total synthesis (*40,41*). The adenophostins have been demonstrated to be full agonists in rat cerebellar microsomes, with potencies 100-fold higher than Ins(1,4,5)P$_3$ [EC$_{50}$ values 1.4 nM **48** and 1.5 nM **49**; *cf.* 170 nM Ins(1,4,5)P$_3$] (*42*); these relative potencies are consistent with binding data (*42*). In another study involving the purified Ins(1,4,5)P$_3$ type 1 receptor, **49** was found to be 10-fold more potent than Ins(1,4,5)P$_3$ [EC$_{50}$ values 11 nM **49**; *cf.* 100 nM Ins(1,4,5)P$_3$] (*43*). In this study, **49** demonstrated a positive cooperativity in binding to the Ins(1,4,5)P$_3$ receptor, not exhibited by Ins(1,4,5)P$_3$. Both **48** and **49** are resistant to the metabolic enzymes 5-phosphatase and 3-kinase, and as expected, produce a sustained Ca^{2+} release (*42*).

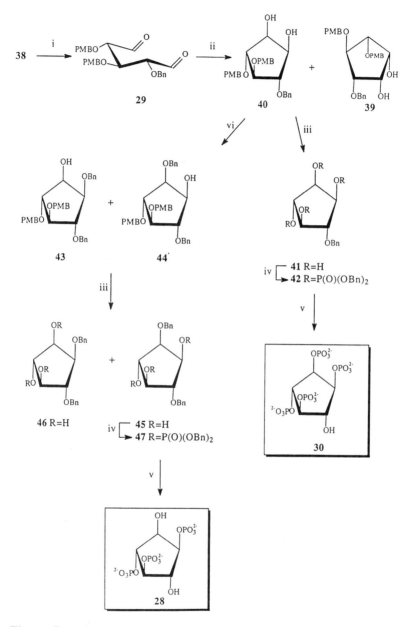

Figure 7. *Reagents and Conditions:* i, (*a*) (COCl)$_2$, DMSO, CH$_2$Cl$_2$, N$_2$, -60°C, 15 min; (*b*) Et$_3$N, room temp., 15 min; (*c*) toluene, reflux (Dean-Stark trap), 1 h; ii, SmI$_2$, *t*-BuOH, N$_2$, -60°C to room temp., 4 h (43% from **17**); iii, MHCl-EtOH (1:2), reflux, 2.5 h (76%); iv (*a*) (BnO)$_2$PN*i*Pr$_2$, 1*H*-tetrazole, CH$_2$Cl$_2$, room temp., 2 h; (*b*) MCPBA, 0 °C, 10 min; v, H$_2$, 10% Pd/C, NaHCO$_3$, MeOH-H$_2$O (4:1), room temp., 48 h; vi, Bu$_2$SnO, BnBr, Bu$_4$NBr, MeCN, 4 Å sieves, reflux, 24 h (87%); x, Ac$_2$O, pyridine, room temp., 2 h.

The high potency of the adenophostins is intriguing, as apart from the presence of a vicinal bisphosphate group and a third phosphate, they bear little apparent resemblance to Ins(1,4,5)P$_3$. However, the structures of **48** and **49** show consistencies with many of the features known to be important for agonism. The glucose 3,4-bisphosphate moiety possesses D-*threo* stereochemistry, and the position 2 hydroxyl group may be regarded as analogous to that of the position 6 of Ins(1,4,5)P$_3$. Indeed, molecular modeling studies (*42,44*) demonstrate similarity of positions 4, 3 and 2 of **48** with positions 4, 5 and 6 respectively of Ins(1,4,5)P$_3$. In addition, the adenophostins possess a third phosphate, which, similarly to Ins(1,4,5)P$_3$, is essential for high potency: 3'-*O*-(α-D-glucopyranosyl)-adenosine-3'',4''-bisphosphate **50**, in which this third phosphate is removed, possessed a 1000-fold lower binding affinity than **48** (*42*). However, although the broad basis for the activity of **48** and **49** is clear, a full structural rationalization for their exceptional potency is lacking. We chose initially to prepare 2-hydroxyethyl-α-D-glucopyranoside 2',3,4-trisphosphate **51**, Glc(2',3,4)P$_3$, to try and establish the relative importance of the adenine and adenosine components of **48**.

Adenophostin A synthesis requires deprotection of a fully protected, phosphorylated intermediate such as **52** (Figure 9), derived from triol **53**. Triol **53** could be obtained from a derivative in which positions 2', 3'' and 4'' are protected with a group removable in the presence of benzyl ethers, benzamides and glycosidic linkages. The *p*-methoxybenzyl ether fulfills this requirement. 6-*N*-Benzoyl-2'-*O*-(*p*-methoxybenzyl)-adenosine **54** is known (*45*) and benzylation of the primary 5'-hydroxyl in favor of the secondary 3'-hydroxyl to give **55** ought to be straightforward. Selective coupling of **55** to position 1 of 2,6-di-*O*-benzyl-3,4-di-*O*-(*p*-methoxybenzyl)-D-glucopyranose **56** to provide the α-anomeric derivative **57** should be possible using trichloroacetimidate (*46*) methodology commonly employed in oligosaccharide synthesis. The β-trichloroacetimidate derivative of **56** could also act as a glycosyl donor to other alcohols, such as 2-(*p*-methoxybenzyloxy)-ethanol, giving a precursor to **51**, or methyl 5-*O*-benzyl-2-*O*-(*p*-methoxybenzyl)-β-D-ribofuranoside, giving a precursor to **58**, Rib(2,3',4')P$_3$. Clearly, compound **56** is an important intermediate and, as part of an effort to synthesize adenophostin A analogs, we first sought a method to prepare **56**.

Synthesis of 2-Hydroxyethyl-α-D-glucopyranoside-2',3,4-tris-phosphate, Glc(2',3,4)P$_3$ 51.

Initially, we chose to synthesize **51**, representing the derivative of adenophostin A in which most of the adenosine moiety has been deleted. Selective protection of positions 2 and 6 of an allyl glycoside as in 2,6-di-*O*-benzyl-3,4-di-*O*-(*p*-methoxybenzyl)-D-glucopyranose **56** was required. As position 2 bears the most reactive secondary hydroxyl group in α-alkyl glycosides, whereas position 3 bears the most reactive in the β-isomers the preparation of allyl α-D-glucopyranoside **59ab** (Figure 10) was required. Using a Fisher glycosidation in which a mixture of D-glucose, allyl alcohol and a strong ion-exchange resin is heated under reflux (*47-49*), an orange syrup was obtained and no crystallization could be induced at this stage. Chromatography of this syrup gave a white solid **59ab**, and ^{1}H NMR spectroscopy revealed a *ca.* 7:3 α:β mixture of allyl glycosides. Fractional crystallization gave pure α-anomer **59a**, but in poor yield. Attention now turned to selective protection of positions 2 and 6. Both esterification and alkylation were attempted using a bis-stannylene approach. When **59ab** was reacted with 2.5 equiv. of dibutyltin oxide, a precipitate formed which could not be redissolved in toluene or dioxane. The stannylation product of **59ab** with 1.05-1.2 equiv. of dibutyltin oxide did not however precipitate on cooling, and treatment with 2.1 equiv. of benzoyl chloride gave a mixture of products by TLC, from which the known allyl 2,6-di-*O*-

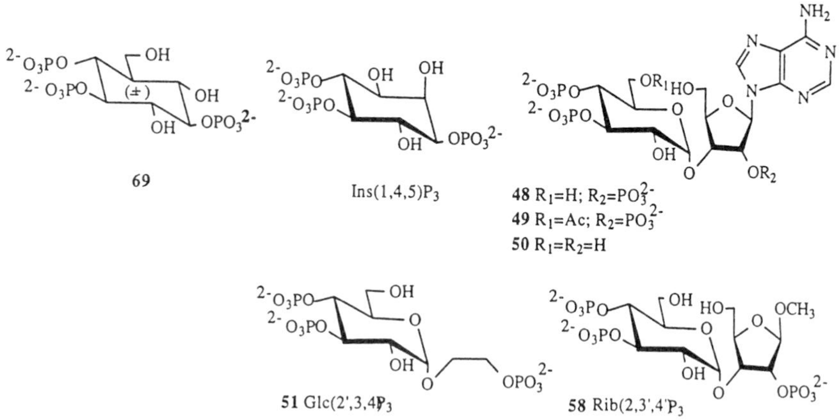

69

Ins(1,4,5)P_3

48 R_1=H; R_2=PO_3^{2-}
49 R_1=Ac; R_2=PO_3^{2-}
50 R_1=R_2=H

51 Glc(2',3,4)P_3

58 Rib(2,3',4')P_3

Figure 8. Comparison of structures of IP_3, adenophostins A and B (**48** and **49**) and their analogs **58** and **59**.

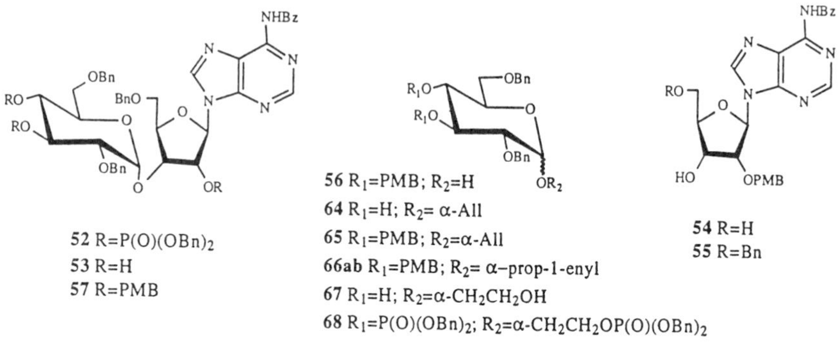

52 R=P(O)(OBn)$_2$
53 R=H
57 R=PMB

56 R_1=PMB; R_2=H
64 R_1=H; R_2= α-All
65 R_1=PMB; R_2=α-All
66ab R_1=PMB; R_2= α–prop-1-enyl
67 R_1=H; R_2=α-CH_2CH_2OH
68 R_1=P(O)(OBn)$_2$; R_2=α-$CH_2CH_2OP(O)(OBn)_2$

54 R=H
55 R=Bn

Figure 9. Structures of protected derivatives of Adenophostin A and its several synthetic precursors.

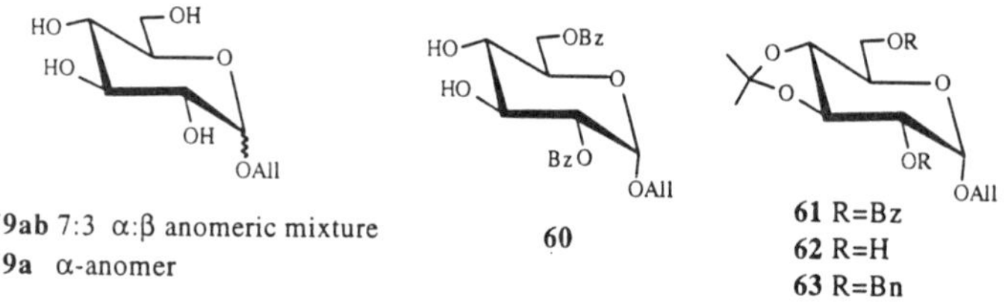

59ab 7:3 α:β anomeric mixture
59a α-anomer

60

61 R=Bz
62 R=H
63 R=Bn

Figure 10. Structures of glucosyl derivatives obtained during synthesis of the adenophostin analog **51**.

benzoyl-α-D-glucopyranoside **60** was isolated in 34% yield by chromatography and crystallization. Presumably, the 2,3-*O*-dibutylstannylene derivative of the α-anomer formed, was soluble in cold toluene and directed substitution at position 2, while the primary hydroxyl was selectively benzoylated over the remaining free secondary hydroxyl at position 4, as would be expected. This two-step preparation of **60** from D-glucose represents an improvement as compared to the five steps of a previous report (*49*). Reaction of **60** with 2-methoxypropene gave the fully protected **61**. Basic methanolysis provided diol **62**, which was benzylated with benzyl bromide to provide allyl 2,6-di-*O*-benzyl-3,4-*O*-isopropylidene-α-D-glucopyranoside **63** (*50*).

For potential direct benzylation, pure α-anomer **59a** was used to test the viability of the method. Treatment of the stannylated product with neat benzyl bromide at 100-110°C, yielded the required allyl 2,6-di-*O*-benzyl-α-D-glucopyranoside **64** in 44% yield, after preparation of its 3,4-dibenzoate. The technique was attempted on **59ab** on a 35 gram scale and was found to be the most convenient overall to produce **64** in multigram quantities. Conversion of **64** to target **56** was achieved in three steps. *p*-Methoxybenzylation of **64** gave fully protected **65**, which was isomerized to the corresponding prop-1-enyl glycoside **66ab** using potassium *t*-butoxide. Finally, acid hydrolysis gave the crystalline **56** in 77% yield.

As both benzyl ethers and isopropylidene acetals should be stable to osmium tetroxide, **63** was chosen as an appropriate selectively protected intermediate for the synthesis of **51**. Reaction of **63** with osmium tetroxide and excess sodium metaperiodate produced a product which was not isolated but reduced with sodium borohydride to furnish 2-hydroxyethyl-2,6-di-*O*-benzyl-α-D-glucopyranoside **67**. Loss of the isopropylidene acetal in this reaction was unexpected, but was attributed to osmic acid, an acid sufficiently strong to remove the labile *trans*-ketal. The loss of the acetal was advantageous, as it directly provided the triol required for phosphorylation. Phosphitylation of **67** and oxidation with MCPBA gave **68** which was deprotected with sodium in liquid ammonia. The required trisphosphate **51** was purified by ion-exchange chromatography (*51*). This compound has also been prepared from D-galactose (*52*).

We examined **51** for Ca^{2+}-mobilizing activity at the platelet Ins(1,4,5)P$_3$ receptor (*53*). It was found to be a full agonist with a potency *ca.* 10-fold lower than Ins(1,4,5)P$_3$ [EC$_{50}$ 0.6 μM; *cf.* Ins(1,4,5)P$_3$ 0.05 μM], a result consistent with binding data and with another study in SH-SY5Y neuroblastoma cells (*44*). A related study is of interest to this discussion. DL-6-Deoxy-6-hydroxymethyl-*scyllo*-inositol-1,2,4-trisphosphate **69** demonstrated a Ca^{2+}-mobilizing potency approximately equal to that of Ins(1,4,5)P$_3$ at the platelet Ins(1,4,5)P$_3$ receptor [EC$_{50}$ 0.1 μM; *cf.* Ins(1,4,5)P$_3$ 0.11 μM] (*54*). Assuming that one enantiomer is inactive, and noting that L-*scyllo*-Ins(1,2,4)P$_3$ has approximately equal Ca^{2+}-mobilizing potency to Ins(1,4,5)P$_3$, implies that the CH$_2$OH component is at least tolerated by the Ins(1,4,5)P$_3$ receptor, and may even give rise to a modest increase in potency. The observation that **51** is less active than either **69** or Ins(1,4,5)P$_3$ suggests that the conformationally rather mobile 2'-phosphate group of **51** is not a good mimic of the 2'-phosphate in adenophostin A, nor of the 1-phosphate of Ins(1,4,5)P$_3$. Thus, all or part of the adenosine moiety in adenophostin A may be necessary to orient the 2'-phosphate group in a particularly favorable way at the receptor binding site.

Disaccharide Polyphosphates Based Upon Adenophostin Activate D-*myo*-Inositol 1,4,5-Trisphosphate Receptors. After showing that the first synthetic carbohydrate-based agonist of the Ins(1,4,5)P$_3$ receptor with the minimal structure 2-hydroxyethyl α-D-glucopyranoside-2',3,4-trisphosphate [Glc(2',3,4)P$_3$, **51**] had ~10-fold reduced potency relative to Ins(1,4,5)P$_3$ in both SH-SY5Y cells (*44*) and platelets (*53*), molecular modeling studies on **51** (*44*) confirmed that the conformationally flexible bimethylene chain did not allow the 2'-phosphate to mimic accurately the positioning of either the 1-phosphate of $Ins(1,4,5)P_3$ or the 2'-phosphate of the adenophostins. We have now synthesized methyl 3-*O*-(α-D-ribofuranoside 2,3',4'-trisphosphate (*55*) [see (c) below]. Biological evaluation of "ribophostin" **58** using permeabilized hepatocytes (*56*) revealed a Ca^{2+}-mobilizing potency 10-fold better than **51** and very close to that of $Ins(1,4,5)P_3$, suggesting that conformational restriction of the 2'-phosphate alone can engender $Ins(1,4,5)P_3$-like, but not adenophostin-like, potency. We have now also synthesized other compounds incorporating the D-glucopyranosyl 3,4-bisphosphate moiety, with an α-glycosidic linkage to a second sugar, containing one or more phosphates. These disaccharide polyphosphates are expected to be conformationally more rigid than $Glc(2',3,4)P_3$, and to place their accessory phosphate group(s) in various positions within the receptor binding site. We have thus synthesized four disaccharide polyphosphates in total: sucrose 3,4,3'-trisphosphate, [$Sucr(3,4,3')P_3$, **70**, Figure 11], α,α'-trehalose 3,4,3',4'-tetrakisphosphate [$Trehal(3,4,3',4')P_4$, **71**] and α,α'-trehalose 2,4,3',4'-tetrakis-phosphate [$Trehal(2,4,3',4')P_4$, **72**], and as above, methyl 3-*O*-(α-D-glucopyranosyl)-β-D-ribofuranoside-2,3',4'-trisphosphate **58**, whose structure is most closely related to adenophostin A (*55*).

Disaccharides offer plenty of scope as frameworks for the synthesis of novel polyphosphates, but the difficulties in selective protection of the hydroxyl groups are often much greater than those encountered with simple inositols. One solution is to protect selected hydroxyl groups in two monosaccharides and then couple the two together. This approach has the advantage that, with sufficient ingenuity, any desired disaccharide polyphosphate can be assembled. Thus, ribophostin, $Rib(2,3'4')P_3$ **58**, as above, was designed to mimic, as closely as possible, the structure of adenophostin A, but with the adenine removed. This approach also allows the glucose and ribose components to be coupled to various other moieties, or a range of monosaccharides may be coupled combinatorially in pairs, thereby generating a number of analogs. The chief disadvantages of this strategy are the time taken to prepare the selectively protected intermediates and the lack of stereospecificity in the coupling reactions. Another solution is simply to choose an existing disaccharide with the required anomeric configuration, selectively protect some of its hydroxyl groups, phosphorylate the unprotected positions and then deprotect. This strategy has the disadvantage that a precise target structure may not be accessible because the required disaccharide is unavailable, but it has the advantage that the polyphosphates may be synthesized much more rapidly from cheap and abundant starting materials, potentially on a large scale. Given the range of available disaccharides, the strategy may have potential for the discovery of lead compounds with novel properties, and for the generation of a range of structurally diverse analogs required to provide data for molecular modeling studies. This method was chosen for compounds **70-72**.

(a) Sucrose 3,4,3'-Trisphosphate, $Sucr(3,4,3')P_3$ 70. Sucrose is the most abundant carbohydrate, and indeed, the cheapest chiral material available (*57*). With its eight free hydroxyl groups, and acid-labile glycosidic linkage, sucrose presents considerable difficulties in selective protection and manipulation. However, a comparison of the structure of sucrose with the disaccharide part of adenophostin A suggested that synthesis of the novel sucrose 3,4,3'-trisphosphate **70** might be worthwhile if a sufficiently concise route could be devised. An appropriately protected triol **73** was obtainable directly from sucrose by regioselective pivaloylation. It was then straightforward to prepare the fully-protected trisphosphate **74** by phosphitylation followed by oxidation. Two-step deprotection, removing first the benzyl groups and then the pivaloyl esters allowed the isolation of $Sucr(3,4,3')P_3$ **70** (*56*).

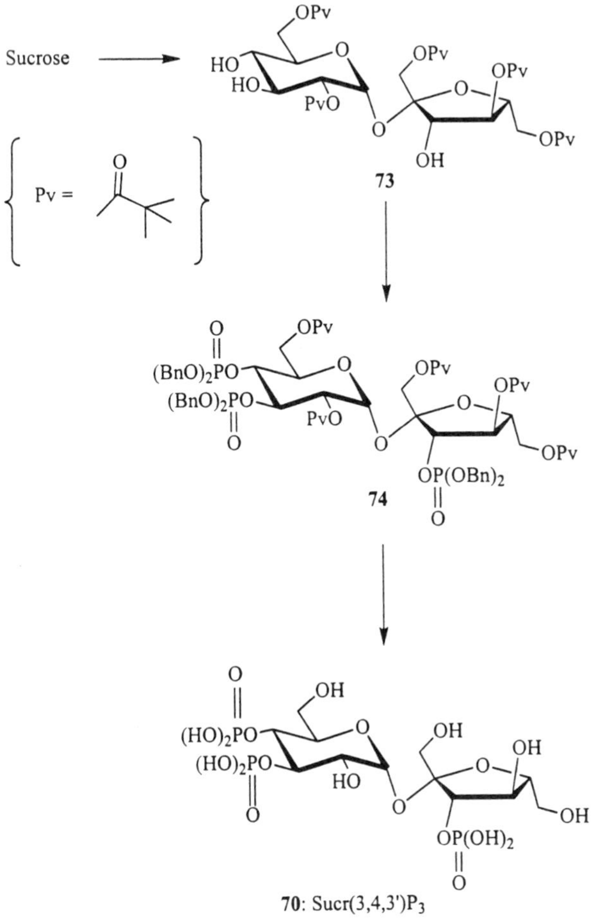

Figure 11. Synthesis of sucrose 3,4,3'-trisphosphate **70** from sucrose. Bn = benzyl, Pv = pivaloyl.

(b) Trehalose Tetrakisphosphates, Trehal(3,4,3',4')P$_4$ 72. Assuming that the basic Ins(1,4,5)P$_3$-like activity of the adenophostins originates in their D-glucose 3,4-bisphosphate component, we reasoned that an analog **71** (Figure 12) consisting simply of two copies of this structure joined by an α-glycosidic linkage might also be active. Either bisphosphate of this C_2-symmetrical molecule could (potentially) be recognized by the anchoring domain of the Ins(1,4,5)P$_3$ receptor binding site, leaving the other phosphates to interact with the accessory domain. Our strategy employed the symmetrically protected trehalose derivative **75**, available from α,α'-trehalose in two steps *via* the regioselective tin-mediated dibenzylation of 4,6:4',6'-di-*O*-benzylidene-α,α'-trehalose (*58*). The by-product of this reaction, the asymmetrically protected **76** would lead to another, potentially interesting, asymmetrical trehalose tetrakisphosphate **72**, in which the position of a single phosphate group on one glucose residue is altered, leaving the other glucose 3,4-bisphosphate component unchanged. A comparison of the biological properties of **71** and **72** enabled us to examine the biological consequences of this slight alteration in structure. The key step in the synthesis of **71** was the simultaneous regioselective reduction of both benzylidene acetals in **77**, leaving the two equivalent hydroxyl groups at positions 4 and 4' exposed for later phosphorylation, and both primary hydroxyls protected as benzyl ethers. This reaction was successfully carried out using sodium cyanoborohydride-hydrogen chloride (*59*). In order to obtain the required regioselectivity it was necessary first to protect positions 3 and 3', adjacent to the acetals. This was easily achieved using benzoyl esters as temporary protecting groups, which were removed after the reduction step to give the symmetrical tetraol **79**. Phosphitylation/oxidation followed by deprotection using sodium in liquid ammonia and ion-exchange chromatography on Q-Sepharose Fast Flow resin gave tetrakisphosphate **71**. Similarly, the symmetry element in the spectra of **75**, **77**, **78** and **79** made identification particularly easy. The same sequence of reactions was then applied with equal success to **76** to yield fully-protected **84**. Deprotection of **84** using catalytic hydrogenation and purification as before gave asymmetrical **72** (*56*).

(c) Methyl 3-*O*-(α-D-Glucopyranosyl)-β-D-ribofuranoside-2,3',4'-tris-phosphate, Rib(2,3',4')P$_3$ 58. We synthesized methyl 3-*O*-(α-D-gluco-pyranosyl)-α-D-ribofuranoside 2,3',4'-trisphosphate **58** (Figure 13), in which the adenine ring of **48** has effectively been deleted, but the third phosphate is held in a ribose ring similarly to **48**, and which should therefore provide information about the relative importance of the conformational restriction of this phosphate for the potency of **48** and **49**. 2,6-Di-*O*-benzyl-3,4-di-*O*-(*p*-methoxybenzyl)-D-glucopyranose **85** was prepared in five steps from D-glucose, involving a regioselective tin-mediated dibenzylation of allyl D-glucopyranoside (*50,51*) followed by *p*-methoxybenzylation and subsequent de-allylation. Reaction of **85** with trichloroacetonitrile in the presence of potassium carbonate (*60*) gave the α-trichloroacetimidate **86** and the crystalline β-anomer **87** in the ratio of 1 : 2.5. With a suitable glycosyl donor in hand, we required methyl 5-*O*-benzyl-2-*O*-(*p*-methoxybenzyl)-β-D-ribofuranoside **88** as an acceptor. D-Ribose was converted into the know (*11*) methyl β-D-ribofuranoside **89**. Reaction of **89** with *p*-methoxybenzaldehyde dimethyl acetal (*61*) in the presence of *p*-toluene-sulfonic acid gave the 2,3-*O*-(*p*-methoxybenzylidene) derivative **90** as a *ca.* 3:2 diastereomeric mixture. Benzylation of **90** gave fully protected **91**. Cleavage of the *p*-methoxybenzylidene acetal with DIBAL-H (*62*) gave **88**, and the more polar isomer **92**, in approximately equal proportions. Although the regioselectivity of acetal cleavage was disappointing, the unwanted isomer **92** was easily reoxidized to **91** (as a 92:8 diastereomeric mixture) using DDQ (*63*). Coupling of **87** and **88** was achieved using trimethylsilyl triflate as promoter (*60*), forming the α-glucopyranosyl compound

173

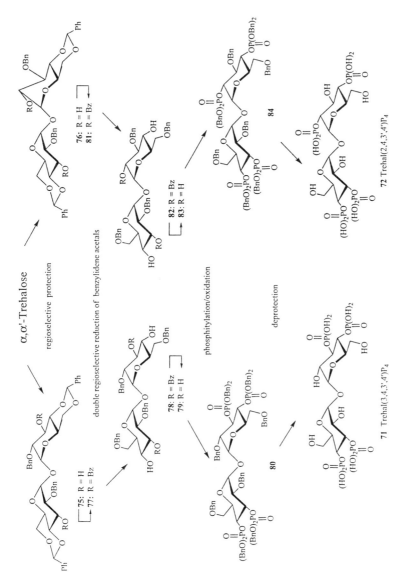

Figure 12. Synthesis of symmetrical **71** and asymmetrical **72** α,α'-trehalose tetrakisphosphates from α,α'-trehalose. Bn = benzyl, Bz = benzoyl, Ph = phenyl.

174

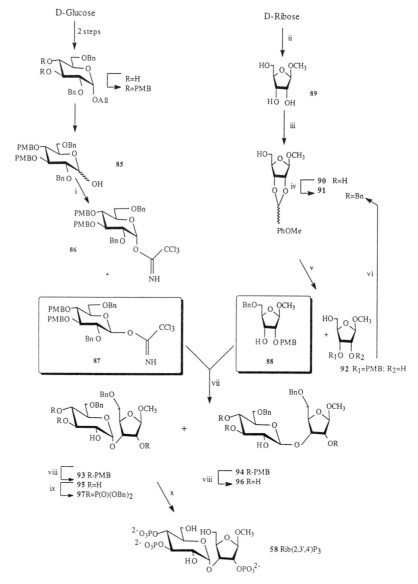

Figure 13. *Reagents and conditions:* i, Cl$_3$CCN, K$_2$CO$_3$, CH$_2$Cl$_2$, N$_2$, rt, 2 h, 65%; ii, MeOH, H$_2$SO$_4$, room temp., 20 h, (β-anomer obtained by crystallization); iii, *p*-MeO-PhCH(OMe)$_2$, PTSA, DMF, 70°C, MeOH (removed *in vacuo*), 4 h, 93%; iv, NaH, BnBr, DMF, 3 h, 82%; v, DIBAL-H, CH$_2$Cl$_2$, room temp., 3 h, 76%; vi, DDQ, CH$_2$Cl$_2$, 3 Å sieves, room temp., 3 h, 71%; vii, Me$_3$SiOSO$_2$CF$_3$, Et$_2$O, 3 Å sieves, room temp., 10 min; viii, DDQ, CH$_2$Cl$_2$-H$_2$O (10:1), room temp., 1 h, 56%; ix, (*a*) (BnO)$_2$PN(*i*-Pr)$_2$, 1*H*-tetrazole, room temp., 30 min, (*b*) MCPBA, -78°C to room temp., 10 min, 82%; x, H$_2$, Pd/C, 40 psi, 16 h, 70%. Ac = acetyl; All = allyl; Bn = benzyl; PMB = *p*-methoxybenzyl.

93 as the major product with the β-glucopyranosyl anomer **94** as a *ca.* 20% contaminant. On treatment of the mixture with DDQ, the crystalline triol **95** could be separated from the β-coupled isomer **96** by chromatography. Phosphitylation of **95** with bis(benzyloxy)-*N,N*-diisopropylaminophosphine (*64*), followed by oxidation of the intermediate trisphosphite triester with MCPBA gave trisphosphate **97**. This was deprotected by hydrogenolysis to give trisphosphate **58**, purified by ion-exchange chromatography. The resulting triethylammonium salt of **58**, "ribophostin" (*55,56*) was converted to the hexapotassium salt for quantification and biological evaluation.

(d) Effects of Adenophostin A on Ins(1,4,5)P₃ Receptors. Ins(1,4,5)P₃ bound to a single class of site on hepatic membranes, and the same sites bound adenophostin A with ~10-fold greater affinity, similar to those obtained with cerebellar membranes in which adenophostin A (K_d = 0.91 nM) bound with ~7-fold greater affinity than Ins(1,4,5)P₃ (K_d = 6.57 nM) (*56*). Thus the predominantly type 2 Ins(1,4,5)P₃ receptors of rat hepatocytes (*65*) and the type 1 Ins(1,4,5)P₃ receptors of cerebellum bind adenophostin A with similar affinity. A maximally effective concentration of Ins(1,4,5)P₃ (10 μM) released 50% of the intracellular Ca²⁺ stores of saponin-permeabilized hepatocytes. The same amount of Ca²⁺ was released by a maximally effective concentration of adenophostin A (1 μM) which evoked 102% of the response to Ins(1,4,5)P₃. Adenophostin A was ~14-fold more potent than Ins(1,4,5)P₃ in evoking Ca²⁺ release. This demonstrates that adenophostin A and Ins(1,4,5)P₃ interact with the same receptors to evoke Ca²⁺ release, but adenophostin A binds with ~10-fold greater affinity. These results with hepatocytes are comparable to those obtained previously from cells expressing largely type 1 Ins(1,4,5)P₃ receptors (*42,43,53*).

(e) Effects of Phosphorylated Carbohydrate Analogs on Ins(1,4,5)P₃ Receptors. The concentration-dependent effects of five phosphorylated carbohydrate analogs on the ⁴⁵Ca²⁺ content of the intracellular Ca²⁺ stores in permeabilized rat hepatocytes were studied (*56*). Each analog released the same fraction of the intracellular Ca²⁺ stores as was released by 10 μM Ins(1,4,5)P₃ (50%), and addition of Ins(1,4,5)P₃ (10 μM) to cells that had previously been stimulated with a maximally effective concentration of each analog failed to evoke further Ca²⁺ release, demonstrating that Ins(1,4,5)P₃ and each of the analogs release the same intracellular Ca²⁺ stores. In binding studies with hepatic membranes, each of the phosphorylated carbohydrate analogs completely displaced specific [³H]-Ins(1,4,5)P₃ binding. The rank order of potency of the analogs [adenophostin A > Ins(1,4,5)P₃ ~ Rib(2,3',4')P₃ > Trehal(2,4,3',4')P₄ > Glc(2',3,4)P₃ ~ Trehal(3,4,3',4')P₄ > Sucr(3,4,3')P₃] was the same in radioligand and functional assays. Both Rib(2,3',4')P₃ and Trehal(2,4,3',4')P₄ bound with significantly greater affinity (~27 and ~3-fold, respectively) than the only active carbohydrate agonist of Ins(1,4,5)P₃ receptors previously examined **51**, [Glc(2',3,4)P₃]. Indeed, the most potent of the carbohydrate analogs, Rib(2,3',4')P₃, was as active as the endogenous agonist, Ins(1,4,5)P₃.

All synthetic analogs contain the same phosphorylated glucose component, identical to that found in the adenophostins. Differences in activity must, therefore, be attributable to the influence of the second component, which bears the accesssory phosphate group(s). Glc(2',3,4)P₃ was designed to place the accessory phosphate group two carbons away from an α-glycosidic oxygen, as in the adenophostins, but the bimethylene chain of Glc(2',3,4)P₃ adopts extended conformations, placing the phosphate group too distant from the ring. The observation that increasing the chain length and/or inversion of anomeric configuration in related xylopyranoside trisphosphates further reduces activity (*31*) supports this hypothesis. Trehal(3,4,3',4')P₄ is unique in that either half of the molecule can mimic the 4,5-

bisphosphate/6-hydroxyl of Ins(1,4,5)P_3. It gives greater rigidity than Glc(2',3,4)P_3, although its potency is similar. It is likely that one glucose bisphosphate residue of Trehal(3,4,3',4')P_4 interacts with the receptor binding site in the same way as does this component in the other analogs, but that neither phosphate group on the second glucose residue is placed close enough to the active bisphosphate to give increased activity over Glc(2',3,4)P_3. Trehal(2,4,3',4')P_4, is an asymmetrical regioisomer of Trehal(3,4,3',4')P_4 in which a single phosphate is relocated to a position two carbons removed from the glycosidic oxygen. This modification leads to a significant gain in activity, suggesting that the 2-phosphate group is better-placed than either the 3 or 4 phosphates. The fact that Trehal(2,4,3',4')P_4 is more potent than Glc(2',3,4)P_3, despite its increased size, validates the general approach of increasing rigidity.

The finding that Rib(2,3',4')P_3 still does not approach the potency of the adenophostins suggests a role for the adenine, either by engaging in its own stabilizing interactions with a nearby area of the Ins(1,4,5)P_3 receptor, and/or by further optimizing the positioning of the 2'-phosphate at the binding site. However, Rib(2,3',4')P_3 is equipotent to Ins(1,4,5)P_3, and much more potent than the other disaccharides. This supports the idea that accurate positioning of the accessory phosphate is important in achieving Ins(1,4,5)P_3-like activity. The weak activity of Sucr(3,4,3')P_3 compared to all the other analogs is surprising, considering its apparent structural resemblance to Rib(2,3',4')P_3. It could be that steric hindrance from one or both hydroxymethyl groups on the fructofuranoside in Sucr(3,4,3')P_3 interferes with binding, or that overlapping anomeric effects (66), not present in Rib(2,3',4')P_3, influence the conformation about the glycosidic linkage. Finally, the presence of the quaternary furanosyl anomeric center in Sucr(3,4,3')P_3, may lead to increased flexibility about the fructofuranosyl linkage (67).

It remains to be established whether adenophostin-like activity can be achieved without the adenine, simply by constraining the phosphate group in the position that it adopts at the receptor binding site. The possibility must be considered that, although Rib(2,3',4')P_3 can attain the necessary conformation, it is still too flexible. Alongside a study of adenine-containing analogs, therefore, it will also be important to investigate conformationally restricted analogs lacking the adenine. Molecular modeling studies may aid the development of these, although simulating the combined influence of highly charged phosphate groups, anomeric effects in disaccharides and the environment at the receptor binding site presents many difficulties.

In summary, we have demonstrated that adenophostin A is a potent agonist of the type 2 Ins(1,4,5)P_3 receptor expressed in hepatocytes and have synthesized four disaccharide polyphosphates, all of which bind to and activate Ins(1,4,5)P_3 receptors. Rib(2,3',4')P_3, whose structure is closely related to adenophostin A, is as potent as Ins(1,4,5)P_3 and more potent than most conventional inositol-based agonists. Thus, phosphorylated disaccharides provide novel opportunities to develop high affinity ligands of Ins(1,4,5)P_3 receptors. We have therefore also shown that radical redesign of the Ins(1,4,5)P_3 molecule is possible, without losing its potent Ca^{2+}-mobilizing activity. Diverse compounds can be designed either by using carbohydrates as starting materials or by incorporating carbohydrate motifs into the final structure. This opens up new ways for chemists to interfere pharmacologically in the polyphosphoinositide pathway of cellular signaling.

Acknowledgments

All colleagues whose names appear on the publications cited herein are acknowledged for their efforts in respect of the work discussed here, especially Drs A. M. Riley, D. J. Jenkins and Miss R. D. Marwood for chemistry and Prof J. Westwick, Drs C. T. Murphy, A. M. Marchant, M. D. Beecroft and C. W. Taylor for pharmacology. We thank Susan Alston for manuscript preparation. This work was supported by the Wellcome Trust (Programme Grant 045491).

Literature Cited

1. Berridge, M. J. *Nature (London)* **1993**, *361*, 315-325.
2. Streb, H.; Irvine, R. F.; Berridge, M. J.; Schulz, I. *Nature (London)* **1983**, *206*, 67-69.
3. Potter, B. V. L.; Lampe, D. *Angew. Chem. Int. Ed. Engl.* **1995**, *34*, 1933-1972.
4. Lampe, D; Mills, S. J.; Potter, B. V. L. *J. Chem. Soc. Perkin Trans 1* **1992**, 2899-2906.
5. Sawyer, D. A.; Potter, B. V. L. *J. Chem. Soc. Perkin Trans 1* **1992**, 923-932.
6. Lampe, D.; Liu, C.; Potter, B;V.L. *J. Med. Chem.* **1994**, *37*, 907-912.
7. Riley, A. M.; Payne, R.; Potter, B. V. L. *J. Med. Chem.* **1994**, *37*, 3918-3927.
8. Batty, I. R.; Nahorski, S. R.; Irvine, R. F. *Biochem. J.* **1985**, *232*, 211-215.
9. Irvine, R. F.; Letcher, A. J.; Heslop, J. P.; Berridge, M. J. *Nature (London)* **1986**, *320*, 631-634.
10. Choi, K. Y.; Kim, H. K.; Lee, S. Y.; Moon, K. H.; Kim, S. S.; Kim, J. W.; Chung, H. K.; Rhee, S. K. *Science* **1990**, *248*, 64-66.
11. Abdullah, M.; Hughes, P. J.; Craxton, A.; Gigg, R.; Desai, T.; Marecek, J. F.; Prestwich, G. D.; Shears, S. J. *J. Biol. Chem.* **1992**, *267*, 22340-22345.
12. Shears, S. B. In *Advances in Second Messenger and Phosphoprotein Research*; Putney J. W., Jr., Ed.; Raven Press Ltd.: New York, 1992, Vol. 26, pp. 63-92.
13. Irvine, R. F. *FEBS Lett.* **1990**, *263*, 5-9.
14. Lückhoff, A.; Clapham, D. E. *Nature (London)* **1992**, *355*, 356-358.
15. Cullen, P. J.; Dawson, A. P.; Irvine, R. F. *Biochem. J.* **1995**, *305*, 139-143.
16. Cullen, P. J.; Chung, S. -K.; Chang, Y. -T.; Dawson, A. P.; Irvine, R. F. *FEBS Lett.* **1995**, *358*, 240-242.
17. Cullen, P. J.; Hsuan, J. J.; Truong, O.; Letcher, A. J.; Jackson, T. R.; Dawson, A. P.; Irvine, R. F. *Nature (London)* **1995**, *376*, 527-530.
18. Riley, A. M.; Jenkins, D. J.; Potter, B. V. L. *J. Am. Chem. Soc.* **1995**, *117*, 3300-3301.
19. Ferrier, R. J.; Middleton, S. *Chem. Rev.* **1993**, *93*, 2779-2831.
20. Ito, H.; Motoki, Y.; Taguchi, L.; Hanzawa, Y. *J. Am. Chem. Soc.* **1993**, *115*, 8835-8836.
21. Hanzawa, Y.; Ito, H.; Taguchi, T. *Synlett* **1995**, 299-305.
22. Tatsuta, K.; Niwata, Y.; Umezawa, K.; Toshima, K.; Nakata, M. *J. Antibiot.* **1991**, *44*, 456-458.
23. Johansson, R.; Samuelsson, B. *J. Chem. Soc. Perkin Trans 1* **1984**, 2371-2374.
24. Joniak, D.; Kosíková, B.; Kosáková, L. *Collect. Czech. Chem. Commun.* **1978**, *43*, 769-773.
25. Jenkins, D. J.; Riley A. M.; Potter, B. V. L. *J. Org. Chem.* **1996**, *61*, 7719-7726.
26. Liu, C.; Potter, B. V. L. *J Org. Chem.* **1997**, *62*, 8335-8340.
27. Chénedé, A.; Pothier, P.; Sollogoub, M.; Fairbanks, A. J.; Sinaÿ, P. *J. Chem. Soc., Chem. Commun.* **1995** 1373-1374.
28. Perrin, E.; Mallet, J. -M.; Sinaÿ, P. *Carbohydr. Lett.* **1995**, *1*, 215-216.
29. Carpintero, M; Fernández-Mayoralas, A.; Jaramillo, C. *J. Org. Chem.* **1997**, *62*, 1916-1917.
30. Jenkins, D. J.; Potter, B. V. L., *J. Chem. Soc. Perkin Trans 1* **1998**, 44-49.

31. Moitessier, N.; Chrétien, F.; Chapleur Y.; Humeau, C. *Tetrahedron Lett.* **1995**, *36*, 8023-8026.
32. Haque, M. E.; Kikuchi, T.; Yoshimoto, K.; Tsuda, Y. *Chem. Pharm. Bull.* **1985**, *33*, 2243-2255.
33. Montchamp, J. -L.; Tian, F.; Hart, M. E.; Frost, J. W. *J. Org. Chem.* **1996**, *61*, 3897-3899.
34. Naleway, J. J.; Raetz, C. R. H.; Anderson, L. *Carbohydr. Res.* **1988**, *179*, 199-209.
35. Rio, S.; Beau, J. -M.; Jacquinet, J. -C. *Carbohydr. Res.* **1991**, *219*, 71-90.
36. Sable, H. Z.; Anderson, T.; Tolbert, B.; Posternak, T. *Helv. Chim. Acta* **1963**, *46*, 1157-1165.
37. Angyal, S. J.; Luttrell, B. M. *Aust. J. Chem.* **1970**, *23*, 1831-1835.
38. Takahashi, M.; Kagasaki, T; Hosoya, T.; Takahashi, S. *J. Antibiot.* **1993**, *46*, 1643-1647.
39. Takahashi, S.; Kinoshita, T.; Takahashi, M. *J. Antibiot.* **1994**, *47*, 95-100.
40. Hotoda, H.; Takahashi, M.; Tanzawa, K.; Takahashi, S.; Kaneko, M. *Tetrahedron Lett.* **1995**, *36*, 5037-5040.
41. van Straten, N. C. R.; van der Marel, G. A.; van Boom, J. H. *Tetrahedron Lett.* **1996**, *37*, 3599-3602.
42. Takahashi, M.; Tanzawa, K.; Takahashi, S. *J. Biol. Chem.* **1994**, *269*, 369-372.
43. Hirota, J.; Michikawa, T.; Miyawaki, A.; Takahashi, M.; Tanzawa, K.; Okura, I.; Furuichi, T.; Mikoshiba, K. *FEBS Lett.* **1995**, *368*, 248-252.
44. Wilcox, R. A.; Erneux, C.; Primrose, W. U.; Gigg, R.; Nahorski, S. R. *Mol. Pharmacol.* **1995**, *47*, 1204-1211.
45. Takaku, H.; Kamaike, K. *Chem. Lett.* **1982**, 189-192.
46. Schmidt, R. R. *Angew. Chem. Int. Ed. Eng.* **1986**, *25*, 212-235.
47. Nepogod'ev, S. A.; Backinowsky, L. V.; Grzeszczyk, B.; Zamojski, A. *Carbohydr. Res.* **1994**, *254*, 43-60.
48. Lee, R. T.; Lee, Y. C. *Carbohydr. Res.* **1974**, *37*, 193-201.
49. Pelyvás, I.; Lindhorst, T.; Thiem, J. *Liebigs Ann. Chem.* **1990**, 761-769.
50. Jenkins, D. J. J.; Potter, B. V. L. *J. Chem. Soc. Chem. Comm.* **1995**, 1169-1170.
51. Jenkins, D. J.; Potter, B. V. L. *Carbohydr. Res.* **1996**, *287*, 169-182.
52. Desai, T.; Gigg, J.; Gigg, R. *Aust. J. Chem.* **1996**, *49*, 305-309.
53. Murphy, C. T.; Riley, A. M.; Lindley, C. J.; Jenkins, D. J.; Westwick, J.; Potter, B. V. L. *Mol. Pharmacol.* **1997**, 741-748.
54. Riley, A. M.; Murphy, C. T.; Lindley, C. J.; Westwick, J.; Potter, B. V. L. *Bioorg. Med. Chem. Lett.* **1996**, *6*, 2197-2200.
55. Jenkins, D. J.; Marwood, R. D.; Potter, B. V. L. *Chem. Commun.* **1997**, 449-450; *Corrigendum* **1997**, 805.
56. Marchant, J. S.; Beecroft, M. D.; Riley, A. M.; Jenkins, D. J.; Marwood, R. D.; Taylor, C. W.; Potter, B. V. L. *Biochemistry* **1997**, *36*, 12780-12790.
57. Jarosz, S. *Polish J. Chem.* **1996**, *70*, 972-987.
58. Dowd, M. K.; Reilly, P. J.; French, A. D. *J. Comp. Chem.* **1992**, *13*, 102-114.
59. Garegg, P. J.; Hultberg, H.; Wallin, S. *Carbohydr. Res.* **1982**, *108*, 97-101.
60. Schmidt, R. R.; Michel, J.; Roos, M. *Liebigs Ann. Chem.* **1984**, 1343
61. Johansson, R.; Samuelsson, B. *J. Chem. Soc. Perkin Trans 1* **1984**, 2371-
62. Evans, D. A.; Kaldor, S. W.; Jones, T. K.; Clardy, J.; Stout, T. J. *J. Am. Chem. Soc.* **1990**, *112*, 7001-7031.
63. Sviridov, A. F.; Ermolenko, M. S.; Yashunsky, D. V.; Borodin, V. S.; Kochetkov, N. K. *Tetrahedron Lett.* **1987**, *28*, 3835-3838.

64. Yu, K. -L.; Fraser-Reid, B. *Tetrahedron Lett.* **1988**, *29*, 979-982.
65. Wojcikiewicz, R. J. H. *J. Biol. Chem.* **1995**, *270*, 11678-11683.
66. French, A. D.; Schafer, L.; Newton, S. Q. *Carbohydr. Res.* **1993**, *239*, 51-60.
67. O'Leary, D. J.; Kishi, Y. *J. Org. Chem.* **1994**, *59*, 6629-6636.

Chapter 11

Stereospecific Syntheses of Inositol Phospholipids and Their Phosphorothioate Analogs

Robert J. Kubiak, Xiangjun Yue, and Karol S. Bruzik[1]

Department of Medicinal Chemistry and Pharmacognosy, University of Illinois at Chicago, Chicago, IL 60612

Following receptor stimulation, phosphatidylinositol phosphates (PIP_n) undergo numerous processes involving cleavage of the phophate group. Earlier results obtained with phosphorothioate analogs of biophosphates and our recent finding of the resistance of the phosphorothioate analog of PI to phosphatidylinositol-specific phospholipase C suggest that analogs of PIP_n could be valuable research tools in studying interdependence of various pathways of inositol metabolism, and could be used as mechanistic probes and inhibitors of such enzymes as PI-PLC, PI kinases and PIP_n phosphatases. This work describes synthesis of a range of phosphorothioate analogs of PIP_n in which sulfur atom substitutes the oxygen atom in the nonbridging position of the phosphodiester and monoester functions, as well as in the bridging position of the phosphodiester function.

Inositol signaling pathways are extremely complex and employ a large number of inositol phosphates and phospholipids (*1*). Due to their multiplicity it is difficult to understand their spatio-temporal relationships unless some signaling branches are eliminated or inhibited. This can be realized by way of gene mutations, where the specific functions of some proteins are abolished, and the effect of the protein modification on the cellular physiology is examined (*2*). Such an effect can also be accomplished by using synthetic analogs of inositol phospholipids or phosphates with properties altered in such a way, as to increase their resistance to enzymatic cleavage, inhibit synthesis or modify cell membrane permeability. These avenues have been explored in many other fields, and should be especially useful in the area of inositol signaling. Interconversion between inositol phosphates and phospholipids following signal stimulation employs almost exclusively reactions involving nucleophilic displacement at phosphorus, such as phosphodiester cleavage at the inositol 1-position and/or addition/removal of phosphomonoester residues at the inositol 3-, 4- and 5-positions (*1, 3, 4*). In all, six phosphatidylinositols with the inositol residue bearing different number of phosphomonoester groups are known (*1,3,4*), five of which (Figure 1, **1-5**) have been recently synthesized in our Laboratory (*5, 6*). A similar progress has also been achieved in other laboratories (*8-16*). Phosphatidylinositol (PI) phosphorylation and dephosphorylation reactions result in vastly different recognition

[1]Corresponding author.

by the receptor or enzymatic proteins. For example, the presence of the phosphate groups at the 4- and 5-positions of inositol makes phosphatidylinositol 4,5-bisphosphate (PI-4,5-P$_2$) a preferred substrate for phosphatidylinositol-specific phospholipase C (PI-PLC) (*17, 18*), while the addition of the phosphate group at the 3-position makes phosphatidylinositol 3,4,5-trisphosphate completely resistant to this enzyme (*19, 20*). The ability to intervene into metabolic phosphate addition/removal steps can therefore be useful in understanding the overall phosphoinositide metabolism.

This chapter describes synthesis of saturated, fatty acid-bearing phosphatidylinositol phosphates **1-5**, as well as their analogs **6-17** (Figure 1) in which phosphate functions have been replaced by the phosphorothioate groups, at either or both the phosphodiester and phosphomonoester positions (*6*). The trademark of these analogs is that the phosphate-phosphorothioate replacement is a rather minor structure modifications, which should ensure the same mode of their interaction with the target proteins as those of the natural compounds. These modifications are expected to impose, however, selective resistance to hydrolytic enzymes involved in the removal of phosphate groups of phosphoinositides. The effect of structure modification by sulfur substitution in the phosphate group (Figure 2) brings about three effects, briefly discussed below.

(i) Most of the differences in the hydrolytic behavior of phosphorothioates, as compared to phosphates, arise from an impaired hydrogen bonding ability of the sulfur atom as compared to oxygen (*21*). This has a general effect of slowing down the reactions catalyzed by phosphoryl transfer enzymes, such as phosphodiesterases and phosphatases, where such hydrogen bonding to the phosphate group is an important activating catalytic factor (*22*). Furthermore, sulfur interacts only weekly with such metal ions as magnesium and calcium, which are also frequent elements of active sites of phosphotransferases (*23*). As a result, sulfur substitution at the nonbridging position of the phosphate group of oligonucleotides renders the phosphodiester group partially or completely resistant to nucleases (*24*). In the area of inositol enzymology, it has been shown that the 5-phosphorothioate analog of inositol 1,4,5-trisphosphate (IP$_3$) is resistant to the 5-phosphatase (*25*). This leads to slower clearance of this analog as compared to IP$_3$, and a prolonged calcium-mobilizing effect of this analog (*26*). We have also shown earlier that sulfur substitution of the pro-S nonbridging oxygen in the phosphate group of PI brings about an almost complete resistance of the Sp-diastereomer (Sp-DPPsI, Sp-**6**) to the cleavage by the bacterial PI-PLC (*27-30*). The loss of the hydrogen bonding to the phosphate is the most likely reason for the very slow cleavage of this analog (see the following chapter by Hondal et al.). The Sp-diastereomer of the phosphorothionate analog of PI is probably the structurally closest, cleavage-resistant analog of PI. These findings have suggested that sulfur modification at various positions in phosphatidylinositol polyphosphates could provide useful research tools in phosphoinositide enzymology.

(ii) Sulfur modification at the nonbridging position creates a stereogenic center at the phosphorus atom, and enables studying steric courses of reactions catalyzed by phosphotransferases (*22*). Application of the P-chiral, sulfur or oxygen-isotope modified analogs, allowed us in the past to determine that the cleavage of the phosphodiester bond in PI by both the bacterial and mammalian phospholipases C occurs by an analogous double-displacement mechanism, with the inositol 1,2-cyclic phosphate as an intermediate (*31-33*).

(iii) Phosphatidylinositol analogs modified by sulfur substitution at the bridging position of the leaving group generate a thiol product upon their cleavage by PI-PLC, instead of the usual diacylglycerol, and hence constitute convenient substrates for quantitation of enzyme activity based on the amount of the thiol released (*34, 35*). Furthermore, the comparison of cleavage kinetics of the natural substrate and

A

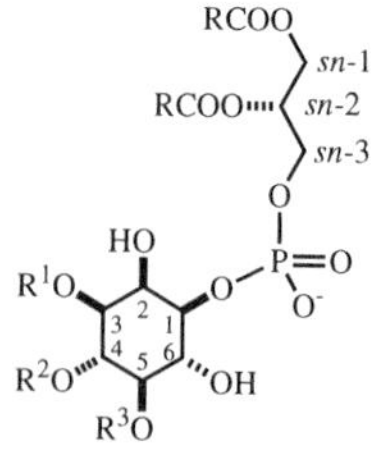

RCOO sn-1
RCOO''''' sn-2
sn-3

HO
R^1O
R^2O''''
R^3O
'''''OH

$R^1 = R^2 = R^3 = H$, PI (**1**)
$R^1 = PO_3^{2-}$; $R^2 = R^3 = H$, PI-3-P (**2**)
$R^1 = R^2 = PO_3^{2-}$; $R^3 = H$, PI-3,4-P$_2$ (**3**)
$R^1 = H$; $R^2 = R^3 = PO_3^{2-}$; PI-4,5-P$_2$ (**4**)
$R^1 = R^2 = R^3 = PO_3^{2-}$; PI-3,4,5-P$_3$ (**5**)

B

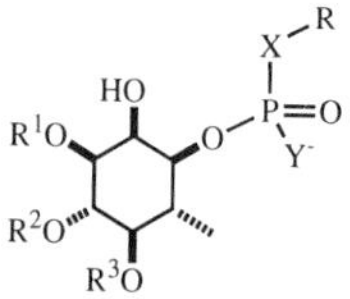

HO
R^1O
R^2O
R^3O''''

R^4COO
$R = R^4COO$'''''

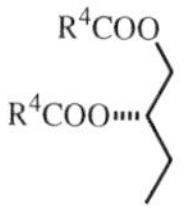

DP: dipalmitoyl
DO: dioctanoyl

$R^1 = R^2 = R^3 = H$, X = O, Y = S; DPPsI (**6**)
$R^1 = R^2 = R^3 = H$, X = S, Y = O; DOsPI (**7**), DPsPI (**8**)
$R^1 = PSO_2^{2-}$; $R^2 = R^3 = H$, X = O, Y = S; DPPI-3-Ps (**9**)
$R^1 = PSO_2^{2-}$; $R^2 = R^3 = H$, X = O, Y = S; DPPsI-3-Ps (**10**)
$R^1 = R^2 = PSO_2^{2-}$, $R^3 = H$, X = O, Y = O; DPPI-3,4-Ps$_2$ (**11**)
$R^1 = H$; $R^2 = R^3 = PO_3^{2-}$, X = S, Y = O; DPsPI-4,5-P$_2$ (**12**)
$R^1 = H$; $R^2 = R^3 = PO_3^{2-}$, X = O, Y = S; DPPsI-4,5-P$_2$ (**13**)
$R^1 = H$; $R^2 = R^3 = PSO_2^{2-}$, X = O, Y = O; DPPI-4,5-Ps$_2$ (**14**)
$R^1 = R^2 = R^3 = PSO_2^{2-}$, X = O, Y = O; DPPI-3,4,5-Ps$_3$ (**15**)
$R^1 = R^2 = R^3 = PSO_2^{2-}$, X = O, Y = S; DPPsI-3,4,5-Ps$_3$ (**16**)

R = (4-nitrophenyl) NO$_2$

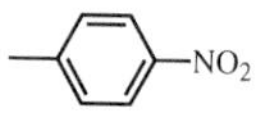

$R^1 = R^2 = R^3 = H$, X = O, Y = S; NPIPs (**17**)

Figure 1. (A) Structures of synthetic phosphatidylinositol phosphates (PIP$_n$) **1**-**5**. All synthesized PIP$_n$ are 1,2-dipalmitoyl (DP) glycerides except the compound **4** which was also synthesized as 1,2-dioctanoyl derivative (DO). (B) Structures of synthesized phosphorothioate analogs of PIP$_n$ **6**-**17**. All analogs were synthesized as 1,2-dipalmitoyl (DP) glycerides, except the analog **7** which was synthesized as 1,2-dioctanoyl (DO) derivative.

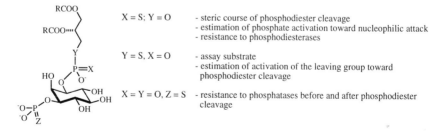

RCOO
RCOO''''

X = S; Y = O
- steric course of phosphodiester cleavage
- estimation of phosphate activation toward nucleophilic attack
- resistance to phosphodiesterases

Y = S, X = O
- assay substrate
- estimation of activation of the leaving group toward phosphodiester cleavage

X = Y = O, Z = S
- resistance to phosphatases before and after phosphodiester cleavage

Figure 2. Summary of potential applications of the synthesized phosphorothioate analogs of phosphatidylinositol phosphates.

phosphorothiolate analogs offers an interesting possibility of estimating the overall contribution made by PI-PLC to stabilize the negative charge on the leaving group in the nucleophilic displacement reaction at the phosphorus center (27, 30).

Applications of the phosphorothioate analogs of PIP_n will vary depending on which phosphate group the sulfur atom is positioned. We have synthesized three types of phosphorothioate analogs: (i) with the sulfur atom in the nonbridging position in the phosphodiester function (analogs **6**, **13** and **17**); (ii) with the sulfur atom in the bridging position of the phosphodiester function (analogs **7**, **8** and **12**); (iii) with the sulfur atom in the nonbridging position in the phosphomonoester function (analogs **9**, **11**, **14** and **15**), and (iv) analogs with sulfur in both the nonbridging diester and monoester positions (analogs **10** an **16**). The analogs **6-17** should be useful in addressing various issues in the metabolic turnover of inositol phospholipids and in studying the mechanisms of such enzymes as PI-specific phospholipases, PI kinases and PI phosphate phosphatases. In addition, the cleavage resistant substrate analogs may prove useful in solving structures of enzyme-substrate complexes. Finally, analogs which selectively inhibit the above enzymes could be potential drug candidates. The two potential enzymatic targets of such analogs could be the 5-phosphatase of phosphatidylinositol 3,4,5-trisphosphate, and β- and γ-isozymes of the mammalian PI-PLCs.

Synthesis of Inositol Precursors: General Strategies

Precursors of PI-4-P and PI-4,5-P₂. Synthesis of precursors of inositol phospholipids and phosphates has been a subject of intense effort in the last decade resulting in the development of numerous efficient pathways. These methods have been a subject of several recent reviews and monographs (36-38). The major synthetic strategies are: (i) the use of diastereomeric derivatives of inositol for separations of enantiomeric forms (38, 39), (ii) the use of enzymic esterification reaction for separation of enantiomers of inositol derivatives (40, 41); (iii) the use of chiral non-cyclitol precursors for synthesis of asymmetrically substituted inositol derivatives (13, 15, 42-45). In this Laboratory we have adopted the first approach, whereby we combine regioselective protection of inositol hydroxyl groups with the chiral separation of enantiomers (46). This is achieved by employing a chiral bornanediyl moiety derived from D-camphor as an acetal protective group for inositol 2- and 3-hydroxyl groups (Scheme 1). This approach has resulted in developing the acetal **18a** as a key starting intermediate for all our syntheses. We then systematically explored ways to achieve regioselective protection of the remaining four hydroxyl groups, by examining a variety of reagents (47). The first important finding was that high regioselectivity (>20:1) of the mono-protection of the 1,4,5,6-tetrol **18a** at the 1-position can be achieved using sterically bulky silyl and acyl groups to give the derivatives such as **19**. This finding was of the paramount importance, in view of the fact that most phosphoinositides have a phosphate group at this position. The disadvantage of using the acyl and silyl groups for protection of vicinal polyalcohols is that these groups tend to migrate under a variety of conditions used to introduce further hydroxyl-protective groups. Of the acyl and silyl protective groups examined, we have found that the tert-butyldiphenylsilyl (TBDPS) group affords the best combination of regioselectivity in its introduction, and sufficient stability during further protection of the 4,5,6-triol moiety (47).

For the reason stated above, further manipulation of the 4,5,6-triol **19** is somewhat restricted, and has to be performed under mild conditions. The second important finding was that the 4,5,6-triol **19** can be highly selectively derivatized at the 4-position using low temperature acylation or silylation (47). The triol **19** can be also regioselectively derivatized at the 4- and 5-positions simultaneously using low

Scheme 1

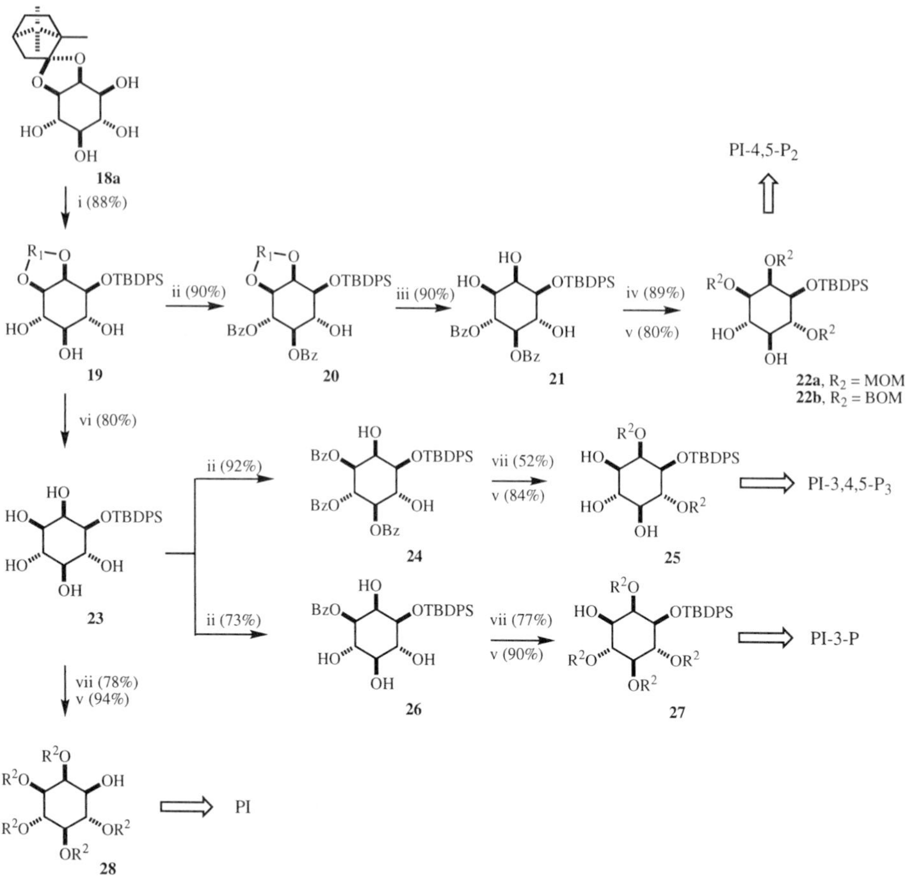

PI-4,5-P$_2$

18a

i (88%)

19 ii (90%) → **20** iii (90%) → **21** iv (89%) / v (80%) →

22a, R$_2$ = MOM
22b, R$_2$ = BOM

vi (80%)

23

ii (92%) → **24** vii (52%) / v (84%) → **25** ⇒ PI-3,4,5-P$_3$

ii (73%) → **26** vii (77%) / v (90%) → **27** ⇒ PI-3-P

vii (78%) / v (94%)

28 ⇒ PI

R^1 = bornanedi-2,2-yl, R^2 = MOM or BOM
i: TBDPS-Cl, imidazole, Py; ii: Bz-Cl/Py; iii: TFA/MeOH, iv: MOM-Cl or BOM-Cl,
iPr$_2$EtN, v: MeONa; vi: BF$_3$/SHCH$_2$CH$_2$OH; vii: MOM-Cl, Pr$_2$EtN; viii: TBAF

temperature benzoylation (-40°C) with two equivalents of benzoyl chloride in pyridine to give the bisbenzoate **20**, or with the bidentate silyl reagent, tetraisopropyldisiloxane dichloride (TIPDS-Cl$_2$) (*47, 48*), to give a 4,5-bissilyl derivative (not shown). This is an important result because it renders an easy access to inositol derivatives differentially protected at the 1- and 4-, or 1- and 4,5-positions, thus affording a convenient route to PI-4-P, PI-4,5-P$_2$, as well as inositol 1,4-bisphosphate (I-1,4-P$_2$) and inositol 1,4,5-trisphosphate (I-1,4,5-P$_3$). In all cases where the protection at the 1-hydroxyl was the TBDPS group (such as in the derivative **19**), we have found that among acyl groups the benzoate group gives the best yields, regioselectivity and product stability toward acyl migration. To complete synthesis of precursors of phosphatidylinositol phosphates, further steps were needed to exhaustively protect the remaining hydroxyl groups and to remove selectively the acyl groups prior to phosphorylation. The fact that the acyl and silyl protective groups are used in the derivative **20**, automatically precludes the use of the alkyl protective groups such as benzyl or allyl groups for the remaining 6-hydroxyl, as they could not be introduced without causing migration of the benzoate and TBDPS groups. Instead, we made an extensive use of the acetal protective groups such as methoxymethylene (MOM), methoxyethoxymethylene (MEM), or benzyloxymethylene (BOM) functions, since their introduction can be achieved under conditions which do not cause acyl or silyl migration.

The obstacle that we encountered with regard to protecting the 6-hydroxyl in the derivative **20**, was that the bornanediyl 2,3-acetal group causes a distortion of the inositol ring into a twisted conformation in which the access to the 6-hydroxyl is very limited. As a result, protection of the 6-hydroxyl under acyl non-migrating conditions could not be satisfactorily performed. In addition, the later steps of phosphorylation of the 4,5-diol in the 2,3-bornanediyl-protected inositol derivatives were also unsuccessful (Bruzik et al., unpublished results). We have found, however, that the treatment of the acetal **20** with trifluoroacetic acid in methanol selectively removes the bornanediyl group to give the triol **21** (Scheme 1), without causing migration of the TBDPS group. Further exhaustive alkylation of the intermediate **21** with MOM or BOM chloride in DMF in the presence of diisopropylethylamine, and the subsequent removal of the benzoyl groups with methylamine or sodium methoxide afforded the 4,5-diols **22a** and **22b**, respectively. Both diols were found to be efficient substrates in the subsequent phosphorylations reactions leading to PI-4,5-P$_2$ and its analogs. In summary, the synthetic strategy described above uses three highly optimized steps: (i) regio- and stereoselective protection of the inositol *cis*-2,3-diol with a chiral bornanediyl acetal group, (ii) regioselective silylation of the 1-hydroxyl group, and (iii) regioselective acylation of the 4- or 4,5-hydroxyls (prior to further deacetalization) to achieve the protection pattern suitable for synthesis of PI-4-P and PI-4,5-P$_2$ and their analogs.

Precursors of PI, PI-3-P and PI-3,4,5-P$_3$. The low reactivity of the 6-hydroxyl, reflected in the selective benzoylations of the triol **19** at the 4- and 5-positions, is most certainly due to the steric hindrance imposed by the large TBDPS group. We have found that it is possible to remove the acetal group in the triol **19** using mercaptoethanol/BF$_3$ reagent to afford a versatile intermediate, 1-TBDPS-inositol (**23**) (*47*). In contrast to the triol **19**, the TBDPS group in the pentol **23** blocks the access to both the 2- and 6-hydroxyl groups, thus leaving an open access to the 3-hydroxyl. As a result, benzoylation of **23** with three and one equivalent of benzoyl chloride in pyridine at -40°C afforded selectively the 3,4,5-trisbenzoyl derivative **24** and the 3-benzoyl derivative **26**, respectively (*5*). These two derivatives are manipulated in the analogous fashion as the compound **21** to give the triol **25** and the alcohol **27**, the precursors of PI-3,4,5-P$_3$ and PI-3-P, respectively. In addition, complete protection of the pentol **23** with MOM or MEM groups, followed by the removal of the TBDPS function gave the alcohol **28**, a useful intermediate in the syntheses of phosphorothionate (**6**) and phosphorothiolate (**7, 8**) analogs of PI. The pentol **23** is thus a third key intermediate used in syntheses of PIP_n.

Precursor of PI-3,4-P_2. Although the yield of the diastereomerically pure acetal **18a**, that we obtain in a single step from inositol, is greater than those of the racemic inositol bisacetals used in many synthetic approaches used by others (*36-39*), still a considerable amount of the mixture of four diastereomers is left behind as a by-product. To improve the overall efficiency of our syntheses, we have elaborated a route aimed at utilization of the mixture of acetals **18a-d**. Thus, the silylation of **18a-d** with TIPDS-Cl_2 afforded the mixture of the corresponding bissilyl derivatives **29a-d** (*47*) (Scheme 2). This mixture can be chromatographically separated into two fractions containing the derivatives **29a,b** and **29c,d**, which differ in their C1-*exo* or C1-*endo* configurations at the C-2' carbon of the bornanediyl moiety. Further bisbenzoylation into **30a,b** and **30c,d** allows chromatographic separation of these pairs into individual diastereomers differing in the inositol configuration. We have then combined pairs of diastereomers having the same inositol configuration, i.e. **30a** with **30c**, and **30b** with **30d**, and subjected the mixture **30a,c** to a sequence of deacetalization and debenzoylation to obtain the 3,4-bissilyl-derivative **31**. Further benzoylation of **31** occurred regioselectively to give 1-benzoate **32**. Finally, exhaustive MOMylation and removal of the benzoate group afforded the 1-alcohol **33**, which served as a precursor of PI-3,4-P_2. This derivative is then phosphorylated from right-to-left as opposed to derivatives **22**, **25** and **27**, which are assembled from the left-to-right (*vide infra*). The mixture **30b,d** was subjected to deacetalization and desilylation to afford 4,5-dibenzoyl-*myo*-inositol (**34**). Further work aiming at utilization of this derivative to synthesis of phosphoinositides is underway.

In summary, regioselective benzoylation of the 1-TBDPS-2,3-acetal **19** gives rise to precursors of PI-4,5-P_2, and the regioselective benzoylation of 1-TBDPS-inositol **23** affords precursors of PI-3-P and PI-3,4,5-P_3. There are several advantages to our approach: (i) all precursors are obtained using the same key intermediates **18a** and **19**, (ii) we apply the same reagents for protection/deprotection schemes in all syntheses; (iii) due to the identical chemistry the syntheses of all PIP_n can be easily mastered. The syntheses of the precursors **22**, **25**, **27** and **28** have been optimized from several perspectives. The use of larger silyl and acyl groups might provide even greater regioselectivity in their introduction, but would most likely hamper further exhaustive protection of the remaining hydroxyl group. For example, protection of the TBDPS-derivatives **22**, **25**, and **27** already takes ca. 24 h for completion. The use of the MOM and similar alkoxymethylene groups is advantageous due to their small size, high reactivity of the alkylating reagents, mild basic conditions required for their introduction (no need to generate alkoxides), as well as due to mild nonaqueous acidic conditions required for their removal. Such conditions are the only ones which are compatible with sulfur modification of the multiple phosphate groups in the adjacent positions, and with the presence of the acyl groups in the diacylglycerol residue of phospholipid analogues. Especially, the last deprotection step is vulnerable to phosphate migration, once the MOM or BOM groups are removed. In all syntheses that we described here, we found no evidence of phosphate migration, as attested by spectroscopic analyses of the final PIP_n products.

Scheme 2

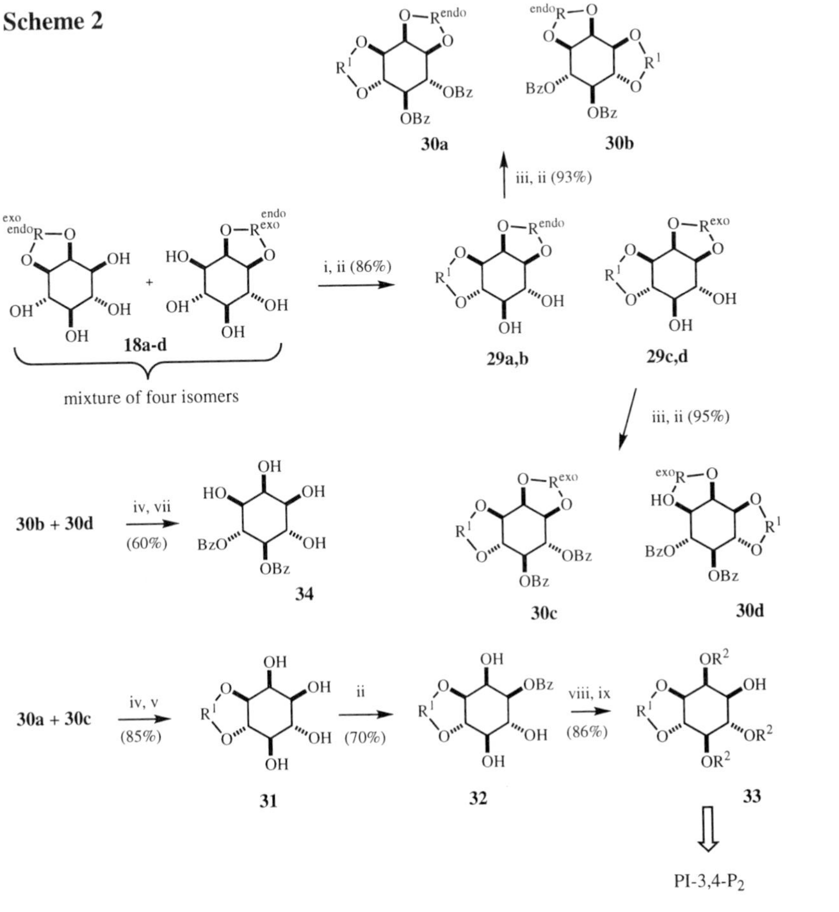

i: TIPDS-Cl$_2$/Py; ii: separation; iii: Bz-Cl/Py; iv: TFA/CHCl$_3$; v: MeNH$_2$/MeOH; vi: TBAF; vii: aq. HF, viii: MOM-Cl, iPr$_2$EtN; ix: MeNH$_2$/MeOH

Synthesis of Inositol Phospholipids

Phosphatidylinositol Phosphates. Most of the precursors described thus far feature the hydroxyl groups at the positions to be phosphorylated into a phosphomonoester, and a TBDPS group at the 1-hydroxyl to be later converted to the phosphodiester function. A typical sequence of reactions involving left-to-right assembly of the fully protected PIP_n is illustrated in Scheme 3 (for a review of phosphorylation methods see ref. *49*). The precursor **22a** was subjected to phosphitylation with *O,O*-dibenzyl-*N,N*-diisopropylphosphoramidite in the presence of tetrazole to give the corresponding bisphosphite **34**. Depending on whether the 4,5-bisphosphate or 4,5-bisphosphorothioate was a target final compound, this derivative was either oxidized with MCPBA or sulfurized with elemental sulfur, to give derivatives **35a** and **35b**, respectively. Both derivatives were next desilylated with tetra-*n*-butylammonium fluoride (TBAF) to give the alcohols **36a** and **36b**, correspondingly, the precursors in the further assembly of the phosphodiester moiety at the 1-position. The alcohols **36a,b** were subsequently phosphitylated at the 1-position with *P*-chloro-*N,N*-diisopropyl-*O*-methylphosphoramidite and then coupled with 1,2-dipalmitoyl-*sn*-glycerol (DPG) in presence of tetrazole to give the phosphites **37a,b**. These derivatives were either oxidized with MCPBA or sulfurized with elemental sulfur to give the neutral triesters **38a-d**. The syntheses of all phosphatidylinositol phosphates and their phosphorothionate analogs described in this account followed this general pathway up to the point of neutral intermediates **38**, however, different deprotection strategies had to be used depending on whether the phosphotriesters at the 3-, 4- and 5-positions were the phosphates or phosphorothioates, as described in the later section. Treatment of the all-oxygen derivative **38a** with neat trimethylamine removed one alkyl group (methyl or benzyl) from each triester functions to give the tris(phosphodiester) **39a**. The analogous reaction also occurs with derivatives **38c-d**, but due to difficulties in further deprotection steps it was not applicable in our synthetic scheme. The subsequent removal of the benzyl groups in **39a** by hydrogenolysis over 10% Pd-catalyst generated the 4,5-bisphosphomonoester **40a**. The final deprotection of the MOM groups was achieved using BF_3/ethanethiol mixture to afford the dipalmitoyl analog of PI-4,5-P_2 (**3**). The crude product was converted to the sodium salt and purified by repeated precipitation with methanol from the chloroform solution. The product was fully characterized by means of ^1H and ^{31}P NMR and electrospray MS. Using the same approach we have also converted precursors **28** into DPPI, **27** into DPPI-3-P, and **25** into DPPI-3,4,5-P_3.

Analogs with Sulfur in Nonbrigding Monoester Positons. The synthetic routes to phosphorothioate analogs of phosphates are well established (*49, 50*), however most of them apply to isolated phosphate groups, such as those in oligonucleotides. In our hands, however, some of the best known methods applied to syntheses of analogs of phosphatidylinositol polyphosphates have failed during the last deprotection steps due to high concentration of functional groups in these compounds, and spatial proximity of the phosphate/phosphorothioate groups. For example, while the hydrogenolytic removal of ester *O*-benzyl groups described in the preceding section was applicable toward synthesis of natural PIP_n, the deprotection of the multiple phosphorothionate groups at the 3-, 4- and 5-positions proved much more problematic. First, application of the BF_3/ethanethiol reagent, known to remove benzyl ester groups, was unsuccessful due to the occurrence of the thiono-thiolo rearrangement (*51*), leading to formation of *S*-benzyl by-products (not shown). Likewise, the application of trimethylsilyl iodide gave significant amounts of side-products resulting from the cleavage of the diacylglycerol residue, thereby precluding

Scheme 3

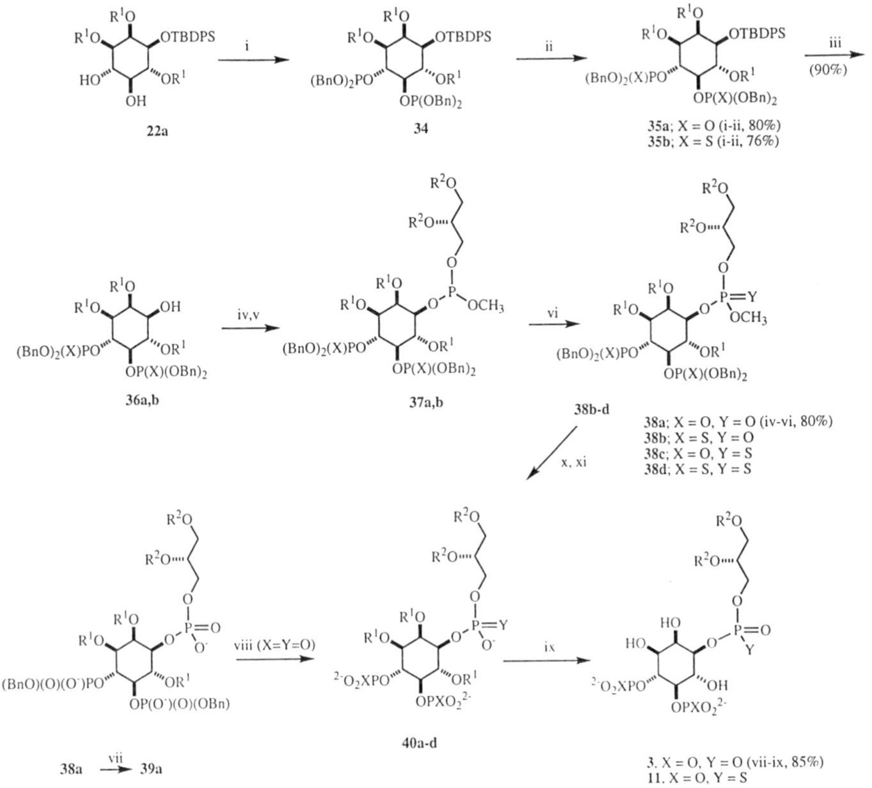

R^1 = MOM; i: $(BnO)_2PNiPr_2$/tetrazole; ii: MCPBA (X = O) or S_8 (X=S); iii: TBAF; iv:
Cl-P(OMe)NiPr$_2$, iPr$_2$EtN; v: DPG/tetrazole; vi: MCPBA (Y = O), S_8 (Y = S); vii: NMe$_3$; viii: H$_2$/Pd;
ix: EtSH, BF$_3$; x: Me$_3$N/TMS-Cl; xi: aqueous buffer, pH=7.0

the use of this reagent. Several other benzyl-removing reagents could not be used due to the presence of sulfur (H_2/Pd could not be employed), or labile acyl groups (Na/NH$_3$ could not be used).

We have eventually solved the problem by applying a novel deprotecting reagent composed of trimethylchlorosilane and trimethylamine (Scheme 4). The rationale for using this reagent is as follows: the dealkylation of the phosphotriesters by such nucleophilic reagents as trimethylamine or the iodide stops abruptly at the stage of the phosphodiester, due to a poor leaving group properties of the negatively charged phosphomonoester, and due to the repulsion between the nucleophile and the negatively charged phosphodiester (52). The second alkyl group can be removed if the negative charge of phosphodiester is neutralized by the protic or Lewis acid. Unfortunately, these conditions resulted in formation of significant amounts of the desulfurized products. The removal of the negative charge can also be accomplished by a transient conversion of the diester into a labile triester, such as TMS-phosphate ester, using the Me$_3$N/TMS chloride mixture. In the presence of TMS chloride, the dealkylation of the triester **41**, obtained analogously as shown in Scheme 3 starting from the triol **25**, is followed by the rapid O-silylation of the diester to form the TMS ester **42**, which is further dealkylated and silylated to form the bis(O-TMS) ester **43**. The course of the deprotection can be conveniently followed by ^{31}P NMR due to the fact that each replacement of the alkyl group by the silyl group results in ca. 10-13 ppm upfield shift of the ^{31}P NMR signal. The obtained TMS esters were readily hydrolyzed by the aqueous buffer at neutral pH to give the MOM-protected phosphorothioate **44**, and the final deprotection of the MOM groups was achieved by means of BF$_3$/ethanethiol, analogously as in the case of PIP$_n$ described above.

Analogs with Sulfur in the Bridging Diester Position.   Synthesis of phosphorothiolate analogs with sulfur in the bridging position was first described by Hendrickson (*34, 53*), followed by the results from this Laboratory (*54*). Our method used a standard phosphoramidite chemistry for the subsequent introduction of both the 2R-2,3-diacyloxypropanethiol (**45**) and penta(methoxymethylene)-*myo*-inositol (**28**) onto the phosphorus atom (*54*) (Scheme 5A). We and others (*55*) have found, however, that the replacement of the amide group in the phosphorothioamidite **46** by an alcohol can lead to several side products, giving an overall low yield of the protected diester derivative **47**. In general, the phosphoramidite method of synthesis of phosphorothiolate analogs of PI showed difficulty in batch-to-batch reproducibility. The alternative pathway which we tested involved alkylation of the phosphorothioate **48** with the diacyliodohydrin **49** obtained in two steps from glycidol (Scheme 5B). In contrast to expectations, the alkylation reaction afforded the S-benzyl ester **50** as a major product (80%), and only a small amount of the desired thiolester **47**. The formation of the diester **50** most likely involves the initial S-alkylation of the diester **48** to give the triester **51** followed by the cleavage of its O-benzyl group with the by-product iodide anion to produce the diester **47** and benzyl iodide (Scheme 5C). Once benzyl iodide is generated, it acts as an alkylating reagent (Scheme 5D), superior to the iodohydrin **49**, and benzylates the substrate **48** to give the triester **52**, which then undergoes debenzylation to give the diester **50** as the main product and benzyl iodide. In an effort to avoid debenzylation of the triester **51** by the iodide anion, we attempted alkylation of **48** by sulfonate esters of diacylglycerol, but have found that the dipalmitoylglycerol triflate was too unstable, whereas the corresponding methylsulfonate was not sufficiently reactive.

Finally, the most efficient and currently favored synthesis of the phosphorothiolate analogs of PI (**7, 8**) and PI-4,5-P$_2$ (**12**) is shown in Scheme 6. Two consecutive displacements of the chloride in O-methylphosphorodichloridite (**56**) by the alcohol **28** or **34** at -78°C and the thiol **45** at 0°C afforded the

Scheme 4

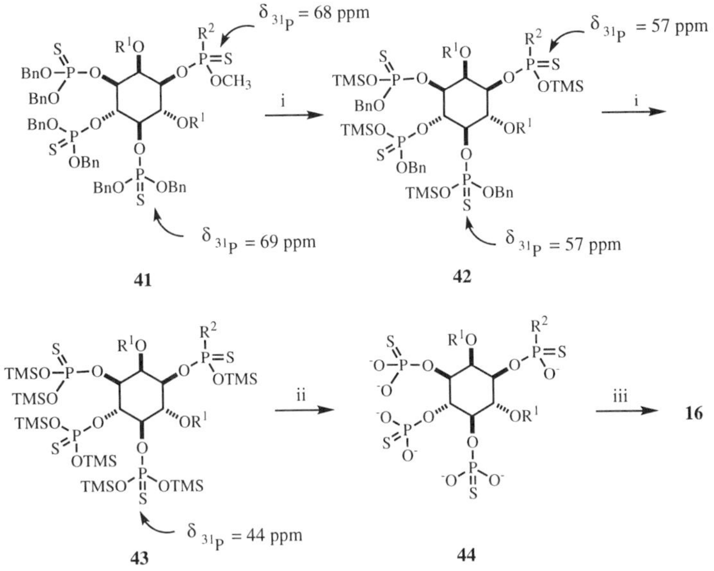

$R^1 = MOM; R^2 = DPG$; i: TMS-Cl/Me$_3$N; ii: aqueous buffer pH 7.0, iii: BF$_3$/EtSH

Scheme 5

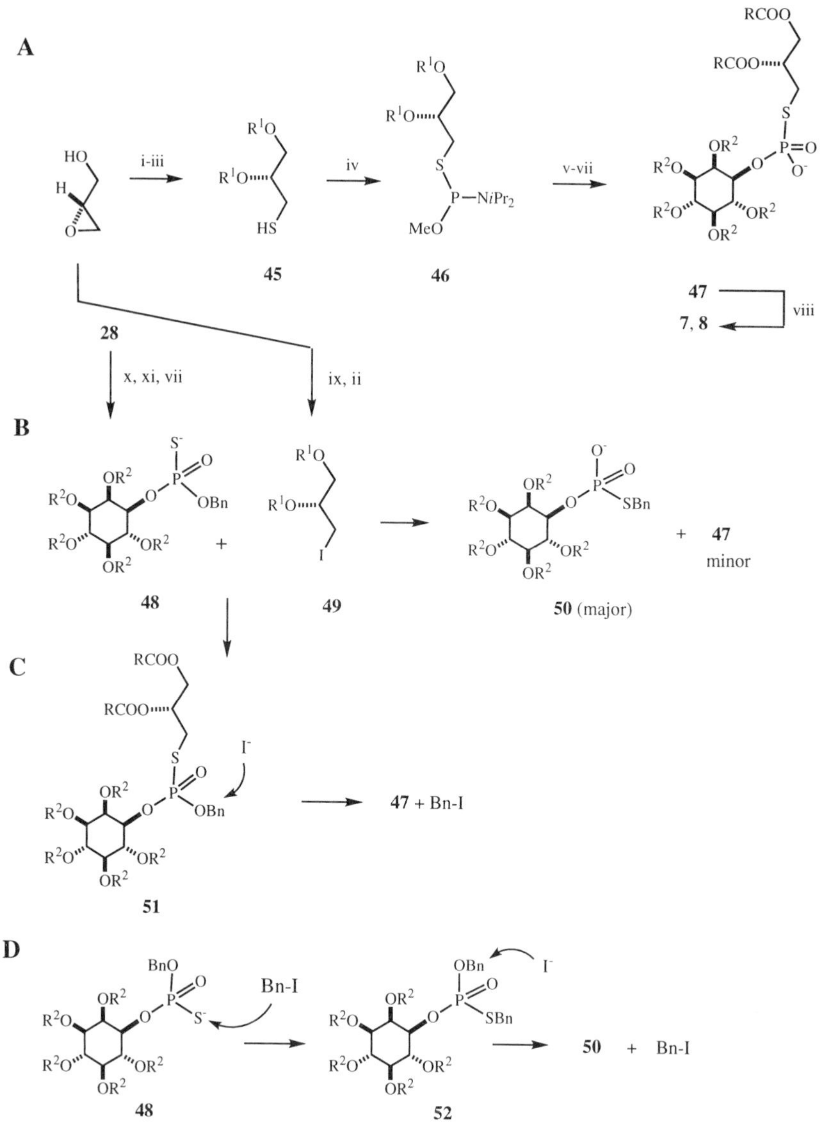

$R^1 = C_7H_{15}CO$ or $C_{15}H_{31}CO$; $R^2 = MOM$ or MEM; i: CH_3COSH; ii: $C_{15}H_{31}COCl/Py$;
iii: $EtOH/AgNO_3$; iv: $Cl\text{-}P(OMe)NiPr_2/iPr_2EtN$; v: alcohol **28**/tetrazole; vi: N^+Bu_4,
IO_4^-; vii: Me_3N; viii: $BF_3/EtSH$; ix: Bu_4N^+, I^-; x: $iPr_2NP(OBn)_2$/tetrazole; xi: S_8

Scheme 6

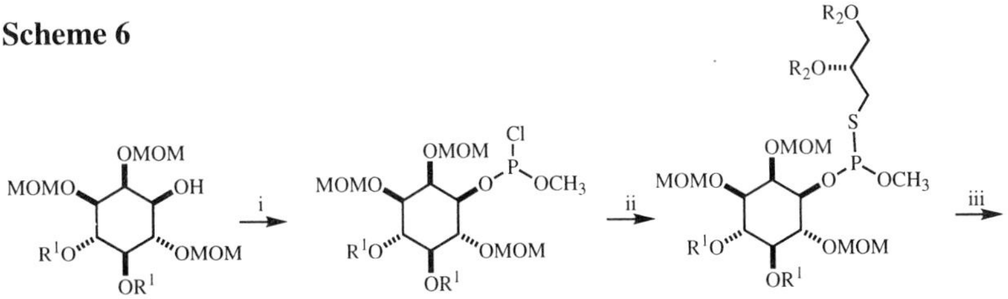

28 (R^1 = MOM)
34 (R^1 = -P(O)(OBn)$_2$)

53 (R^1 = MOM), (i-iii, 85%)
54 (R^1 = -P(O)(OBn)$_2$, (i-iii, 65%)

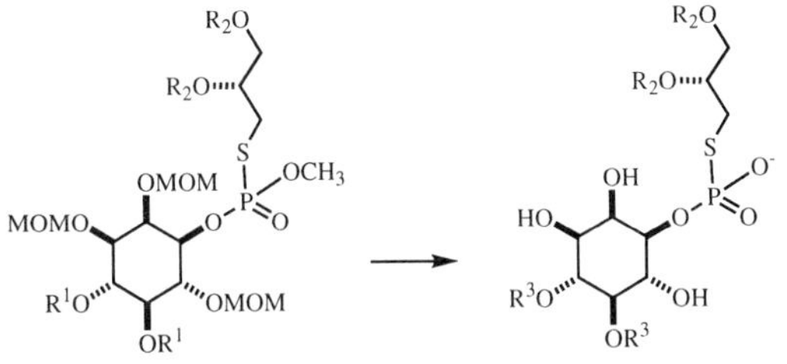

55 (R^1 = MOM) $\xrightarrow{\text{iv,v}}$ **8** (R^3 = H), (iv-v, 78%)

56 (R^1 = -P(O)(OBn)$_2$) $\xrightarrow{\text{iv,vi,v}}$ **12** (R^3 = -PO$_3$$^{2-}$), (iv-vi, 83%)

R^2 = C$_7$H$_{15}$CO or C$_{15}$H$_{31}$CO; i: Cl$_2$POCH$_3$/iPr$_2$EtN; ii: thiol **45**/iPr$_2$EtN; iii: Bu$_4$N$^+$, IO$_4^-$; iv: Me$_3$N; v: BF$_3$/EtSH, vi: H$_2$/Pd

phosphorothioites **53** and **54**, respectively, in high yields. The subsequent oxidation of these thiophosphites with tetra-*n*-butylammonium periodate gave the fully protected phosphorothiolates **55** and **56**, correspondingly. Ultimately, sequential deprotection of the triester **55** with trimethylamine and ethanethiol/BF$_3$-etherate gave the phosphorothiolate analog of PI (**8**, R = H). The deprotection of **56** was completed analogously, except that the complete removal of the benzyl group from the phosphates was achieved by hydrogenolysis over Pd-charcoal catalyst. Application of the analog **7** to studies of the mechanism of the bacterial PI-PLC has been reported earlier (*27*), and use of the analog **12** for elucidation of the mechanism of the mammalian enzymes will be reported elsewhere.

Summary

In conclusion, we have accomplished a general, systematic synthesis of almost all naturally occurring phosphatidylinositols, and many of their phosphorothioate analogs which offer selective resistance towards enzymes involved in inositol phospholipid metabolism. Application of the phosphorothioate analogs to study the mechanism of the bacterial phospholipase C resulted in obtaining results which would have been difficult to obtain otherwise. We are currently studying the behavior of analogs of PI-4,5-P2 with the mammalian phospholipase C-δ_1, and application of other analogs to studies of PI kinases and PIP$_n$ phosphatases will be reported elsewhere.

Acknowledgements: This work was supported by the grant from the National Institutes of Health, GM30327.

References

1. Berridge, M. J. *Nature (London)* **1993**, *361*, 315.
2. Spivakkroizman, T.; Mohammadi, M.; Hu, P.; Jaye, M.; Schlessinger, J.; Lax, I. *J. Biol. Chem.* **1994**, *269*, 14419.
3. Stephens, L. R.; Jackson, T. R.; Hawkins, P. T. *Biochim. Biophys. Acta* **1993**, *1179*, 27.
4. Duckworth, B. C.; Cantley, L. C. *PI 3-Kinase and Receptor-Linked Signal Transduction,* Bell, R. M.; Exton, J. H.; Prescott, S. M. Eds., *Handbook of Lipid Research,* Plenum Press, New York, 1996, Vol. 8, p. 125.
5. Bruzik, K. S.; Kubiak, R. J. *Tetrahedron Lett.* **1995**, *36*, 2415.
6. Kubiak, R. J.; Bruzik, K. S. *Bioorg. Med. Chem. Lett.* **1997**, *7*, 1231.
7. Falck, J. R.; Abdali, A. in *Inositol Phosphates and Derivatives. Synthesis, Biochemistry, and Therapeutic Potential*, Reitz, A. B. Ed., *ACS Symp. Ser.* **1991**, *463*, 145.
8. Gou, D.-M.; Chen, C.-S. *J. Chem. Soc. Chem. Commun.* **1994**, 2125.
9. Watanabe, Y.; Hirofuji, H.; Ozaki, S. *Tetrahedron Lett.* **1994**, *35*, 123.
10. Watanabe, Y.; Tomioka, M.; Ozaki, S. *Tetrahedron* **1995**, *33*, 8969.
11. Reddy, K. K.; Saady, M.; Falck, J. R.; Whited, G. *J. Org. Chem.* **1995**, *60*, 3385.
12. Desai, T.; Gigg, J.; Gigg, R.; Martin-Zamora, E. In *Synthesis in Lipid Chemistry*, Tyman, J. H. P.; Ed.; Royal Society of Chemistry, London, 1996.
13. Prestwich, G. D. *Acc. Chem. Res.* **1996**, *29*, 503.
14. Wang, D.-S.; Chen, C.-S. *J. Org. Chem.* **1996**, *61*, 5905.
15. Chen, J.; Profit, A. A.; Prestwich, G. D. *J. Org. Chem.* **1996**, *91*, 6305.
16. Aneja, S. G.; Parra, A.; Stonescu, C.; Xia, W.; Aneja, R. *Tetrahedron Lett.* **1997**, *38*, 803.
17. Rhee, S. G.; Suh, P.-G.; Ryu, S.-H.; Lee, S. Y. *Science* **1989**, *244*, 546.

18. Bruzik, K. S.; Tsai, M.-D. *Bioorg. Med. Chem.* **1994**, *2*, 49.
19. Lips, D. L.; Majerus, P. W.; Gorga, F. R.; Young, A. T.; Benjamin, T. L. *J. Biol. Chem.* **1989**, *264*, 8759.
20. Serunian, L. A.; Haber, M. T.; Fukui, T.; Kim, J. W.; Rhee, S. G.; Lowenstein, J. M.; Cantley, L. C. *J. Biol. Chem.* **1989**, *264*, 17809.
21. Hinrichs, W.; Steifa, M.; Saenger, W.; Eckstein, F. *Nucleic Acids Res.* **1987**, *15*, 4945.
22. Eckstein, F. *Annu. Rev. Biochem.* **1985**, *54*, 367.
23. Jaffe, E. K.; Cohn, M. *J. Biol. Chem.* **1979**, *254*, 10839.
24. Cummins, L.; Graff, D.; Beaton, G.; Marshall, W. S.; Caruthers, M. H. *Biochemistry* **1996**, *35*, 8734.
25. Safrany, S. T.; Wojcikiewicz, R. J. H.; Strupish, J.; McBain, J.; Cooke, A. M.; Potter, B. V. L.; Nahorski, S. R. *Mol. Pharmacol.* **1991**, *39*, 754.
26. Potter, B. V. L.; Lampe, D. *Angew. Chem. Int. Ed. Engl.* **1995**, *34*, 1933.
27. Hondal, R. J.; Bruzik, K. S.; Tsai, M.-D. *J. Am. Chem. Soc.* **1997**, *119*, 5477.
28. Hondal, R. J.; Riddle, S. R.; Kravchuk, A. V.; Zhao, Z.; Bruzik, K. S.; Tsai, M.-D. *Biochemistry* **1997**, *36*, 6633.
29. Hondal, R. J.; Zhao, Z.; Riddle, S. R.; Bruzik, K. S.; Tsai, M.-D. *J. Am. Chem. Soc.* **1997**, *119*, 9933.
30. Hondal, R. J.; Zhao, Z.; Kravchuk, A. V.; Liao, H.; Riddle, S. R.; Bruzik, K. S.; Tsai, M.-D. *Biochemistry* **1998**, *37*, 4568.
31. Lin, G.; Bennett, C. F.; Tsai, M.-D. *Biochemistry* **1990**, *29*, 2747.
32. Bruzik, K. S.; Lin, G.; Tsai, M.-D. *ACS Symp. Ser.* **1991**, *463*, 172.
33. Bruzik, K. S.; Morocho, A. M.; Jhon, D.-Y.; Rhee S. G.; Tsai, M.-D. *Biochemistry* **1992**, *31*, 5183.
34. Hendrickson, E. K.; Johnson, J. L.; Hendrickson, H. S. *Bioorg. Med. Chem. Lett.* **1991**, *1*, 615.
35. Hendrickson, E. K.; Hendrickson, H. S.; Johnson, J. L.; Khan, T. H.; Chial, H. J. *Biochemistry* **1992**, *31*, 12169.
36. Billington, D. C. (1993) *"The Inositol Phosphates. Chemical Synthesis and Biological Significance"* pp. 95-138, VCH, Weinheim.
37. Potter, B. L. V. *Nat. Prod. Rep.* **1990**, 1.
38. Billington, D. C. *Chem. Soc. Rev.* **1989**, *18*, 83.
39. Stepanov, A. E.; Shvets, V. I. *Chem. Phys. Lipids* **1979**, *25*, 247.
40. Liu, Y.-C.; Chen, C.-S. *Tetrahedron Lett.* **1989**, *30*, 1617.
41. Ozaki, S.; Lei, L. In *Carbohydrates in Drug Design* Witczak, Z. J.; Nieforth, K. A. Eds., Marcell Dekker, New York, 1997, pp. 343.
42. Bender, S. L.; Budhu, R. J. *J. Am. Chem. Soc.* **1991**, *113*, 9883.
43. Chiara, J. L.; Martin Lomas, M. *Tetrahedron Lett.* **1994**, *35*, 2969.
44. Ley, S. V.; Sternfeld, F. *Tetrahedron* **1989**, *45*, 3463.
45. Nguyen, B. V.; York, C.; Hudlicky, T. *Tetrahedron* **1997**, *53*, 8807.
46. Bruzik, K .S.; Salamonczyk, G. M.; Stec, W. J. *J. Org. Chem.* **1986**, *51*, 2368.
47. Bruzik, K. S.; Tsai, M.-D. *J. Am. Chem. Soc.* **1992**, *104*, 6361.
48. Bruzik, K. S.; Myers, J; Tsai, M.-D. *Tetrahedron Lett.* **1992**, *33*, 1009.
49. Beaucage, S. L.; Iyer, R. P. *Tetrahedron* **1993**, *49*, 10441.
50. Beaucage, S. L.; Iyer, R. P. *Tetrahedron* **1993**, *49*, 6123.
51. Bruzik; K. S.; Stec, W. J. *J. Org. Chem.* **1981**, *46*, 1618.
52. Chandler, A. J.; Hollfelder, F.; Kirby, A. J.; O'Carroll, F.; Stromberg, R. *J. Chem. Soc. Pertin Trans 2* **1994**, 327.
53. Bushnev, A. S.; Hendrickson, E. K.; Shvets, V. I.; Hendrickson, H. S. *Bioorg. Med. Chem.* **1994**, *2*, 147.

54. Mihai, C.; Mataka, J.; Riddle, S. R.; Tsai, M.-D.; Bruzik, K. S. *Bioorg. Med. Chem. Lett.* **1997**, *7*, 1235.
55. Cosstick, R.; Vyle, J. S. *Nucl. Acids Res.* **1990**, *18*, 829.
56. Martin, S. F.; Josey, J. A.; Wong, Y.-L.; Dean, D. W. *J. Org. Chem.* **1994**, *59*, 4805.

Chapter 12

Chemical Synthesis of Phosphatidylinositol Phosphates and Their Use as Biological Tools

Yutaka Watanabe

Department of Applied Chemistry, Faculty of Engineering, Ehime University, Matsuyama 790-77, Japan

Phosphorylated inositol phospholipids, such as phosphatidylinositol 4,5-bisphosphate, phosphatidylinositol 3,4,5-trisphosphate and 3,4-bisphosphate have received much attention due to their biological role in cellular signal transduction systems.   In this report we describe synthesis of both unsaturated (oleoyl and linolenoyl) and saturated (stearoyl) fatty acid-containing polyphosphorylated phosphatidyl-inositols.   New synthetic methodologies involving phosphorylation based on phosphite chemistry, and novel phosphate and alcohol protecting   groups   [9-fluorenylmethyl   and o-(2-O-substituted-oxyethyl)benzoyl, respectively], have been developed to accomplish this task. Optically active inositol intermediates have been obtained by kinetic resolution of 1,2-diol derivatives, enantioselective enzymatic acetylation,   and   chromatographic   separation   of   diastereomeric menthoxyacetyl and O-acetylmandelyl derivatives.

The metabolic map of inositol phospholipids is briefly illustrated in Figure 1 (*1*) . The old pathway generating two second messengers, D-*myo*-inositol 1,4,5-trisphosphate [Ins(1,4,5)P$_3$] and diacylglycerol (DAG) has been well established. The discovery of these signal transduction pathways stimulated the progress in both inositol chemistry and biochemistry. More recent synthetic efforts have been mainly directed toward synthesis of phosphatidylinositols. This is due to discovery of a newer signaling pathway involving activation of PI 3-kinase, resulting in formation of 3-phosphorylated phosphatidylinositols (*2*).   In addition, new results indicate that PI(4,5)P$_2$ is not only a precursor of a second messenger inositol 1,4,5-trisphosphate, but is itself a cellular modulator due to its interactions with cytoskeletal and signaling proteins. These 3-phosphorylated PIs have also been a focus of our most recent synthetic efforts as described below.

Methodologies for Synthesis of Phosphoinositides

To accomplish this goal we have developed several synthetic methodologies as shown in Figure 2.* The salient feature of our approach is application of the phosphite-pyridinium tribromide reagent for regioselective phosphorylation of 1,2-diol

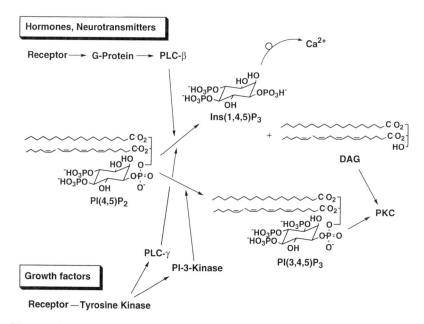

Figure 1. Schematic representation of major signalic pathways involving phosphatidylinositol phosphates.

derivatives of *myo*-inositol (*3*). This is important since 1,2-diols are common synthetic intermediates, and have been found to undergo regioselective acylation, alkylation and silylation at the 1-position, however, selective phosphorylation at this position has proven difficult so far (*4*). Similar phosphite chemistry is the basis for a new glycosylation procedure using a glycosyl phosphite as a glycosyl donor in the presence of a Lewis acid (*5,6*). These methods have been successfully used in the regioselective synthesis of dimannosyl phosphatidylinositol (PIM$_2$) from mycobacteria (*7,8*).

PAC Hydroxy Protecting Groups. We have previously reported the reaction of *meso*-tribenzoyl inositol **1** (Figure 3) with tartaric acid monomethyl ester **D-3** in the presence of MsCl as a condensing reagent to give high enantioselective acylation product **2**. This reaction has also been be applied for diastereoselective derivatization of 1,2-diol **4** to give fairly high diastereoselective acylation products **5** (*9*). In this reaction, the presence of the benzoate groups at the 5- and 6-positions was essential, as the corresponding acetate derivative did not provide good results.

The benzoyl group can be modified such that it can be removed chemoselectively in the presence of other acyl functions in the same molecule. Such modified benzoyl group could function not only as a hydroxyl protecting group, but also as an auxiliary for kinetic resolution in the synthesis of phosphatidylinositols. Along these lines, we designed three benzoyl groups featuring oxyethyl function at the phenyl ortho-position. We designate these group as PAC protective groups (*Proximity-Assisted Cleavable*). The structures and designations of further structural variations of these protective groups are given below (Figure 4). What distinguishes PAC groups from benzoate is that they can be readily deprotected due to lactonization of the hydroxyethyl benzoate generated after removal of the R groups. The cyclization reaction allowed chemoselective deprotection without affecting the diacylglycerol moiety, as has been demonstrated in the synthesis of PI(3,4)P$_2$ and PI(4,5)P$_2$ discussed below.

In order to obtain optically pure synthetic intermediates for the synthesis of phosphatidylinositol 3,4- and 4,5-bisphosphate, the inositol derivatives bearing PAC$_{Lev}$ groups, **DL-6** and **DL-8** have been subjected to kinetic resolution as shown in Figure 5, to yield the corresponding **D-6** and **D-8** 1,2-diols with high optical purities along with 1-*O*-tartrates of the opposite inositol enantiomers, **L-7** and **L-9**.

The Use of Fluorenylmethyl Group for Phosphate Protection. Phosphatidylinositol phosphates which have been prepared so far carried saturated fatty acid chains. In these cases, hydrogenolysis was employed to remove benzyl and related phosphate protecting groups. In order to deprotect phosphate functions without affecting unsaturated fatty acid chain, we first examined β-elimination type protecting groups, such as β-cyanoethyl group, *p*-nitrophenylethyl and trimethylsilylethyl, but found that they were unsuitable for our purposes. Eventually, we found that 9-fluorenylmethyl group (Fm) was very promising. Thus, di(fluorenylmethyl)phosphoramidite **10** (Figure 6) reacted smoothly with alcohols in the presence of tetrazole, and the subsequent oxidation with mCPBA gave the corresponding phosphoric triesters **11** in high yields. The deprotection of the esters was accomplished by the reaction with DBU or triethylamine to generate phosphoric monoesters **12** (Watanabe, Y.; Nakamura, T.; Mitsumoto, H. *Tetrahedron Lett.*, in press). The above protection strategy was successfully applied to the synthesis of phosphoinositides having unsaturated fatty acid chains as described in the next section.

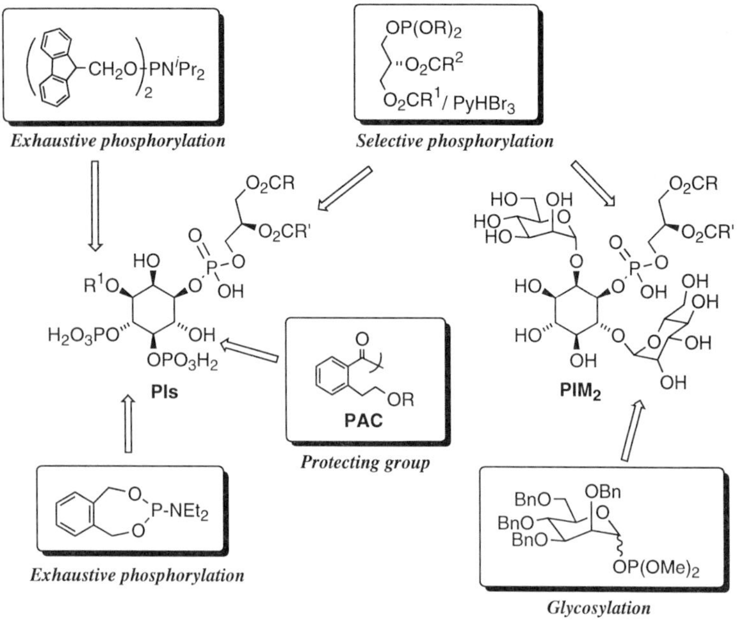

Figure 2. Critical elements involved in the design of synthetic methods leading to phosphatidylinositol polyphosphates.

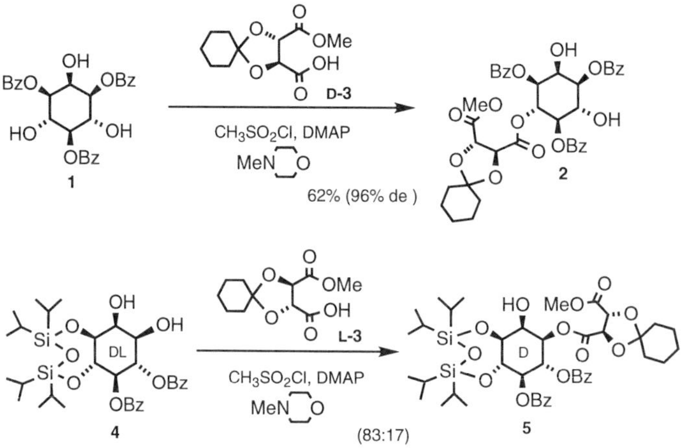

Figure 3. Enantiospecific desymmetrization of the prochiral inositol triol and kinetic resolution of the racemic 1,2-diol, as two major approaches to enantiomerically pure inositol precursors.

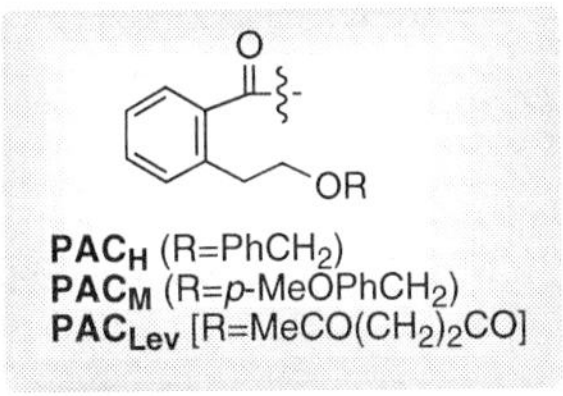

Functions:
1. Protection of a hydroxy group
2. Chemoselective deprotection
3. Auxiliary for kinetic resolution

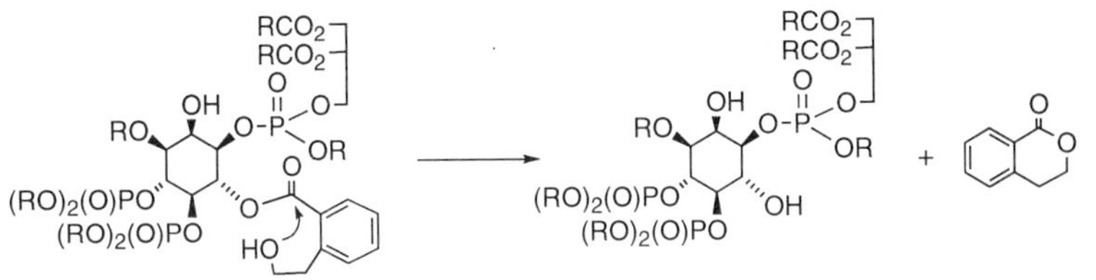

Figure 4. Structure and chemoselective deprotection of PAC groups.

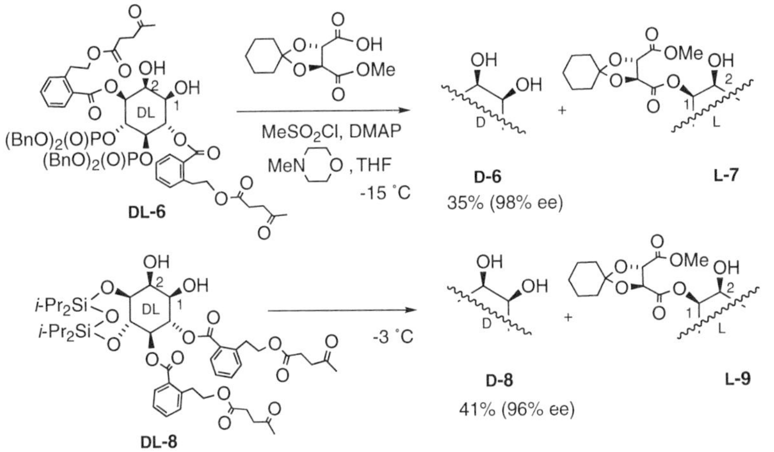

Figure 5. Kinetic resolution of enantiomers of the precursors used in synthesis of $PI(4,5)P_2$ and $PI(3,4)P_2$.

Synthesis of Phosphoinositides

Among various phosphoinositides, 3-O-phosphoryl phosphatidylinositol derivatives have received much synthetic interest, mainly because they are difficult to obtain from natural sources. In our strategies, we have employed preferably 1,2-cyclohexylidene-3,4-disiloxanediyl-*myo*-inositol **15** as a pivotal intermediate for syntheses of various phosphorylated inositol derivatives as shown in Figure 7 (*11*). The bissilyl derivative **15** was readily prepared from *myo*-inositol **13** via **14** in a regioselective manner. In the derivative **15** the 5-hydroxyl group is sterically hindered by the adjacent bulky silyl group. As a result, the 6-hydroxyl can be preferentially derivatized by acylation and glycosylation.

Saturated analog of PI(3,4)P$_2$. The synthesis of saturated PI(3,4)P$_2$ analog was accomplished using the PAC$_{Lev}$ group for protection and kinetic resolution (Figure 8). The PAC$_{Lev}$ groups were introduced at the 5 and 6 positions in the intermediate **15** by the reaction with PAC$_{Lev}$-OH in the presence of DCC and DMAP to give the derivative **19**. After removal of the cyclohexylidene group, the 1,2-diol **20** was regioselectively phosphorylated at the 1-position by the reaction with O,O-dibenzyl 1,2-isopropylideneglycerol phosphite in the presence of pyridinium tribromide. The resultant phosphate **21** was levulinoylated and desilylated to afford the 3,4-diol **23**, which was then phosphorylated via phosphoramidite method using the cyclic phosphoramidite XEPA (*12,13*) to give **24**. Careful removal of the isopropylidene group in **24** followed by stearoylation gave the fully protected product **26**, which was then subjected to hydrogenolysis on Pd-C in the presence of NaHCO$_3$, resulting in the formation of the phosphate-deprotected derivative. The use of NaHCO$_3$ is important, as the ketone groups were partially reduced to yield secondary alcohols in its absence. The hydrogenolysis product was then sequentially treated with hydrazine in pyridine-acetic acid and *t*-BuOK to form the desired final product PI(3,4)P$_2$. The latter steps did not affect the diacylglycerol residue.

Analog of PI(4,5)P$_2$. The PAC$_{Lev}$ group was also applied in the synthesis of a saturated PI(4,5)P$_2$ analog in a manner analogous to that of PI(3,4)P$_2$ (Figure 9). The diPAC$_{Lev}$ derivative **28** obtained from dicyclohexylideneinositol **26** was transformed to 1,2-diol **6** by two transacetalization steps and phosphorylation. The further regioselective phosphorylation of **6** with phosphite-tribromide in the presence of the bulky pyridine base completed the assembly of the fully protected product. The final deprotection stage was accomplished by sequential hydrogenolysis, hydrazinolysis, and finally lactonization in quantitative yields, giving the desired dibenzoyl-PI(4,5)P$_2$ (Y. Watanabe and T. Nakamura, *Natural Product Lett.* in press).

Unsaturated analog of PI(4,5)P$_2$. The strategy used for synthesis of saturated PI(4,5)P$_2$ was applied in synthesis of the unsaturated analog. For this purpose, however, the benzyl protecting group on the phosphate was replaced by the Fm group. Fluorenylmethyl phosphate **31** (Figure 10) was obtained in the high yield analogously as described above. The diol **32** generated by removal of the cyclohexylidene group was smoothly phosphorylated at OH-1 with complete regioselectivity giving **33** in 90% yield, much higher than phosphorylation of the derivative **6** under similar conditions. The final deprotection of **33** was accomplished by the successive procedures shown in Figure 10 to give dioleoyl PI(4,5)P$_2$.

Benzyl vs. Fluorenylmethyl Protecting Groups. We observed the difference in reactivity in phosphorylation of the 1,2-diol depending on the phosphate protecting group. These results are summarized in Figure 11. When the derivative **6** was treated

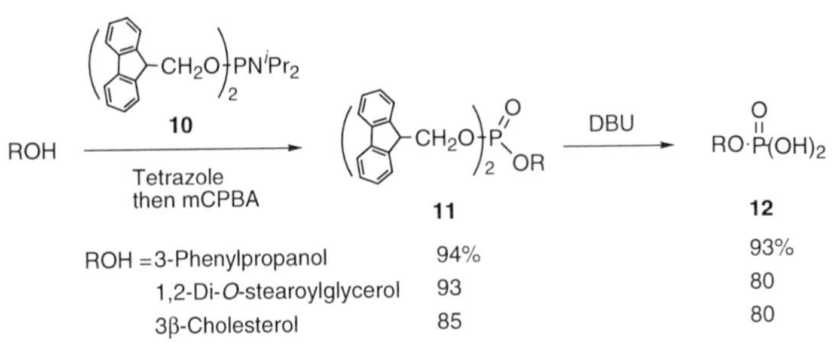

Figure 6. Application of fluorenylmethylene protective group for synthesis of phosphate monoesters.

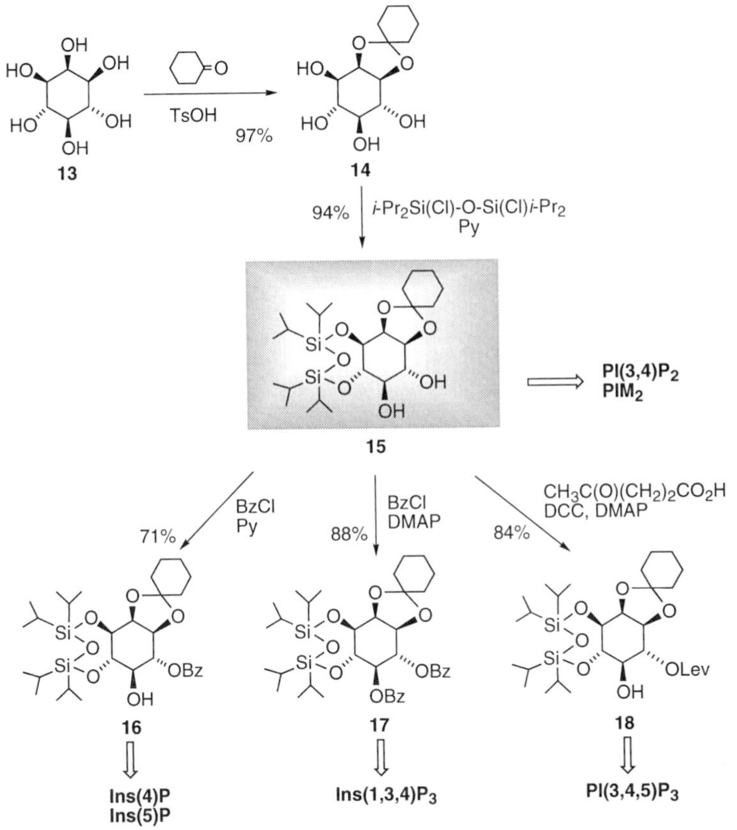

Figure 7. Usefulness of disiloxanediyl ether as a synthetic intermediate.

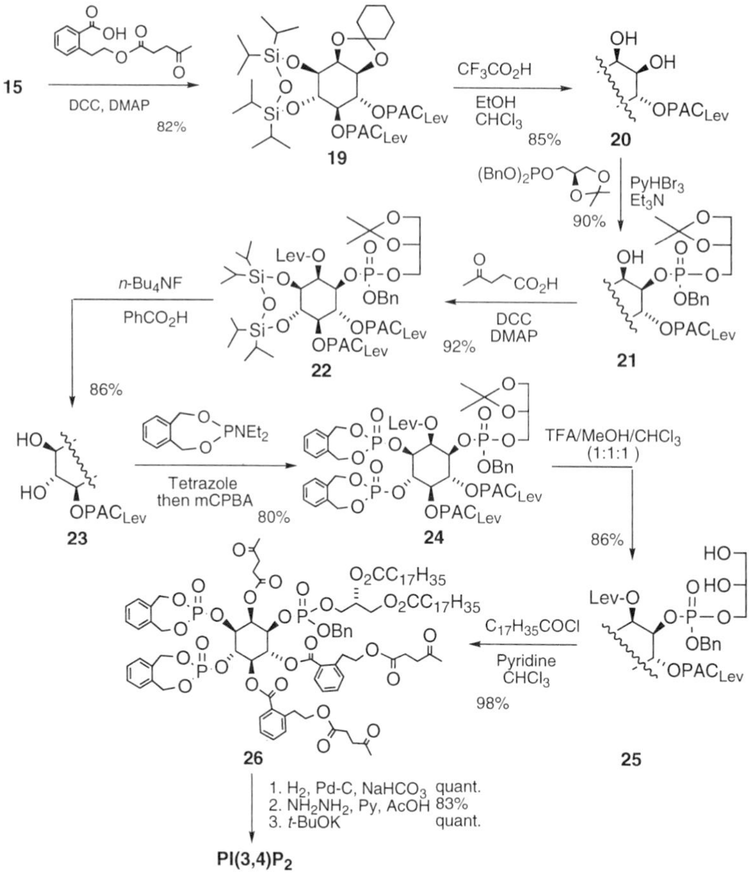

Figure 8. Synthesis of saturated PI(3,4)P$_2$ analog.

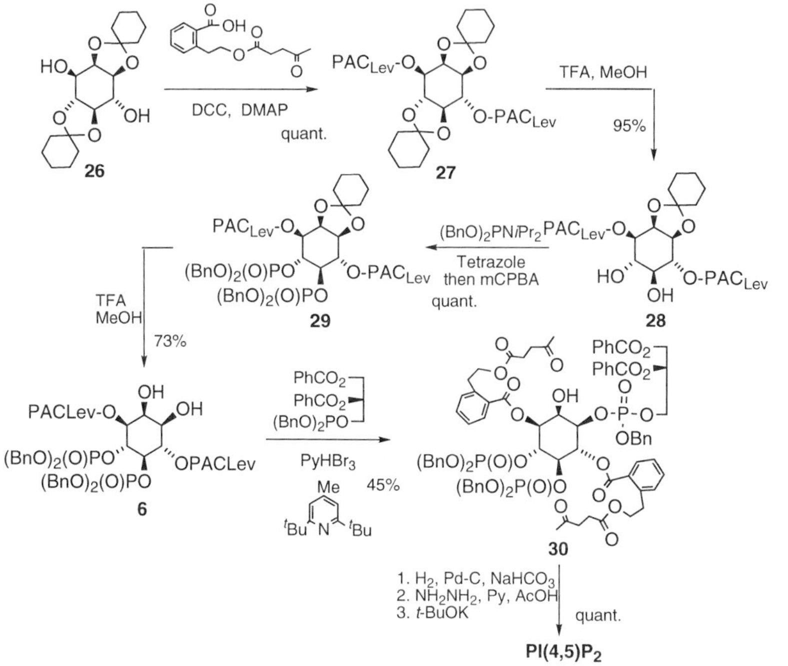

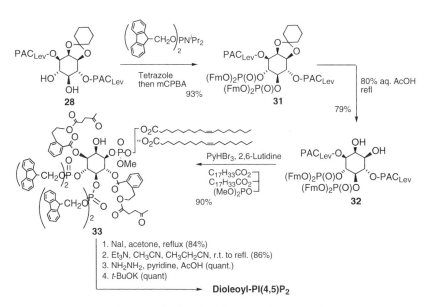

Figure 9. Synthesis of 1,2-dibenzoylglycerol analog of PI(4,5)P$_2$.

Figure 10. Synthesis of dioleoyl analog of PI(4,5)P$_2$.

with 1,2-di-O-benzoylglyceryl O,O-dibenzyl phosphite and pyridinium tribromide in the presence of triethylamine or lutidine, the 1-O-phosphorylated product **36** was accompanied by a significant amount of 2-phosphate, most likely formed via 1,2-phosphate migration. In contrast, the fluorenylmethyl phosphate derivative **32** was very smoothly converted to 1-phosphate (R=oleoyl, R=methyl) **36** in 90% yield. On the other hand, in the synthesis of a PI(3,4,5)P$_3$ analog, the benzyl derivative **34** was phosphorylated at OH-1, while from fluorenylmethyl derivative **35** the desired 1-phosphate could not be obtained at all by the identical phosphorylation procedures. The reasons for the observed differences in reactivities are unclear.

Synthesis of PI(3,4,5)P$_3$

We reported synthesis of saturated racemic PI(3,4,5)P$_3$ in 1994 as shown in Figure 12 (*14*). During the same year, several other papers appeared (*15-19*), one of which reported the synthesis of short chain water soluble analogs of PI(3,4,5)P$_3$ and PI(4,5)P$_2$ (*16*). Synthesis of photoactivatable analogs of PI(3,4,5)P$_3$ was recently reported by the Prestwich's group (*20,21*).

Enantiomerically pure saturated PI(3,4,5)P$_3$. In order to obtain a precursor of an optically active PI(3,4,5)P$_3$ analog, we applied the enzymatic method which involved acetylation of 1,2-cyclohexylidene-*myo*-inositol **14** using Lipase CES from *Pseudomonas sp.* (Amano Pharmaceutical Co. Ltd.) to afford 3-acetate **40** with 98% optical purity. This acetate was converted to enantiomerically pure PI(3,4,5)P$_3$ according to the procedures shown in Figure 13 (*22*) .

Unsaturated PI(3,4,5)P$_3$. Preparation of polyunsaturated PI(3,4,5)P$_3$ was accomplished in a manner analogous to that used for the saturated PIP$_3$ (Figure 14). Thus, the triol **37** was phosphorylated by successive treatment with difluorenylmethyl phosphoramidite **10** and with mCPBA to afford **44** in 90% yield. The phosphate **44** was converted to 1,2-diol **35**, however, as described above, all attempts at regioselective phosphorylation at the 1-position failed. Therefore, the diol **35** was transformed to 2-chloroacetyl derivative **45** by a three-step procedure. On the way to this compound we attempted first to perform regioselective silylation at 1-position, however, we were obtaining large amounts of the 2-silyl ether. Eventually, the reaction of the diol **35** with silyl triflate in the presence of a bulky base gave regioselectively the desired silyl derivative which was then chloroacetylated and desilylated to give the alcohol **45**. Phosphitylation of **45** with linolenoylglyceryl phosphoramidite **46** followed by oxidation with *t*-butyl hydroperoxide produced the fully protected product **47**. *t*-Butyl hydroperoxide was employed instead of mCPBA to avoid epoxidation of the double bonds by the excess of the peracid. Deprotection of the four phosphates in **47** bearing fluorenylmethyl and β-cyanoethyl groups was accomplished simultaneously by treatment with triethylamine for 15 h at room temperature, while the analogous deprotection during synthesis of PI(4,5)P$_2$, the second fluorenylmethyl groups at each phosphate at the 4- and 5-positions were removed under refluxing conditions. Finally, the levulinoyl and the chloroacetyl groups were removed at the same time by the reaction with hydrazine dithiocarbonate to form polyunsaturated analog of PI(3,4,5)P$_3$. Both deprotection steps proceeded quite smoothly, and were as convenient as that using hydrogenolysis.

In addition to enzymatic resolution of the inositol intermediate **40**, we achieved chromatographic separation of diastereomeric 6-O-menthoxyacetyl or O-acetylmandelyl 5-O-triethylsilylated derivatives **48** (Figure 15). Removal of the acyl functions in **49** by treatment with hydrazine or Grignard reagent afforded the 6-alcohol

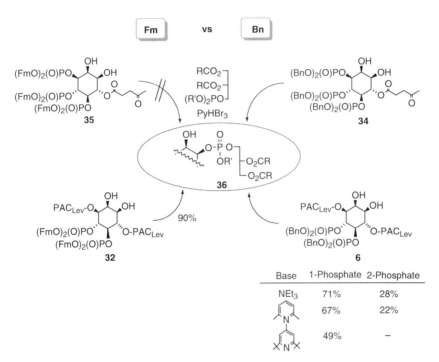

Figure 11. Comparison of reactivities of *O,O*-dibenzyl and *O,O*-di(flurenylmethylene) phosphate derivatives of the inositol 1,2-diol in pyridine tribromide-catalyzed phosphitylation.

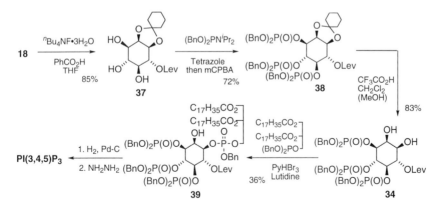

Figure 12. Synthesis of the distearoyl analog of PI(3,4,5)P₃.

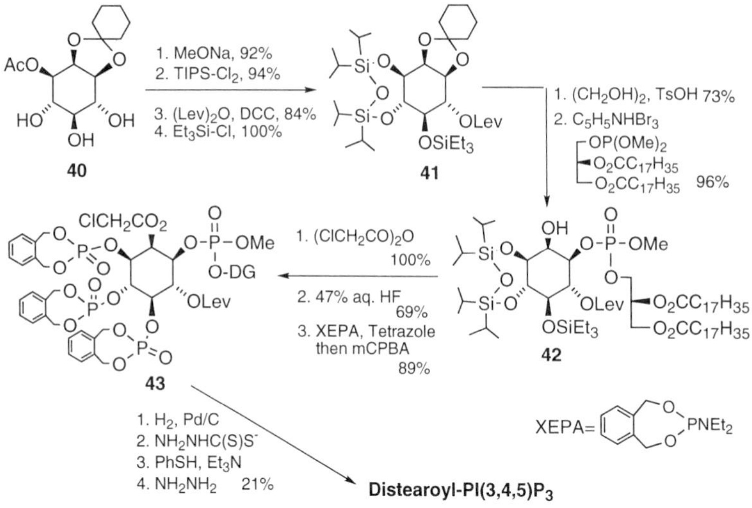

208

Figure 13. Synthesis of the enantiomerically pure distearoyl analog of PI(3,4,5)P$_3$.

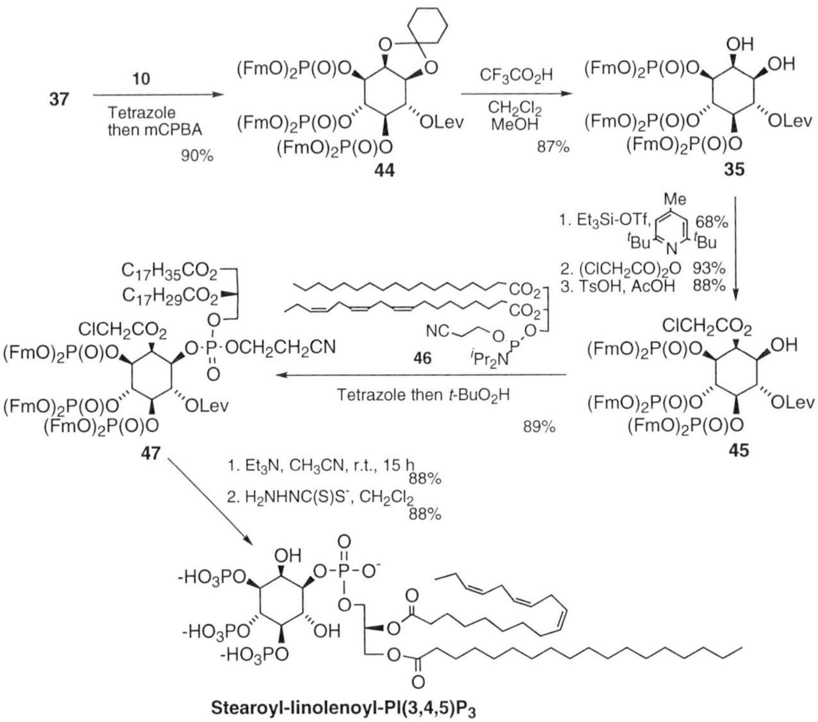

Figure 14. Synthesis of the polyunsturated analog of PI(3,4,5)P$_3$.

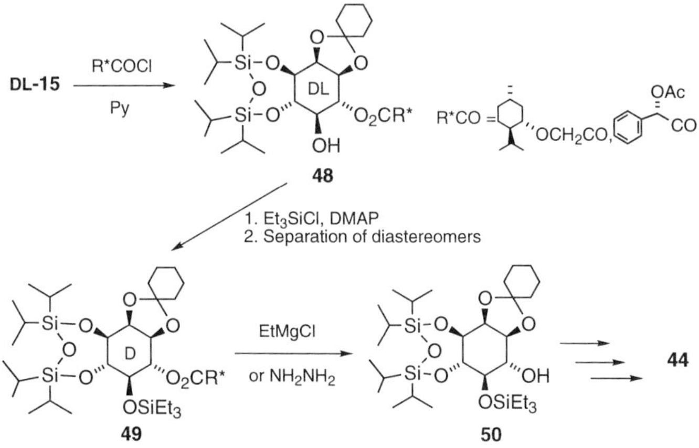

Figure 15. Separation of enntiomers of the 5,6-diol **15** via formation of diastereomeric esters.

50 without migration of the silyl group. This derivative was converted to **44** (Scheme 14) by consecutive levulinoylation, exhaustive and phosphorylation.

Synthetic and biological aspects of unsaturated phosphatidylinositols

In summary, we have developed synthetic methods leading to polyunsaturated phosphatidylinositols, including natural PIP_3 and PIP_2. Very recently, Reese's group has completed the synthesis of natural $PI(3,4,5)P_3$, with stearoyl and arachidonoyl groups at the sn-1- and sn-2- positions of the glycerol moiety, respectively. Although saturated distearoyl $PI(3,4,5)P_3$ was found to activate protein kinase C (*23*), it did not show other expected activities in various biological systems. The reason for the above is not clear, however, long chain saturated molecules tends to aggregate strongly, therefore, interaction with a PI-binding protein may be affected. In contrast, the synthetic polyunsaturated PIP_3 displayed better solubility and showed well-resolved NMR spectra, indicating decreased aggregation properties. Therefore, polyunsaturated phosphatidylinositol phosphates should be able to mimic the properties of natural $PI(3,4,5)P_3$ better than the saturated analogs.

Acknowledgments: This work was financially supported in part by the Grant-in-Aid for Scientific Research (No. 08680631) and Grant-in-Aid for Scientific Research on Priority Areas (No. 06240105) from the Ministry of Education, Science and Culture, Japan. The work was also funded by CIBA-GEIGI Foundation (Japan) for the Promotion of Science.

Literature Cited

1. Berridge, M. J. *Nature* **1993**, *361*, 315-325.
2. Duckworth, B. C.; Cantley, L. C. *PI 3-Kinase and Receptor-Linked Signal Transduction,* Bell, R. M.; Exton, J. H.; Prescott, S. M. Ed.; *Handbook of Lipid Research,* Plenum Press: New York; 1996, Vol. 8; pp 125-175.
3. Watanabe, Y.; Inada, E.; Jinno, M.; Ozaki, S. *Tetrahedron Lett.* **1993**, *34*, 497-500.
4. Watanabe, Y. *Selective Reactions and Total Synthesis of Inositol Phosphates;* Rahman, A. Ed.; Studies in Natural Products Chemistry; Elsevier: Amsterdam, 1996, Vol. 18, Stereoselective Synthesis (Part K); pp 391-456.
5. Watanabe, Y.; Nakamoto, C.; Ozaki, S. *Synlett* **1993**, 115-116.
6. Watanabe, Y.; Nakamoto, C.; Yamamoto, T.; Ozaki, S. *Tetrahedron* **1994**, *50*, 6523-6536.
7. Watanabe, Y.; Yamamoto, T.; Ozaki, S. *J. Org. Chem.* **1996**, *61*, 14-15.
8. Watanabe, Y.; Yamamoto, T.; Okazaki, T. *Tetrahedron* **1997**, *53*, 903-918.
9. Watanabe, Y.; Oka, A.; Shimizu, Y.; Ozaki, S. *Tetrahedron Lett.* **1990**, *31*, 2613-2616.
10. Watanabe, Y.; Ishimaru, M.; Ozaki, S. *Chemistry Lett.* **1994**, 2163-2166.
11. Watanabe, Y.; Mitani, M.; Morita, T.; Ozaki, S. *J. Chem. Soc., Chem. Commun.* **1989**, 482-483.
12. Watanabe, Y.; Komoda, Y.; Ebisuya, K.; Ozaki, S. *Tetrahedron Lett.* **1990**, *31*, 255-256.
13. Watanabe, Y.; Komoda, Y.; Ozaki, S. *Tetrahedron Lett.* **1992**, *33*, 1313-1316.
14. Watanabe, Y.; Hajimu, H.; Ozaki, S. *Tetrahedron Lett.* **1994**, *35*, 123-124.
15. Gou, D.-M.; Chen, C.-S. *J. Chem. Soc., Chem. Commun.* **1994**, 2125-2126.
16. Toker, A.; Meyer, M.; Reddy, K. K.; Falck, J. R.; Aneja, R.; Aneja, S.; Parra, A.; Burns, D. J.; Ballas, L. M.; Cantley, L. C. *J. Biol. Chem.* **1994**, *269*, 32358-32367.

17. Bruzik, K. S.; Kubiak, R. J. *Tetrahedron Lett.* **1995**, *36*, 2415-2418.
18. Wang, D. S.; Chen, C. S. *J. Org. Chem.* **1996**, *61*, 5905-5910.
19. Aneja, S. G.; Parra, A.; Stoenescu, C.; Xia, W.; Aneja, R. *Tetrahedron Lett.* **1997**, *38*, 803-806.
20. Chen, J.; Profit, A. A.; Prestwich, G. D. *J. Org. Chem.* **1996**, *61*, 6305-6312.
21. Gu, Q.-M.; Prestwich, G. D. *J. Org. Chem.* **1996**, *61*, 8642-8647.
22. Watanabe, Y.; Tomioka, M.; Ozaki, S. *Tetrahedron* **1995**, *51*, 8969-8976.
23. Moriya, S.; Kazlauskas, A.; Akimoto, K.; Hirai, S.; Mizuno, K.; Takenawa, T.; Fukui, Y.; Watanabe, Y.; Ozaki, S.; Ohno, S. *Proc. Nat. Acad. Sci. USA* **1996**, *93*, 151-155.

Chapter 13

Synthesis of Phosphatidyl-*myo*-inositol Polyphosphates and Derivatives

Shoichiro Ozaki[1], Yutaka Watanabe[2], Takehiro Maekawa[2], Yuichi Higaki[2], Tomio Ogasawara[2], and Xiang Zheng Kong[3]

[1]Department of Medicinal Chemistry, The University of Utah, Salt Lake City, UT 84112
[2]Department of Applied Chemistry, Faculty of Engineering, Ehime University, Matsuyama 790, Japan
[3]Department of Chemistry, Shandong University, Jinan, China, 250100

This paper describes synthesis of a variety of new molecular probes for investigation of signal transduction events involving inositol phosphates. The synthesized molecular probes include: the affinity resin bearing phosphatidyl-*myo*-inositol 3,4-bisphosphate ligand, unsaturated fatty acid-bearing inositol polyphosphates, such as 1-dioleoylphosphatidyl-*myo*-inositol 3,4,5-trisphosphate, 1-(1-stearoyl-2-arachidonoyl-glycerophosphoglycolyl)-*myo*-inositol 4,5-bisphosphate, membrane-permeant inositol polyphosphates, such as acyl-*myo*-inositol polyphosphate and acylinositol polyphosphate acyloxymethyl ester. In addition we have demonstrated utility of a new phosphitylating agent, di-*O*,*O*-*p*-methoxybenzyl *N*,*N*-diisopropylphosphoramidite.

Synthesis of Affinity Column Bearing the Phosphatidylinositol 3,4-Bisphosphate Ligand

In order to identify proteins that bind inositol polyphosphates (IP$_n$), we have synthesized in the past the affinity columns bearing inositol 1,4,5-trisphosphate (I-1,4,5-P$_3$) (*1*), I-1,3,4,5-P$_4$ (*2*), and I-1,3,4,5,6-P$_5$ (*2*). This work describes synthesis of affinity column, which instead of an inositol polyphosphate ligand bears an inositol phospholipid, such as phosphatidylinositol 3,4-bisphosphate (PI-3,4-P$_2$) as an affinity ligand (*3*). The synthetic route to such affinity resin is shown in Scheme 1.

Alkylation of 1,2:5,6-di-*O*-cyclohexylidene-*myo*-inositol (**1**) with *p*-methoxybenzyl chloride in the presence of sodium hydride gave 1,2:5,6-di-*O*-cyclohexylidene-3,4-di-*O*-(*p*-methoxybenzyl)-*myo*-inositol (**2**). The *trans*-5,6-*O*-cyclohexylidene group was cleaved with ethylene glycol in the presence of catalytic amounts of *p*-toluenesulfonic acid to give 1,2-*O*-cyclohexylidene-3,4-*O*-di-(*p*-methoxybenzyl)-*myo*-inositol (**3**). The diol **3** was exhaustively alkylated with benzyl chloride to give 5,6-di-*O*-benzyl-1,2-cyclohexylidene-3,4-di-*O*-(*p*-methoxybenzyl)-*myo*-inositol (**4**). The subsequent cleavage of the *cis*-acetal group in **4** with trifluoroacetic acid at room temperature afforded 5,6-di-*O*-benzyl-3,4-*O*-di(*p*-methoxybenzyl)-*myo*-inositol (**5**). This diol was first regioselectively silylated at the 1-position with triethylsilyl chloride in pyridine to give 2-alcohol **6**, and further acylated with *p*-nitrobenzoyl chloride in the presence of dimethylaminopyridine to

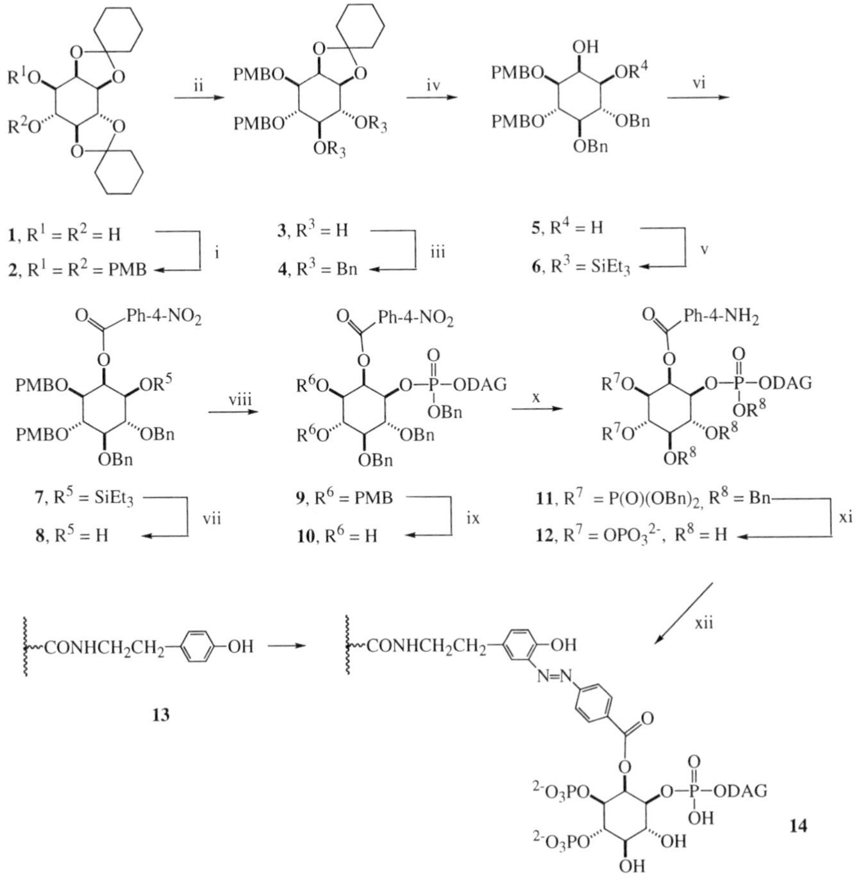

Scheme 1. Synthesis of affinity resin with PI-3,4-P_2 ligand. i: *p*-methoxybenzyl chloride (PMB-Cl), NaH, DMF, 6 min at 0°C then 14 h at rt, 45%; ii: ethylene glycol, TsOH, DMF, rt, 2 h, 58%; iii: BnCl, NaH, DMF, 0°C → rt, 2 h, 81%; iv: CF_3COOH, CH_2Cl_2, rt, 2 h, 71%; v: $SiEt_3Cl$, pyridine, rt, 14 h, 92%; vi: *p*-$NO_2PhCOCl$, DMAP, pyridine, rt, 14 h, 89%; vii: 80% AcOH, TsOH, rt, 6 h, 95%; viii: *i*$Pr_2NP(OBn)DAG$, tetrazole, CH_2Cl_2, rt, 0.5 h, MCPBA, -40°C, 0.5 h, rt, 1 h, 68%; ix: DDQ, H_2O, CH_2Cl_2, rt, 4 h, 89%; x: *i*$Pr_2NP(OBn)_2$, tetrazole, CH_2Cl_2, rt, 5 h, MCPBA, -40°C, 0.2 h, rt, 1 h, 98%; xi: H_2, Pd/C, CH_3OH, rt, 19 h, 88%; xii: $NaNO_2$, HCl.

produce the fully protected derivative **7**. The triethylsilyl group of **7** was selectively cleaved with 80% acetic acid during 6 h at room temperature yielding the 1-hydroxy derivative **8**. This product features a protection pattern suitable for differential phosphorylations at the 1- and 3,4-positions.

Thus, the reaction of **8** with *O*-benzyl-(1,2-di-*O*-stearoyl-*sn*-glycero)-*N*,*N*-diisopropylphosphoramidite in the presence of tetrazole and subsequent oxidation with *t*-butyl hydroperoxide gave the phosphate **9**. In order to liberate the hydroxyl groups at the 3- and 4-positions, the compound **9** was treated with 2,3-dichloro-5,6-dicyanobenzoquinone (DDQ) in the presence of water to give the 3,4-diol **10**. Phosphorylation of **10** with *O*,*O*-dibenzyl-*N*,*N*-diisopropylphosphoramidite in the presence of tetrazole and subsequent oxidation with *m*-chloroperbenzoic acid gave the 3,4-bisphosphotriester **11**. This compound was deprotected via hydrogenolysis in the presence of Pd/C, with the simultaneous reduction of the nitro function of the *p*-nitrobenzoyl group to afford the analog of PI-3,4-P$_2$ (**12**).

Since we found out that the amino group of **12** did not react with CH-Shepharose 4B directly, CH-Cepharose was first coupled with tyramine to give the derivatized solid support **13**. The *p*-aminobenzoyl group of **12** was then diazotized, and the diazo group was coupled with the pendant tyramine of **13** to give the PI-3,4-P$_2$-derivatized affinity resin **14**. This PI-3,4-P$_2$ affinity resin should be an effective tool for identification, isolation and purification of PI-3,4-P$_2$ binding proteins, such as PI-3,4-P$_2$ receptor, PI-3,4-P$_2$ 5-kinase and PI-3,4-P$_2$ 3-phosphatase.

Synthesis of Unsaturated Phosphatidylinositol 3,4,5-Trisphosphate (*4*)

The chemical synthesis of saturated fatty acid analogs of phosphatidyl-*myo*-inositol 3,4,5-trisphosphate has been reported previously (*5-7*). In contrast, natural PI-3,4,5-P$_3$ contain unsaturated fatty acids such as arachidonic acid, and therefore, the saturated analogs might not be effective in reproducing all biological effects of the natural PIP$_n$. We therefore deemed synthesis of PIP$_n$ containing unsaturated fatty acids as necessary. The synthetic problem is that most approaches employed so far used benzyl group for protection of the hydroxyl and phosphate groups. As these benzyl groups are finally cleaved by hydrogenolysis, the unsaturated acyl groups would also be reduced to saturated fatty acid group in this process. Synthesis of unsaturated PIP$_n$ clearly necessitates a different approach, and hence, we devised a new strategy to dealing with this problem.

We used the levulinoyl (4-oxopentanoyl) group instead of benzyl group for hydroxyl protection and ethyl instead of benzyl group for phosphate protection. The synthetic route for the synthesis of unsaturated 1-(1,2-di-*O*-oleoyl-3-*sn*-phosphatidyl)-*myo*-inositol 3,4,5-trisphosphate (**25**) is shown in Scheme 2. The reaction of 1,2-*O*-cyclohexylidene-3,4-*O*-(tetraisopropyldisiloxane-1,3-diyl)-6-*O*-(4-oxopentanoyl)-*myo*-inositol (**15**), obtained as described previously (*6a*), with *O*,*O*-diethyl phosphorochloridite in the presence of diisopropylethylamine in methylene chloride at room temperature, followed by oxidation of the phosphite intermediate with hydrogen peroxide afforded the 5-phosphoryl derivative **16** in quantitative yield. The 1,2-cyclohexylidene group of **16** was cleaved by treatment with trifluoroacetic acid in methanol at room temperature to produce the diol **17**. The reaction of **17** with triethylsilyl chloride in pyridine at room temperature gave regioselectively the 1-silyl ether **18** which was further acylated with levulinic acid in the presence of dicyclohexylcarbodiimide (in CH$_2$Cl$_2$ / pyridine as solvent) to yield the 2-ester **19**. The final step in precursor synthesis constituted the removal of the silyl group using *p*-toluenesulfonic acid in 80% acetic acid to give the 1-alcohol **20**. This precursor was coupled with the unsaturated phosphatidyl group synthon, *O*-ethyl-3-(1,2-di-*O*-oleoyl-*sn*-glycero)-*N*,*N*-diisopropylphosphoramidite in the presence of tetrazole in methylene

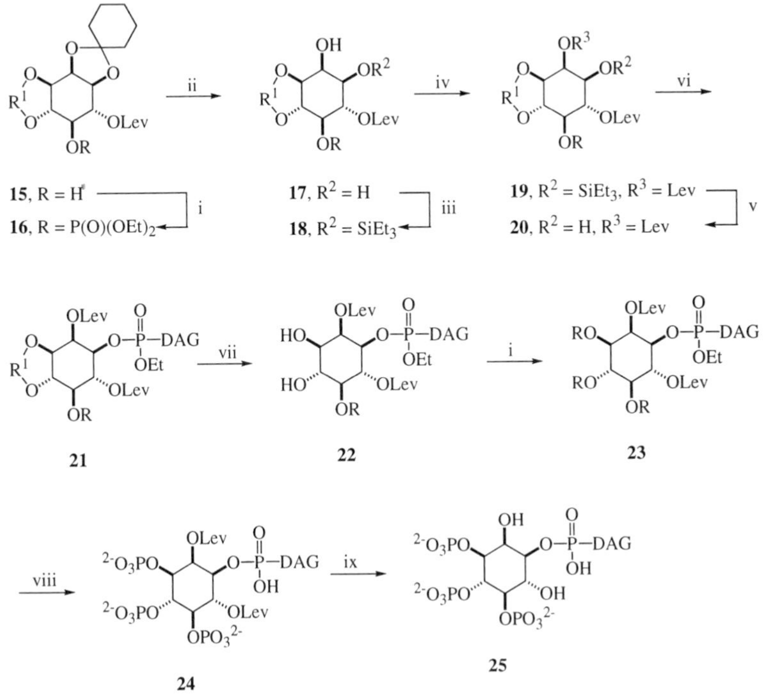

Scheme 2. Synthesis of 1-(1,2-dioleoylphosphatidyl)-*myo*-inositol. i: (EtO)$_2$PCl,*i*Pr$_2$EtN, CH$_2$Cl$_2$, rt, 14 h, 70%; H$_2$O$_2$, -10°C → 0°C, 1h, 90% (**15 → 16**), 40% (**22 → 23**); ii: CF$_3$COOH, CH$_2$Cl$_2$, 0°C, 14 h, 54%; iii: Et$_3$SiCl, pyridine, rt, 14h, 92%; iv: levulinic acid, DCC, CH$_2$Cl$_2$, pyridine, rt, 14 h, 56%; v: 80% AcOH, TsOH, CHCl$_3$, rt, 14 h, 54%; vi: *i*Pr$_2$NP(OEt)DAG, tetrazole, CH$_2$Cl$_2$, rt, 5 h, 70% H$_2$O$_2$, 0°C, 5 h, 92%; vii: Bu$_4$N$^+$F$^-$,THF, rt, 2h, 90%; viii: H$_2$NNH$_2$, pyridine, acetic acid, rt, 1 h, 54%; R^1 = -Si(*i*Pr$_2$)$_2$-O-Si(*i*Pr$_2$)$_2$-.

chloride at room temperature, and the resulting PIII intermediate was oxidized with H_2O_2 at 0°C to produce the phosphotriester **21**. The last part of the synthesis consisted of introducing phosphate groups at the 3- and 4-positions. Thus, the cleavage of the tetraisopropyldisiloxane-1,3-diyl group in **21** was carried out by tetrabutylammonium fluoride and benzoic acid in dry THF at room temperature to give the 3,4-diol **22**, which was further phosphorylated by diethyl chlorophosphite in the presence of diisopropylethylamine at room temperature, and oxidized with 70% H_2O_2 at 0°C for 5 min, and at room temperature for 30 min, to give the fully protected precursor of PI-3,4,5-P$_3$ (**23**). Finally, seven ethyl groups were cleaved by triethylsilyl bromide in CHCl$_3$ at room temperature to give **24**, and the two levulinoyl groups were cleaved by hydrazine hydrate in pyridine-acetic acid to afford unsaturated PI-3,4,5-P$_3$ **25**. Biological testing of compound **25** is now in progress.

Synthesis of 1-(1-Stearoyl-2-arachidonoyl-*sn*-glycerophosphoglycolyl)-*myo*-inositol 4,5-Bisphosphate

Synthesis of analogs of phosphatidyl-*myo*-inositol is of interest in view of its role as a substrate of phosphatidylinositol-specific phospholipase C. Since the P-O bond between inositol and diacylglycerol is cleaved by PLC, its alteration could produce much needed inhibitors and ligands for this enzyme. Here we describe synthesis of compound **32**, which in which the COCH$_2$O group is inserted in place of the P-O bond. The compound **32** was synthesized by the route shown in Scheme 3.

The 1,2-ketal function of the precursor **26** was hydrolyzed with 80% aqueous acetic acid to give the 1,2-diol **27**. Further regioselective acylation of **27** with *t*-butyldimethylsilylglycolic acid chloride gave the monoester **28**. The acetylation of 2-hydroxyl in compound **28** gave the acetate **29**, from which the *t*-butyldimethylsilylgroup was cleaved using tetra-*n*-butylammonium fluoride to afford 2-acetyl-3,6-dibenzyl-1-glycolyl-4,5-dibenzylphosphoryl-*myo*-inositol **30**. This product was first exhaustively deprotected by hydrogenolysis with hydrogen gas over Pd/C catalysts to give 1-glycolyl 4,5-bisphosphate **31**. Compound **31** feature the hydroxyl group in the glycolyl moiety for the purpose of its coupling with the phosphatidyl group. This was accomplished by means of phospholipase D-catalyzed transphosphatidylation (8) (45°C, 12 h) of phosphatidylcholine in the biphasic solvent system: acetate buffer - chloroform to give the analog **32** in 21% yield.

Synthesis of Membrane-Permeant Inositol Polyphosphates

Inositol polyphosphates are membrane-impermeant, because of multiple anionic charges of phosphate groups. The permeability of inositol phosphate can be substantially increased in two ways: (i) inositol polyphosphates can be derivatized with bulky acyl groups, expecting that two or more bulky groups might cover anionic charge of phosphate groups, thus facilitating membrane anchoring and penetration; (ii) the phosphate groups themselves are converted into charge-devoid labile triesters such as acyloxymethyl esters.

Synthesis of acyl inositol polyphosphates. Synthesis of this type of derivatives is shown in Scheme 4. 1,2:4,5-di-*O*-cyclohexylidene-*myo*-inositol **33** was acylated to give 3,6-diacyl derivative **34**. The cyclohexylidene group was cleaved with trifluoroacetic acid in methanol to give 1,2-diol **35**, followed by phosphorylation with *O,O*-dibenzyl-*N,N*-diisopropylphosphoramidite, oxidation with *m*-CPBA, and subsequent hydrogenolysis to afford 1,2,4,5-tetrakisphosphate derivative **37**. To synthesize acyl derivative of I-1,4,5-P$_3$, 2,3-*O*-cyclohexylidene-*myo*-inositol (**38**) was acetylated to give 6-*O*-acetyl derivative **39**. The hydroxy groups at 1-, 4- and 5-

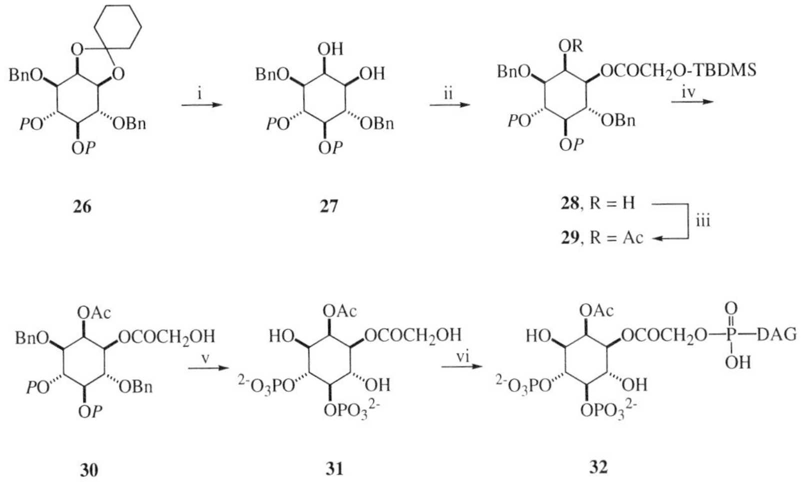

Scheme 3. Chemo-enzymatic synthesis of 2-acetyl-1-(1-stearoyl-2-arachidonoyl-sn-glycerophosphoglycolyl)-myo-inositol 1,4,5-bisphosphate. i: 80% AcOH, CH₂Cl₂, reflux, 1 h, 75%; ii: tBuMe₂SiOCH₂COCl, DMAP, CH₂Cl₂, rt, 12 h, 80%; iii: Ac₂O, pyridine, rt, 1 h, 61%; iv: Bu₄N⁺F⁻,THF, rt, 12 h, 95%; v: H₂, Pd/C, CH₃OH, rt, 14 h, 96%; vi: phosphatidylcholine, phospholipase D, acetate buffer, 45°C, 14 h, 21%, silica gel column, R_f 0.2, CHCl₃ : CH₃COCH₃ : CH₃OH : CH₃COOH : H₂O (40:15:15:12:8); P = P(O)(OBn)₂.

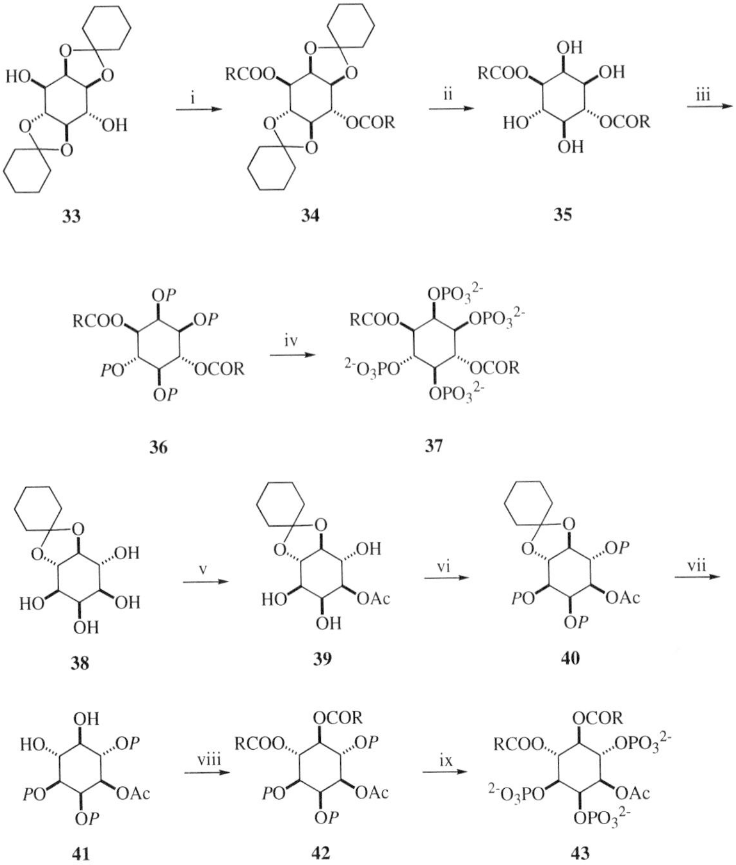

Scheme 4. Synthesis of acylated I-1,2,4,5-P$_4$ and acylated I-1,4,5-P$_3$. I: RCOCl, DMAP, pyridine, 1h at 0°C then 1 h at rt, 90% (R = CH$_3$), 91%, (R = C$_3$H$_7$); ii: CF$_3$COOH, CH$_2$Cl$_2$, MeOH (1:1:1), rt, 2 h, 83% (R = CH$_3$), 91% (R=C$_3$H$_7$); iii: (BnO)$_2$PNiPr$_2$, tetrazole, CH$_2$Cl$_2$, 0°C, 1 h, MCPBA, -40°C, 6 min then rt, 2 h, 80% (R = CH$_3$), 88% (R = C$_3$H$_7$); iv: H$_2$, Pd/C; CH$_3$OH, rt, 7 h, 65% (R = CH$_3$), 55%(R = C$_3$H$_7$); v: Ac$_2$O, molecular sieves 4Å,DMA, rt, 48 h, mixture of 4-acetyl- and 5-acetyl- compounds; vi: (BnO)$_2$PNiPr$_2$, tetrazole, CH$_2$Cl$_2$, 0°C, 1.5 h, MCPBA, -40°C, 6 min, separation by silica gel chromatography, 35% (4-acetyl), 39%(5-acetyl); vii: CF$_3$COOH, CH$_2$Cl$_2$, 0°C, 4 h, 77% (R = CH$_3$), 72% (R = C$_3$H$_7$); viii: RCOCl, DMAP, pyridine, 0°C 1 h, rt, 5 h, 70% (R = CH$_3$), 66% (R = C$_3$H$_7$); ix: H$_2$, Pd/C; CH$_3$OH, rt, 14 h, 69% (R = CH$_3$), 76% (R = C$_3$H$_7$); P = P(O)(OBn)$_2$.

positions were phosphorylated to give the trisphosphotriester **40**, in which the cyclohexylidene group was cleaved by trifluoroacetic acid and the remaining two hydroxy group were acylated. The final deprotection of the phosphate groups by hydrogenolysis afforded 6-*O*-acetyl-2,3-*O*-diacyl-*myo*-inositol 1,4,5-trisphosphate (**43**).

Synthesis of acetoxymethyl esters of inositol polyphosphates. Application of the acetoxymethyl group for modification of nucleotides (*9*) and inositol-3,4,5,6-tetrakisphosphate acetoxymethyl ester (*10,11*) was previously reported. In this work, we have synthesized 3,6-di-*O*-acetyl-*myo*-inositol 1,2,4,5-tetrakisphosphate octakis(acetoxymethyl) ester (**44**), 3,6-*O*-dibutyryl-*myo*-inositol 1,2,4,5-tetrakisphosphate octakis(acetoxymethyl) ester (**45**), 6-*O*-acetyl-*myo*-inositol 1,2,4,5-tetrakisphosphate octakis(acethoxymethyl) ester (**46**), *myo*-inositol 1,2,4,5-tetrakisphosphate octakis(acetoxymethyl) ester (**48**), and *myo*-inositol 1,4,5-trisphosphate hexakis(acetoxymethyl) ester (**50**) starting from 3,6-di-*O*-acyl-*myo*-inositol 1,2,4,5-tetrakisphosphate **37a** and **37b**, 2,3,6-tri-*O*-acyl-*myo*-inositol 1,4,5-trisphosphate (**43b**), *myo*-inositol-1,2,4,5-tetrakisphosphate **47**, and *myo*-inositol 1,4,5-trisphosphate (**49**), as shown in Scheme 5. The reactions were carried out by treatment of inositol phosphates with the mixture of acetoxymethyl bromide and diisopropylethylamine in acetonitrile or DMF as solvents at room temperature (25-36°C) for 2-10 days. The products were extracted with toluene, the extracts were evaporated to dryness under vacuum and the solid residues were chromatographed on silica gel.

Ca^{2+} mobilizing tests of these compounds were carried out in Prof. Hirata's Laboratory. No Ca^{2+} increase was observed in Cos-1 cells when highly acylated IP$_3$, IP$_4$ and acetoxymethyl esters of IP$_3$ and IP$_4$ were incubated at 37°C for 30 min at a high concentration. It is assumed that the lack of Ca^{2+} mobilization resulted from a strong 5-phosphatase activity, resulting in the cleavage of the 5-phosphate group. Alternatively, the acetoxy groups were not sufficiently lipophilic and the polarity of the IP$_n$ derivatives was still to high, or the large molecular size of the derivatives prevented their fast penetration through the membrane. It is also possible that one or more acetoxy groups are cleaved within the time required to pass through the membrane, resulting in the charged and thus impermeable derivatives. To remedy this problem, it is necessary to know metabolism of these molecules, or to synthesize derivatives with more bulky and stable acyloxy groups.

Di-*O*,*O*-*p*-methoxybenzyl-*N*,*N*-diisopropylphosphoramidite as a New Phosphitylating Reagent

In many syntheses it would be beneficial not to use hydrogenolytic conditions for the cleavage of the phosphotriester groups. These cases include, among others, syntheses of unsaturated fatty acid esters, as shown above, and phosphorothioate analogs of phosphates. In such syntheses the use of a popular benzyl group (*12*) is not indicated. In contrast to benzyl group, the *p*-methoxybenzyl group is cleaved under very mild conditions by oxidation with 2,3-dicyano-5,6-dichlorobenzoquinone (DDQ) in the presence of water, or with ammonium ceric nitrate. In order to accomplish synthesis of some unsaturated PIP$_n$ we tested the utility of a new phosphitylating reagent, di-*O*,*O*-*p*-methoxybenzyl-*N*,*N*-diisopropylphosphoramidite (**51**) (*13*).

Compound **51** was obtained by treatment of *N*,*N*-diisopropylphosphoramido-dichloridite with *p*-methoxybenzyl alcohol in the presence of diisopropylethylamine as shown in Scheme 6. The amidite **51** reacted with many alcohols **52** in the presence of tetrazole to give di-*O*,*O*-(*p*-methoxybenzyl)alkyl- (or aryl-) phosphites **53**. The

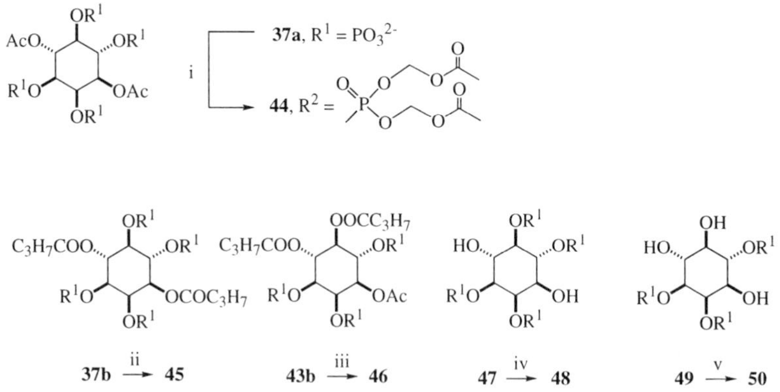

Scheme 5. Synthesis of acetoxymethyl esters of inositol polyphosphates. i: CH_3COOCH_2Br, iPr_2EtN, CH_3CN, rt, 2 days, 29%; ii: CH_3COOCH_2Br, iPr_2EtN, CH_3CN, rt, 5 days, 35%; iii: CH_3COOCH_2Br, iPr_2EtN, CH3CN, rt, 5 days, 45%; iv: CH_3COOCH_2Br, iPr_2EtN, CH_3CN, DMF, rt, 10 days, 10%; v: CH_3COOCH_2Br, iPr_2EtN, CH_3CN, DMF, 10 days, 16%.

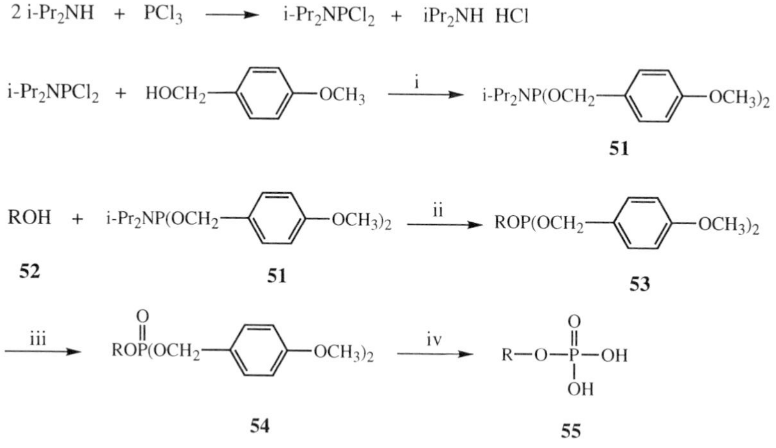

Scheme 6. Synthesis and application of di-O,O-p-methoxybenzyl-N,N-diisopropylphosphoramidite (*51*) as a new phosphitylating reagent. i: iPr_2EtN, THF, 0°C, 6 min, then 1 h at rt, 42%; ii: tetrazole, CH_2Cl_2, 6 min at -40°C then 1 h at rt; iii: MCPBA, CH_2Cl_2, 6 min at -40°C then 1 h at rt, 75% (R = 3,6-dibutyryl-*myo*-inositol); iv: DDQ, H_2O, CH_2Cl_2, rt, 15 h, 32%(R = 3,6-dibutyryl-*myo*-inositol).

phosphites **53** were oxidized with a peracid reagent to give the corresponding alkyl (or aryl) phosphates **54** which were then were deprotected to alkyl (or aryl) phosphates **55** by oxidation with DDQ or ammonium ceric nitrate. We found that the *p*-methoxybenzyl group can be cleaved without affecting the alkene and benzyl group. This new phosphorylating reagent **51** should be useful in syntheses of phosphatidylinositols bearing unsaturated fatty acid components.

Acknowledgments: The authors wish to thank Prof. Glenn D. Prestwich for his advise and support. We also thank Prof. Hirata for conducting Ca^{2+} mobilizing tests.

Literature Cited

1. Hirata, M.; Watanabe, Y.; Ishimatsu, T.; Yanaga, F.; Koga,T.; Ozaki, S. *Biochem. Biophys. Res. Commun.* **1990**, *168*, 379.
2. Ozaki, S.; Koga, Y.; Ling, L.; Watanabe, Y.; Kimura, Y.; Hirata, M. *Bull. Chem. Soc. Jpn.* **1994**, *67*, 1058.
3. Ozaki, S.; Kong, X. Z.; Watanabe,Y.; Ogasawara, T. *Chinese J. Chem.* **1997**, *15*, 556.
4. Ozaki, S.; Kong, X. Z.; Watanabe, Y.; Ogasawara, T. *Chinese J. Chem.* in press.
5   *Inositol Phosphates and Derivatives: Synthesis, Biochemistry, and Therapeutic Potential*, Reitz, A. B. Ed., American Chemical Society, Washington, D.C. 1991, p. 143-154.
6. (a) Watanabe,Y.; Hirofuzi, H.; Ozaki, S. *Tetrahedron Lett.* **1994**, *35*, 123. (b) Watanabe, Y.; Hirofuji, H.; Ozaki, S. *Tetrahedron* **1995**, *51*, 8969.
7. Bruzik, K. S.; Kubiak, R. J. *Tetrahedron Lett.*, **1995**, *36*, 2415.
8. Shuto,S.; Ito, H.; Ueda, S.; Imamura, S.; Furukawa, K.; Tsujino, M.; Matsuda, A.; Ueda, T. *Chem. Pharm. Bull. Jpn* **1988**,*36*, 209.
9. Srivastva, D. N.; Farquhar, D. *Bioorg. Chem.* **1984**, *12*, 118.
10. Vajanaphanich, M.; Scholz, C.; Rudolf, M. T.; Wasserman, M. *Nature* **1994**, *371*, 711.
11. Roemr, S.; Stadler, C.; Rudolf, M.; Jastorff, B.; Schultz, C. *J. Chem. Soc. Perkin Trans 1*, **1996**, 1683.
12. Ya, K. L.; Fraser-Reid B. *Tetrahedron Lett.* **1988**, *29* , 979.
13. Ozaki, S.; Zheng, D.X.; Hao, A.Y.; Kong, X. Z. submitted to *Chinese J. Chem.*

Chapter 14

Syntheses of 2-Modified Phosphatidylinositol 4, 5-Bisphosphates: Putative Probes of Intracellular Signaling

R. Aneja and S. G. Aneja

Nutrimed Biotech, Cornell University Research Park, Langmuir Laboratory, Ithaca, NY 14850

The first syntheses of phosphatidyl-*myo*-inositol 4,5-bisphosphates (PtdIns-4,5-P_2) modified at the 2-OH are described. Complementary procedures are presented for modification of 2-OH by replacement or derivatization, exemplified by the 2-deoxy-2-fluoro epimer and the 2-*O*-acetyl derivative, respectively. The products optionally incorporate an ω-aminoalkyl-type residue for conjugation to photoaffinity or other reporter group or solid matrix. These analogues retain the core PtdIns-4,5-P_2 structure and stereochemistry, but lack the nucleophilic 2-OH deemed essential for substrate hydrolysis by the mammalian phosphoinositide-specific phospholipase C (PI-PLC). Therefore, the 2-modified analogues are putative structure- and mechanism-based competitive inhibitors of PI-PLC, suitable as comparative probes of PtdIns-4,5-P_2-binding to PI-PLC and cellular regulatory proteins.

Multifarious roles emerging for cellular phosphoinositides as vital participants in intracellular signaling and allied processes, created a need for new probes of enzymes and regulatory proteins involved in phosphoinositide metabolism (*1*). For instance, PtdIns-4,5-P_2 (**1**) functions as the preferred substrate of PI-PLC (*2*) and phosphoinositide 3-kinase (PI 3-kinase) (*1*) enzyme families, and as allosteric activating factor of cellular regulatory proteins with and without pleckstrin homology (PH) domains (*3*). The action of PI-PLC on PtdIns-4,5-P_2 causes its hydrolysis to the two intracellular second messengers *myo*-inositol-1,4,5-trisphosphate (Ins-1,4,5-P_3) and *sn*-1,2-diacylglycerol (DAG) (*4*). Evidence is accumulating that a critical early step in catalyzed hydrolysis involves intramolecular nucleophilic attack by the 2-OH on the 1-phosphodiester phosphorus, resulting in concomitant formation of the inositol 1,2-cyclic phosphate intermediate and DAG, followed by further hydrolysis of the cyclic phosphate to Ins-1,4,5-P_3 (*5*). However, with some isoforms of PI-PLC a small proportion of the cyclic phosphate survives. The two-step mechanism is supported, and the mode of binding of the *myo*-inositol-phosphate residue at the catalytic site is revealed, by x-ray crystal structure analyses of the deletion mutant of PI-PLC-δ_1 in ternary complexes with calcium ions and Ins-1,4,5-P_3 as a substrate-mimic (*6*), and DL-*myo*-inositol-2-methylene-1,2-cyclic-monophosphate as an analogue of the putative

R¹COO—CH₂
$|$
R²COO►CH
$|$
CH₂—O

1: X = OH, Y = H;
2a: X = H, Y = F;
2b: X = OAc, Y = H.

cyclic phosphate intermediate (7). New probes incorporating both the inositol phosphate and the glycerolipid moieties could provide a comprehensive view and understanding of the interactions between the phosphoinositides, and, the enzymes and regulatory proteins involved in signaling.

In this chapter, we describe the syntheses of 2-modified PtdIns-4,5-P_2s as putative probes of intracellular signaling mediated by PI-PLCs and allied regulatory proteins. The molecular design of these PtdIns-4,5-P_2 analogs is based on the tenet that the participation of the axial 2-OH as an intramolecular nucleophile is essential for catalyzed hydrolysis; conversely, the intramolecular nucleophilic action is precluded in analogues lacking the 2-OH. The core PtdIns-4,5-P_2 structure and absolute stereochemistry are retained to ensure efficient interaction with the catalytic as well as the non-catalytic PH domain binding sites. Thus, the PtdIns-4,5-P_2 analogues modified at the 2-position by replacement or derivatization of the axial 2-OH are potential competitive inhibitors of PI-PLC enzymes, suitable as comparative probes of protein-binding and enzyme action. An essential caveat is that the modifying groups be small, preferably isosteric with OH, for a good fit at the catalytic site. In this context, we selected 2-deoxy-2-fluoro epimers **2a** and 2-O-acetyl derivatives **2b** as examples of modification at 2-OH by replacement and derivatization, respectively.

Retrosynthetic Analysis and Strategy for Synthesis

The strategy for synthesis is based on the retrosynthetic disconnection of the 2-modified PtdIns-4,5-P_2 structure into sn-3-phosphatidic acid (sn-3-PA) and protected chiral myo-inositol phosphate fragments shown in Scheme I. Thus, the essential stages entail (i) preparation of the key materials: an optically resolved O-protected myo-inositol-4,5-bisphosphate with a free 1-OH as the inositol synthon, and 1,2-di-O-fatty-acyl-sn-glycero-3-phosphoric acid (sn-3-PA) as the lipid synthon; (ii) coupling of the inositol 1-OH and the lipid phosphoric acid by phosphodiester condensation; (iii) deprotection of the condensation product to generate the target 2-modified PtdIns-4,5-P_2. Two tactically distinct options, comprising modification of the 2-OH either prior to, or after the phosphodiester condensation stage, are illustrated in the syntheses of the 2-deoxy-2-fluoro **2a** and the 2-O-acetyl **2b** series.

Inositol and Lipid Starting Materials and their Coupling

Chiral natural products have been chemically modified and converted to 1D-myo-inositol derivatives, however, myo-inositol is an inexpensive and competitive starting material for synthesis. Its use mandates an optical resolution, but resolution via diastereomeric esters with chiral acids is facile and efficient for several selectively O-protected derivatives suitable as starting materials for syntheses (8).

The choice of sn-PA as the lipid synthon and method of its incorporation into the target structure distinguish our approach from related syntheses, albeit of unmodified phosphoinositides, which all utilize DAG as the lipid synthon (9). This is advantageous because the *sn*-3-PA is stable in contrast with DAG which has a strong propensity to isomerize and racemize via 1,2- and 2,3-acyl migration.

Chiral *myo*-Inositol Derivatives. In an earlier synthesis of PtdIns-4,5-P_2, we employed 3-*O*-benzyl-1,2:4,5-di-*O*-cyclohexylidene-*myo*-inositol (*10*). This method is inefficient because regioselective 3-*O*-benzylation of 1,2:4,5-di-*O*-cyclohexylidene-*myo*-inositol concomitantly produces 6-*O*-benzyl and 3,6-di-*O*-benzyl derivatives as by-products. In the present syntheses, we employed 1D-1,2:4,5-di-*O*-cyclohexylidene-*myo*-inositol as the chiral inositol starting material. The compound had been described earlier, albeit in inadequate optical purity, and the 1D- absolute configurations were assigned to both the (-)- and the (+)- enantiomers. We developed protocols for its preparation in an optically pure form, and by correlation with a reference of configuration, established unequivocally that (-)-1,2:4,5-di-*O*-cyclohexylidene-*myo*-inositol belongs to the 1D- absolute configuration series (*11*). In our experience, the enantiomeric 1,2:4,5-di-*O*-cyclohexylidene-*myo*-inositols are versatile starting materials for synthesis of the phosphoinositides. Specifically, complete benzylation of the (-)-enantiomer gave a quantitative yield of 1D-3,6-di-*O*-benzyl-1,2:4,5-di-*O*-cyclohexylidene-*myo*-inositol **3** which was utilized as the key chiral synthon as shown in Schemes II and IV.

***sn*-3-Glycerolipids.** Natural *sn*-3-phosphatidylcholine, the derived *sn*-glycero-3-phosphocholine and 1-acyl-*sn*-glycero-3-phosphocholine have integral chiral glycerol-3-phosphate, and are convenient starting materials for glycerophospholipid synthesis (*12*). These were utilized as the main starting materials for the lipid synthons exemplified by *sn*-3-PAs with identical or different fatty acyls at the *sn*-1 and *sn*-2 positions. Some *sn*-PAs are available from commercial sources, and were used in appropriate cases. An alternative material, 3-*O*-benzyl-*sn*-glycerol, was employed for synthesis of the diether analogues.

Phosphodiester Coupling of the Key *myo*-Inositol and Lipid Synthons. The phosphodiester condensation procedure for coupling a phosphatidic acid with an alcohol was previously developed by us as a general method for the synthesis of glycerophospholipids (*13*). This method was modified and adapted in the present syntheses for conjugating the key *myo*-inositol and lipid synthons. In the modified general protocol, the appropriately protected *myo*-inositol, *sn*-3-PA and triisopropylbenzenesulfonyl chloride (TPSCl) in the molar ratio 1:1:2, were allowed to react in anhydrous pyridine solution at room temperature for 1 to 3 hr. The reaction mixture was treated with water to decompose excess TPSCl and activated phosphate species. The crude product was obtained by evaporation under reduced pressure, and was further purified by chromatography.

Deprotection of Condensation Products. The temporary protection of alcohol and phosphate ester functions was provided by benzyl, and that of aminoalkyl by benzyloxycarbonyl (Cbz) groups. The simultaneous removal of all protecting groups by Pd-catalyzed hydrogenolysis of the condensation products formed the target PtdIns-4,5-P_2 analogues.

Synthesis of the Target Analogues

The 2-deoxy-2-fluoro analogues **2a** ($R^1 = R^2 = C_{15}H_{31}$) formally belonging to 1D-*scyllo* configuration series are accessible more easily than the 1D-*myo* series, and the preparation of **13** serves to illustrate our synthetic approach (Schemes II and IV).

1D-1-[1,2-Di-*O*-hexadecanoyl-*sn*-glycero-3-phospho]-2-deoxy-2-fluoro-*scyllo*-inositol-4,5-bisphosphate (**13**). In this example, the replacement of the 2-OH with inversion by the isosteric and isopolar fluoro residue is carried out in a key inositol synthon prior to phosphodiester condensation.

The chiral *myo*-inositol **3**, selected as the general starting material, was converted into the key *myo*-inositol synthon 1D-3,6-di-*O*-benzyl-2-deoxy-2-fluoro-*scyllo*-inositol-4,5-bis(di-*O*,*O*-benzylphosphate) (**10**) as outlined in Scheme II. Complete deketalization of **3** by hydrolysis with acetic acid-water (90:10) at 95°C gave 1D-3,6-di-*O*-benzyl-*myo*-inositol (**4**). The treatment of **4** in DMSO with cyclohexanone dimethylketal catalyzed by p-toluenesulfonic acid at 40-45°C under reduced pressure (used to distill out methanol) provided kinetic control resulting in 3:1 selective reaction at the 4,5- versus the 1,2-OHs to produce an acceptable yield of the critical novel synthon 1D-3,6-di-*O*-benzyl-4,5-*O*-cyclohexylidene-*myo*-inositol (**5**). Reaction of **5** with Bu$_2$SnO in toluene with azeotropic removal of H$_2$O, rotary evaporation, solvent change to DMF and treatment with 4-methoxybenzyl chloride/CsF (*14*) at 80°C for 2 hr provided high regioselectivity and gave, after purification   by HPLC, pure equatorial 1-*O*- substituted derivative **6**. Mixing **6** in CH$_2$Cl$_2$ with DAST (*15*) at 0-5°C followed by reaction at 35-40°C gave the 2-deoxy-2-fluoro derivative **7** formed by substitution with inversion. The structure of a second, albeit minor, product is being ascertained by independent syntheses via fluorination of the 2-epimer of **6** and related structures.   Transketalization of **7** with ethylene glycol/catalytic *p*-toluenesulfonic acid at 35-40°C yielded the 4,5-diol **8**. Dibenzylphosphorylation (*16*) of **8** using dibenzyl diisopropylphosphoramidite and 1*H*-tetrazole in CH$_2$Cl$_2$ followed by 3-chloroperoxybenzoic acid yielded the 4,5-bis-*O*-(dibenzylphosphate) derivative **9**. The oxidation of **9** in CH$_2$Cl$_2$ solution with DDQ at room temperature removed the methoxybenzyl group and gave **10**.

The structure of the inositol synthon **10** may be varied by replacing reaction of **6** with DAST in Scheme II by other reagents to produce 2-deoxy, *O*-acyl, *O*-alkyl, deoxyhalo or deoxydihalo analogues. As a special case, benzylation of **6** yielded the 2-*O*-benzyl analogue of **7**. Subsequent transformations exactly as in Scheme II gave 1D-2,3,6-tri-*O*-benzyl-*myo*-inositol-4,5-bis(dibenzylphosphate), the inositol synthon for unmodified PtdIns-4,5-P$_2$s.

1,2-Di-*O*-hexadecanoyl-*sn*-glycero-3-phosphoric acid (**11**, Scheme III) was prepared by partial synthesis from natural *sn*-3-phosphatidylcholine by methods developed previously (*12*). In early stage exploratory experiments, a reagent grade material from a commercial source was employed. The preparation of *sn*-PA incorporating useful variation in structure is illustrated later. The phosphodiester condensation reaction of the protected 2-deoxy-2-fluoro derivative **10** and the *sn*-3-PA **11** by the general protocol, yielded 1D-1-[1,2-di-*O*-hexadecanoyl-*sn*-glycero-3-phospho]-2-deoxy-2-fluoro-3,6-di-*O*-benzyl-*scyllo*-inositol   4,5-bis(*O*,*O*-dibenzyl-phosphate) (**12**).   Hydrogenolysis using H$_2$/Pd gave the target 2-deoxy-2-fluoro PtdIns-4,5-P$_2$ **13**.

1D-1-[1,2-Di-*O*-hexadecanoyl-*sn*-glycero-3-phospho]-2-*O*-acetyl-*myo*-inositol-4,5-bisphosphate (**18**).   This analogue illustrates an example of modification wherein the 2-OH is derivatized by esterification, specifically acetylation, subsequent to the phosphodiester condensation.

The chiral *myo*-inositol **3** used in Scheme II was converted in 2 steps into the

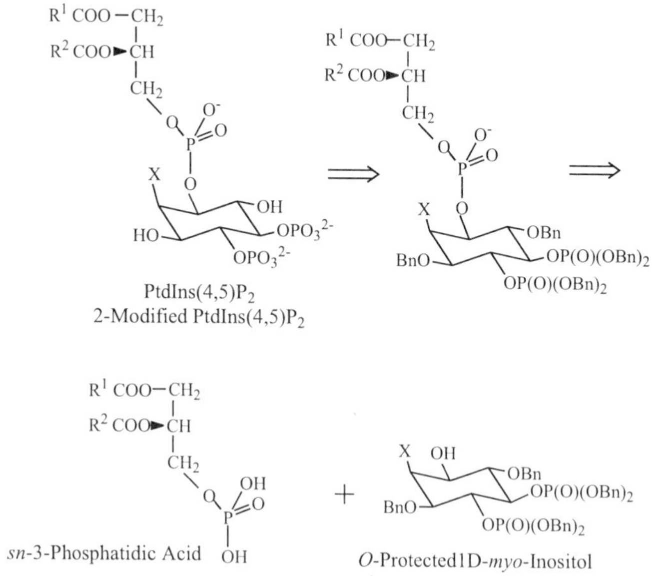

Scheme I. Retrosynthetic analysis of 2-modified PtdIns(4,5)P$_2$

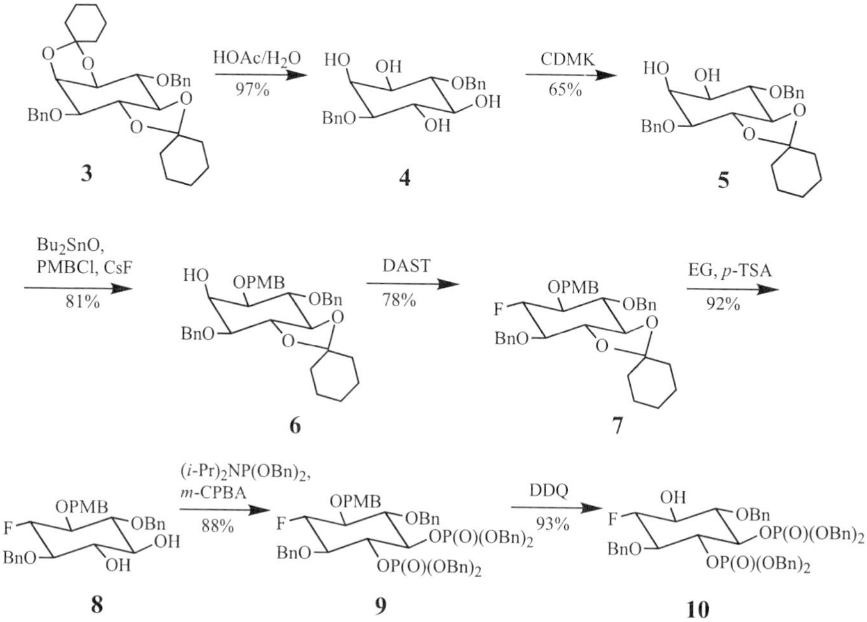

Scheme II. Synthesis of 1D-3,6-di-*O*-benzyl-2-deoxy-2-fluoro-*scyllo*-inositol 4,5-bisdibenzylphosphate (**10**).

key *myo*-inositol synthon, 1D-3,6-di-*O*-benzyl-*myo*-inositol-4,5-bis(*O,O*-dibenzyl-phosphate) (**16**) (Scheme IV). Transketalization under kinetic control by reaction of **3** (Scheme II) with ethylene glycol/catalytic *p*-toluenesulfonic acid in CH_2Cl_2 at room temperature for 3 h effected selective removal of the 4,5-cyclohexylidene group and gave 1D-1,2-*O*-cyclohexylidene-3,6-di-*O*-benzyl-*myo*-inositol (**14**). Bisdibenzyl-phosphorylation of **14** using dibenzyl diisopropylphosphoramidite and 1*H*-tetrazole in CH_2Cl_2 followed by *m*-chloroperbenzoic acid yielded 1D-3,6-di-*O*-benzyl-1,2-*O*-cyclohexylidene-*myo*-inositol 4,5-bis(*O,O*-dibenzylphosphate) (**15**). The 1,2-*O*-cyclohexylidene protecting group in **15** was removed by hot aqueous acetic acid resulting in the 1,2-diol **16** (*10*).

The 1,2-di-*O*-hexadecanoyl-*sn*-glycero-3-phosphoric acid (**11**) prepared for the synthesis of the 2-deoxy-2-fluoro analogue was employed as the lipid synthon. The condensation of the *myo*-inositol synthon **16** and *sn*-3-PA **11** occurred smoothly by the general protocol. Both possible isomeric products were formed, the 1-phosphatidyl derivative **17** (Scheme V) was the predominant product obtained by chromatography in pure state in 55% yield. The observed regioselectivity is consistent with the greater reactivity of the equatorial 1-OH compared to the axial 2-OH flanked by two *cis*-neighbors. For post-condensation modification of the 2-OH, reaction between Ac_2O/DMAP and **17**, without or with DCC, gave a low yield of the corresponding 2-*O*-acetate derivative. A phosphoric-acetic mixed anhydride of **17**, and a strained 5-cyclic phosphotriester are plausible consecutive side-intermediates transformed by hydrolytic work-up into unacetylated starting **17** and its 2-phosphatidyl isomer which were isolated as by-products. Much improved yield of the 2-*O*-acetate derivative was obtained by acetylation using Ac_2O/DMAP with an excess of NaOAc added to suppress the displacement of acetate from phosphoric-acetic mixed anhydride and the consequent formation of the 5-cyclic side-intermediate. Hydrogenolysis gave the title 2-*O*-Ac PtdIns-4,5-P_2 analogue **18**.

Matched Pairs of Normal and 2-Modified PtdIns-4,5-P_2 Analogues

For comparative evaluation of the normal unmodified PtdIns-4,5-P_2 and the 2-modified analogues, matched pairs with identical fatty acyl residues are essential. Both approaches to the 2-modified series are suitable equally for the synthesis of unmodified PtdIns-4,5-P_2s, and hence of the said matched pairs. Thus, condensation of *sn*-3-PA **11** and 1D-2,3,6-tri-*O*-benzyl-*myo*-inositol 4,5-bis-(*O,O*-dibenzylphosphate), prepared by benzylation of **6** and subsequent transformations as described for **7** in Scheme II, followed by hydrogenolysis, gave 1D-1-[1,2-di-*O*-hexadecanoyl-*sn*-glycero-3-phospho]-*myo*-inositol-4,5-bisphosphate to be paired with **13** or **18**. Alternatively, the same dihexadecanoyl PtdIns-4,5-P_2 was obtained by hydrogenolysis of **17** prepared as an intermediate in the synthesis of **18**.

Analogues with hexadecanoyl and other long chain fattyacyl residues are akin to the cellular PtdIns-4,5-P_2 (**1**, R^1CO = stearoyl, R^2CO = arachidonyl) in so far as these form multimolecular aggregates on hydration. Analogues with short chain fatty acyls form clear solutions in water and are monomeric (*8,10*). Representative water soluble analogues of normal and 2-modified PtdIns-4,5-P_2s were synthesized by reaction of 1,2-dihexanoyl-*sn*-glycero-3-phosphoric acid with appropriate inositol synthons.

The broad utility of the analogues as probes and reagents is enhanced by the incorporation of an ω-aminoalkanoyl residue into the analogue structure to obtain conjugands suitable for linking to fluorescent and other reporter groups, and to solid

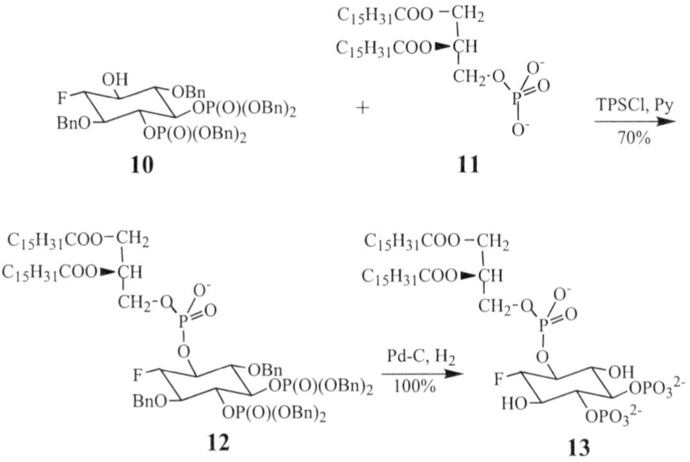

228

Scheme III. Synthesis of 1D-1-[1,2-di-*O*-hexadecanoyl-*sn*-glycero-3-phospho]-2-deoxy-2-fluoro-*scyllo*-inositol 4,5-bisphosphate (**13**).

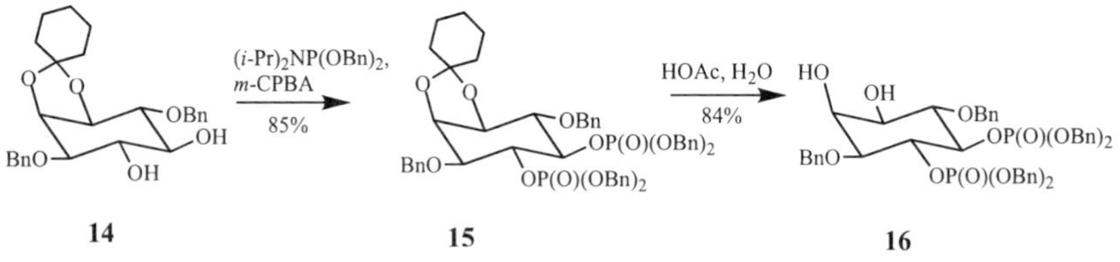

Scheme IV. Synthesis of 1D-3,6-di-*O*-benzyl-*myo*-inositol 4,5-*O*-bis(dibenzylphosphate) (**16**).

matrices such as gold or chromatographic media. As an example, for the preparation of 1D-1-[1-*O*-hexanoyl-2-*O*-(ω-aminobutanoyl)-*sn*-glycero-3-phospho]-*myo*-inositol-4,5-bisphosphate (**23**), 1-*O*-hexanoyl-2-*O*-(ω-carboxybenzylaminobutanoyl)-*sn*-glycero-3-phosphoric acid (**22**) was prepared from 1,2-di-*O*-hexanoyl-*sn*-glycero-3-phosphocholine (**19**) by the chemoenzymatic synthesis outlined in Scheme VI. Hydrolysis of **19** in ether-NaOAc buffer pH 8.5 at 37°C catalyzed by phospholipase A$_2$ (PLA$_2$) from *Crotalus adamanteus* snake venom removed the ester at the *sn*-2 position. The resulting 1-*O*-hexanoyl-*sn*-glycero-3-phosphocholine (**20**), esterified using an excess of ω-Cbz-aminobutanoic acid, DCC and DMAP in anhydrous ethanol-free CHCl$_3$ at room temperature for 12 h, gave 1-*O*-hexanoyl-2-*O*-(ω-Cbz-aminobutanoyl)-*sn*-glycero-3-phosphocholine (**21**). Hydrolysis of **21** in sodium acetate buffer pH 8.5 at 37°C catalyzed by phospholipase D (PLD) from *Streptomyces chromofuscus* yielded **22**. Condensation of **22** with the key inositol synthon **16** followed by hydrogenolysis yielded (**23**). Reaction between **23** and *N*-hydroxysuccinimidyl-4-azidosalicylic acid gave the 4-azidosalicyl photoaffinity-labelled analogue **24**.

The above described methods were also employed for the preparation of other normal and 2-modified matched pairs. Condensation of 1,2-di-*O*-*n*-butyl-*sn*-glycero-3-phosphoric acid (**25**), with the inositols **10** and **16**, respectively, followed by hydrogenolysis, yielded the 2-modified analogue 1D-1-[1,2-di-*O*-*n*-butyl-*sn*-glycero-3-phospho]-2-deoxy-2-fluoro-*scyllo*-inositol-4,5-bisphosphate (**26**), and 1D-1-[1,2-di-*O*-*n*-butyl-*sn*-glycero-3-phospho]-*myo*-inositol-4,5-bisphosphate.

Selected Applications

Compound **26** has been designed as a water-soluble analogue, stable to non-specific chemical hydrolysis, and applicable in the preparation of co-crystallizates with PI-PLC suitable for X-ray crystal structure analysis. Studies of PI-PLC-δ$_1$ have given a clear picture of the mode of binding of Ins-1,4,5-P$_3$, calcium and activated water at the active site, but provided no direct information about binding of the lipid residue because the probes employed were inositol phosphates rather than phosphatidylinositol phosphates (*6,7*). The complete phosphatidylinositol phosphate structure is retained in **26** and related probes, so these may help elucidate the contribution of lipid residue of PtdIns-4,5-P$_2$ to binding in the catalytic domain, as well as the non-catalytic PH-domain. The water soluble monomeric **26** and its 1,2-dihexanoyl analogue, together with corresponding unmodified PtdIns-4,5-P$_2$s, are suitable probes for studying interactions with PI-PLCs by NMR. From a medicinal chemistry perspective, the analogues described herein are important as prototypes for the design of isozyme-specific inhibitors with potential as therapeutics in aberrant signal transduction via PI-PLC. Further, as substrate analogues, these are potential probes for of PI 3-kinase enzyme family.

Summary

We have validated a general approach to synthesis of the 2-modified PtdIns-4,5-P$_2$s by preparation of analogues with long as well as short chain fatty acyl and alkyl-ether type glycerolipid moieties. The approach is suitable also for the unmodified PtdIns-4,5-P$_2$ analogue series. The products are designed as structure- and mechanism-based competitive inhibitors of PI-PLC. These analogues are being applied as comparative probes of substrate-binding to the active site of this enzyme family and to regulatory

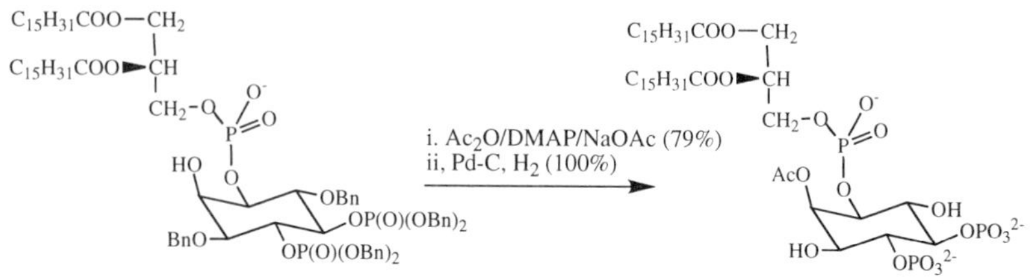

17 18

Scheme V. Synthesis of 1D-1-[1,2-di-*O*-hexadecanoyl-*sn*-glycero-3-phospho]-2-*O*-acetyl-*myo*-inositol 4,5-bisphosphate (**18**).

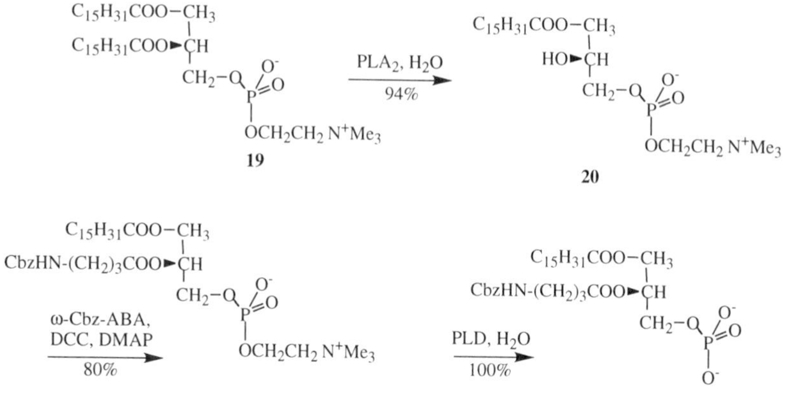

19 20 21 22

Scheme VI. Synthesis of 1-*O*-hexanoyl-2-*O*-(ω-Cbz-aminobutanoyl)-*sn*-glycero-3-phosphoric acid (**22**).

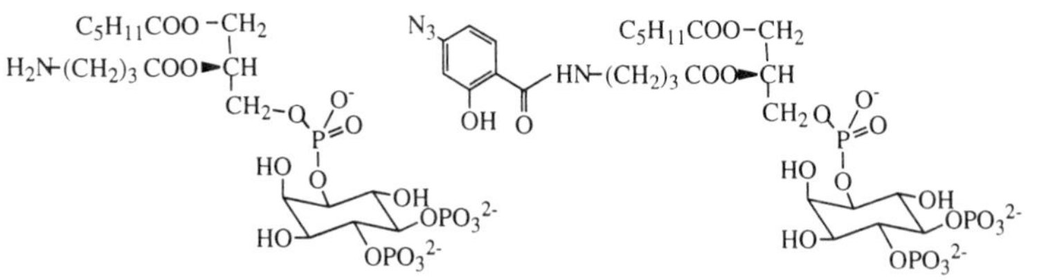

23 **24**

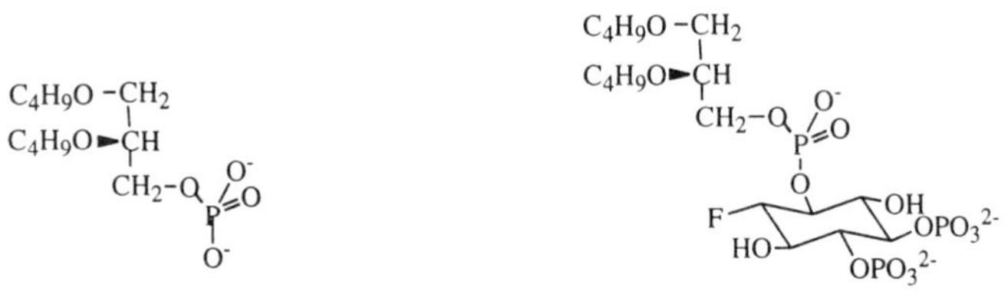

25 **26**

proteins involved in intracellular signaling. Experiments are underway on applications to x-ray crystallography of PI-PLC-δ_1, and on activation of ADP-ribosylation factor GTPase-activating proteins (Arf GAPs).

Note on Safety. The target compounds and intermediates are known putative signaling transducers and modulators. These potent biological agents and precursors need to be handled with care.

Acknowledgments. This work was supported by PHS/NIH grants GM49594 and GM51138. We thank P. T. Ivanova, P. Klosinski, A. Parra, C. Stoenescu, W. Xia, D. Fuller and J. L. Peng for their contributions to the experimental work.

Literature Cited

1. Duckworth, B. C.; Cantley, L. C. *Lipid Second Messengers - Handbook of Lipid Research*; Plenum Press: New York, NY 1996, Vol 8, pp 125-175.
2. Lee, S. B.; Rhee, S. G. *Curr. Opin. Cell Biol.* **1995**, 183.
3. Terui, K.; Kahn, R. A.; Randazzo, P. A. *J. Biol. Chem.* **1994**, *269*, 28130.
4. Berridge, M. J. *Nature* 1993, 361, 315.
5. Bruzik, K. S.; Tsai, M.-D. *Bioorg. Med. Chem.* **1994**, *2*, 49, and references therein.
6. Essen, L.-O.; Perisic, O.; Cheung, R.; Katan, M.; Williams, R. L. *Nature* **1996**, *380*, 595.
7. Essen, L.-O; Perisic, O.; Katan, M.; Wu, Y.; Roberts, M. F.; Williams, R. L. *Biochemistry* **1997**, *36*, 1704.
8. Aneja, R.; Aneja, S. G.; Parra, A. *Tetrahedron Lett.* **1996**, *37*, 5081.
9. Aneja, S. G.; Parra, A.; Stoenescu, C.; Xia. W.; Aneja, R. *Tetrahedron Lett.* **1997**, *38*, 803, and references therein.
10. Toker, A.; Meyer, M.; Reddy, K.; Falck, J. R.; Aneja, R.; Aneja, S.; Parra, A.; Burns, D. J.; Cantley, L. M. *J. Biol. Chem.* **1994**, *269*, 32358.
11. Aneja, R.; Aneja, S. G.; Parra, A. *Tetrahedron Asymmetry* **1995**, *6*, 17.
12. Aneja, R. *Biochem. Soc. Trans.* **1974**, *2*, 38.
13. Aneja, R.; Chadha, J. S.; Davies, A. P. *Biochim. Biophys. Acta* **1970**, *218*, 102.
14. Nagashima, N.; Ohno, M. *Chem. Lett.* **1987**, 141.
15. Posner, G. H.; Haines, S. R. *Tetrahedron Lett.* **1985**, *26*, 5.
16. Yu, K.-L.; Fraser-Reid, B. *Tetrahedron Lett.* **1988**, *29*, 979.

Chapter 15

Membrane-Permeant, Bioactivatable Derivatives of Inositol Polyphosphates and Phosphoinositides

Carsten Schultz[1], Marco T. Rudolf[1], Hartmut H. Gillandt[1], and
Alexis E. Traynor-Kaplan[2,3]

[1]Institut für Organische Chemie, Abt. Bioorganische Chemie, Universität Bremen,
UFT, 28359 Bremen, Germany
[2]Department of Medicine, The Whittier Institute, University of California at San
Diego, La Jolla, CA 92093

The extracellular application of membrane-permeant, bioactivatable
derivatives of myo-inositol 1,4,5-trisphosphate and phosphatidyl-myo-
inositol 3,4,5-trisphosphate to living cells mimicked the intracellular
action of these second messengers, such as elevation of intracellular
Ca^{2+}-levels in PC12 cells and inhibition of calcium-mediated chloride
secretion in T_{84} cells, respectively. 3,6-Di-O-butyryl-myo-inositol
1,2,4,5-tetrakisphosphate hexakis(acetoxymethyl) ester and 1',2'-
dipalmitoyl-6-O-butyryl-phosphatidyl-myo-inositol 3,4,5-trisphosphate
heptakis(acetoxymethyl) ester were prepared by total synthesis.

The field of intracellular signal transduction is certainly one of the most rapidly
expanding areas in the life sciences to date. To better understand the complex
signaling pathways there is an urgent need for agents to specifically modulate
intracellular levels of signaling molecules. While in the past those compounds were
mostly agonists or antagonists directed towards the extracellular moiety of
membrane-located receptors, many of the more recent research tools are
extracellularly applicable compounds, lipophilic enough to pass through cellular
membranes. Some of these tools generated a tremendous boost in particular research
areas: for instance the discovery of the ATPase inhibitor thapsigargin (*1*) for the
calcium field, or wortmannin as an inhibitor of phosphatidylinositol 3-kinase (PtdIns
3-kinase) in tyrosine kinase-mediated signaling (*2*). However, a more direct
manipulation of intracellular messenger levels is generally difficult because many of
the known second messengers carry negative charges in the form of phosphates,
which results in membrane-impermeability. This is especially true of the highly
charged inositol polyphosphates of which some 20 are abundant in living cells. Not
surprisingly, up to not long ago, for only one of them, Ins(1,4,5)P$_3$ (*1*), a clear
physiological function was assigned. Therefore, membrane-permeant derivatives of
the naturally occurring inositol phosphates should be highly beneficial to serve as
tools for studying the functions of inositol phosphates. Similarly attractive appears
the possibility of synthesizing chemically modified derivatives of second messengers
and converting them into membrane-permeant derivatives. These compounds should
allow investigation of pharmacological properties of these derivatives in the natural
environment of the living cell. This methodology should also be applicable when the

[3]Current address: Inologic Inc., 652 First Avenue South, Suite 601, Seattle, WA.

binding proteins still remain unknown, generating the intriguing possibility to map proteins prior to their first isolation. The information extractable from these experiments might subsequently help in designing the affinity resins, photoaffinity probes or even antagonists with certain physiological functions leading to potential intracellularly acting pharmaceuticals.

Masking Hydrophilic Groups - A Prodrug Approach

The introduction of masking groups for phosphates, hydroxyl or amino groups is meant to increase lipophilicity, thus allowing passive transport across plasma membranes. On one hand, the bioactivatable protecting groups should exhibit sufficient stability in extracellular media to allow application for longer periods of time. On the other hand and most importantly, they need to be substrates for intracellular endogenous enzymes like esterases or lipases to ensure a rapid cleavage of protecting groups, and hence a fast delivery of the biologically active compound. Although the products should be impermeant again after the cleavage and might therefore accumulate inside cells, the ratio of the rates of delivery and metabolism of the deprotected product is crucial for the amount of biologically active compound being generated at a particular time point.

Butyrates masking hydroxyl and amino groups have been used by Posternak *et al.* as early as 1962 to increase the lipophilicity of the intracellular messenger cAMP (*3*). Butyrates were also used to mask hydroxyl groups in inositol polyphosphates. Additionally, these groups could serve as protecting groups during total synthesis, and should prevent phosphate migration during the unmasking procedure (see below).

Masking negative charges is expected to have the largest effect on increasing permeability. The most widely used group for this purpose is certainly the acetoxymethyl (AM) ester group, originally employed to convert carboxylates like penicillin (*4*) or ion indicators (*5,6*) into uncharged bioactivatable derivatives. This strategy has subsequently been applied to a variety of organic phosphates (*7,8*) and phosphonates (*9,10*) including cyclic phosphate nucleotides like cAMP, cGMP and their derivatives (*11-13*). The challenging task of converting inositol polyphosphates into biologically active membrane-permeant derivatives was first successfully comleted in 1994. In combining butyrate masking groups for the hydroxyls and AM-esters for the phosphates, 1,2-di-*O*-butyryl-*myo*-inositol 3,4,5,6-tetrakis-phosphate octakis(acetoxymethyl) ester (Bt$_2$Ins(3,4,5,6)P$_4$/AM) was prepared by total synthesis, and helped to establish the messenger function of Ins(3,4,5,6)P$_4$ as a negative regulator of transepithelial ion fluxes in colon epithelial cells (*14*). In the meantime, a variety of modified derivatives of Ins(3,4,5,6)P$_4$ have been used in a similar way (*15*), and the compounds are currently under investigation in *in-vivo*-mapping experiments as mentioned above (*16,17*).

Membrane-Permeant Derivatives of Ins(1,4,5)P$_3$

The potential of the AM-ester methodology can be best demonstrated on the most prominent inositol phosphate, Ins(1,4,5)P$_3$ (**1**), with its well-defined function in regulating intracellular calcium release (*18*). First attempts to prepare a membrane-permeant derivative of Ins(1,4,5)P$_3$ resulted in the synthesis of racemic 2,3,6-tri-*O*-butyryl-*myo*-inositol 1,4,5-trisphosphate hexakis(acetoxymethyl) ester (**2**) (*19*). However, the initial success (*20*) in elevating intracellular Ca^{2+} levels ([Ca^{2+}]$_i$) proved to be unreliable, probably because the generated Ins(1,4,5)P$_3$, or one of its active metabolites, were metabolized faster than the respective compound could be delivered to the cytosol. The cleavage of the butyrates was suspected to be the rate-limiting step. Therefore, a derivative without butyrates, *myo*-inositol 1,4,5-trisphosphate hexakis(acetoxymethyl) ester, was prepared but proved inactive, this time probably due to its lack in exhibiting sufficient lipophilicity. Tsien and co-workers finally increased the membrane-permeability by using propionoxymethyl and butyroxymethyl esters, respectively, and the resulting derivatives were very active in elevating $[Ca^{2+}]_i$ when applied extracellularly to astrocytoma cells. However, there was evidence that some phosphate migration occurred (*19*).

Metabolically Stable, Membrane-Permeant Ins(1,4,5)P₃ Derivative

All previous attempts were aimed at artificial elevation of the intracellular levels of authentic Ins(1,4,5)P₃. Most valuable should be such a derivative which is able to trigger intracellular calcium release, but is no substrate for the dominant metabolic enzymes responsible for the rapid decay of Ins(1,4,5)P₃. This kind of tool would generate a Ins(1,4,5)P₃ signal free of potential effects of metabolites like Ins(1,3,4,5)P₄ and Ins(1,3,4)P₃. Recently, Potter and co-workers presented the synthesis of Ins(1,2,4,5)P₄ (**3**) and showed that this compound represents a very potent agonist of intracellular calcium release (*21*). In addition, Ins(1,2,4,5)P₄ was neither a substrate for the 5-phosphatase nor the 3-kinase. This made Ins(1,2,4,5)P₄ a promising candidate for conversion into a metabolically stable membrane-permeant AM-ester derivative. To avoid phosphate migration we relied on butyrate functions covering the hydroxyl groups. Therefore, the desired compound was 2,6-di-*O*-butyryl-*myo*-inositol 1,2,4,5-octakis(acetoxymethyl) ester (Bt₂Ins(1,2,4,5)P₄/AM, **4**) (Figure 1).

Synthesis. The synthesis has been described recently (*22*). In brief, it starts from the well-known 1,2:4,5-di-*O*-cyclohexylidene-*myo*-inositol (*rac*-**5**) as is summarized in Figure 2. Exhaustive butyrylation and subsequent cleavage of the ketals gave racemic 3,6-di-*O*-butyryl-*myo*-inositol which was phosphitylated using a classic phosphite approach, followed by oxidation to the fully protected 1,2,4,5-tetrakis(dibenzyl)phosphate (*rac*-**6**). At this stage, the enantiomers **6** and *ent*-**6** were separated by preparative HPLC on a chiral stationary phase (Chiradex, 10 μm, Merck). It should be mentioned here that many inositol poly(dibenzyl)phosphates can be separated on this phase (unpublished results). The absolute configurations were verified by preparing the dicamphanates of **5**, separating the diastereomers, and then following the synthetic pathway to the fully protected tetrakisphosphate **6**. The latter and its enantiomer were deprotected, and the resulting free acids were alkylated with acetoxymethyl bromide in the presence of diisopropylethylamine in dry acetonitrile. Purification of the octakis(acetoxymethyl) esters Bt₂Ins(1,2,4,5)P₄/AM (**4**) and Bt₂Ins(2,3,5,6)P₄/AM (*ent*-**4**) by preparative reverse phase HPLC yielded more than 0.5 g of each enantiomer.

Measurement of intracellular Ca^{2+}-levels in PC12 cells. PC12 cells were loaded with the fluorescent calcium indicator fura-2 and transferred into a cuvette for ratio measurements with excitation at 340 / 380 nm, and emission at 505 nm as described before (*23*). Bt₂Ins(1,2,4,5)P₄/AM in the stock solution containing DMSO and pluronic 127 (1% and 0.04 % final concentration, respectively) was added to the gently stirred cell suspension. The addition of 1.3 mM Bt₂Ins(1,2,4,5)P₄/AM (**4**) led to a smooth rise in $[Ca^{2+}]_i$ within about 30 min to reach a plateau of about 700 nM $[Ca^{2+}]_i$ as is shown in Figure 3. The subsequent addition of the natural agonist bradykinin (2 μM) gave no further effect. Smaller doses (0.9 mM) exhibited half-maximal responses with a sharp transient after a dose of bradykinin. Control experiments with similar doses of the enantiomer Bt₂Ins(2,3,5,6)P₄/AM (*ent*-**4**) had

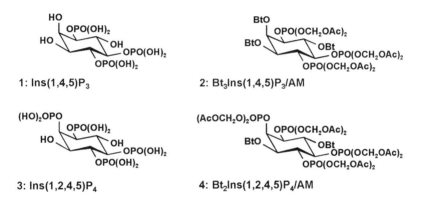

Figure 1. Membrane-permeant derivatives of Ins(1,4,5)P$_3$ and Ins(1,2,4,5)P$_4$.

236

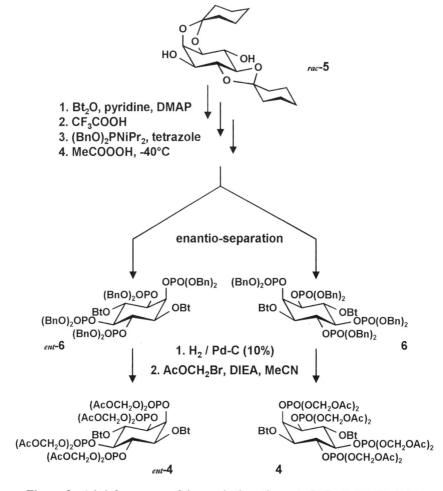

Figure 2. A brief summary of the synthetic pathway to Bt$_2$Ins(1,2,4,5)P$_4$/AM.

no effect on $[Ca^{2+}]_i$, nor was the response to a subsequent dose of bradykinin altered. The results showed that $Bt_2Ins(1,2,4,5)P_4/AM$ was able to enter the cells, and that intracellular endogenous enzymes deprotected the compound to give $Ins(1,2,4,5)P_4$ or a biologically active derivative thereof. The constant rise in $[Ca^{2+}]_i$ suggested that $Ins(1,2,4,5)P_4$ was metabolically stable inside the cells, as was verified by HPLC analysis of cell extracts (24).

Membrane-Permeant Derivative of PtdIns(3,4,5)P_3

Phosphatidyl-*myo*-inositol 3,4,5-trisphosphate (PtdIns(3,4,5)P_3, **7**) has attracted enormous attention since its first discovery in 1988 (25). Nevertheless, its actual second messenger function still remains to be shown directly, because most of the knowledge about its potential effects results from studies with the enzyme which forms PtdIns(3,4,5)P_3: phosphatidylinositol 3-kinase (PI 3-kinase, 26). We therefore set out to prepare a membrane-permeant derivative which would allow to deliver PtdIns(3,4,5)P_3 to the interior of the cell without disrupting the plasma membrane, and circumventing the tyrosine kinase receptors usually involved in activating PI 3-kinase. This analog would generate a „clean" PtdIns(3,4,5)P_3 signal and would offer the advantage to study the down-stream effect of PtdIns(3,4,5)P_3 without the interference of other signaling pathways.

Design. Membrane-permeability was intended to be provided by masking the phosphates with acetoxymethyl esters as was successfully shown for inositol polyphosphates (see above). Ideally, the two hydroxy groups should be converted to butyrates to avoid phosphate migration during enzymatic AM-ester hydrolysis. However, this pathway would require alkylation of 2,6-di-*O*-butyryl-PtdIns(3,4,5)P_3 with acetoxymethyl bromide. We suspected that this precursor would be insoluble in moderately polar solvents (such as MeCN) necessary for the reaction. Alternatively, the AM-esters could be first generated from 2,6-di-*O*-butyryl-Ins(3,4,5)P_3, and the product then coupled with an activated diacylglycerol derivative, e.g. an activated phosphite. Unfortunately, these coupling reactions usually need the help of nucleophilic bases, and we did not trust the bis(acetoxymethyl) triester groups to survive such treatment. We therefore developed an alternative route which combines the lipid coupling and the AM-alkylation in one-pot reaction. The starting point for this pathway was our assumption that during formation of inositol phosphate acetoxymethyl ester, cyclic 5-membered-ring phosphate triesters will be formed in high yields when an unprotected hydroxy group is *cis*-vicinally located to the phosphate (C. Stadler, C. Schultz, unpublished data). We have shown before (27) that these cyclic phosphate are readily opened by alcohols to give predominantly the thermodynamically more stable equatorial phosphate triester (Figure 5). Therefore, we aimed for a precursor which would first allow the intermediate formation of a 1,2-cyclic phosphate acetoxymethyl ester, and could subsequently be ring-opened by diacylglycerol to give the phospholipid linkage. We chose dipalmitoylglycerol for this purpose to finally yield 1',2'-dipalmitoyl-6-*O*-butyryl-phosphatidyl-*myo*-inositol 3,4,5-trisphosphate heptakis-(acetoxymethyl) ester, DiC_{16}-6-Bt-PtdIns(3,4,5)P_3/AM (**8**, Figure 4).

Synthesis. The synthetic pathway (Figure 6) started from the well-described racemic 3-*O*-benzyl-1,2:5,6-di-*O*-cyclohexylidene-*myo*-inositol (**9**). After butyrylation of the 6-hydroxy group, the ketals were removed by acidic hydrolysis. The resulting tetrol **10** was phosphorylated employing *O,O*-dibenzyl-*N,N*-diisopropylphosphoramidite followed by oxidation with peracetic acid at -40°C to give the fully protected 2,3,4,5-tetrakisphosphate (**11**). Hydrogenolysis with Pd on carbon in glacial acetic acid

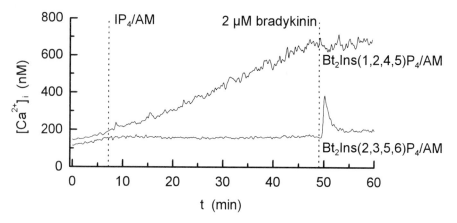

Figure 3. Bt$_2$Ins(1,2,4,5)P$_4$/AM, but not its enantiomer elevates Ca^{2+}-levels in PC12 cells.

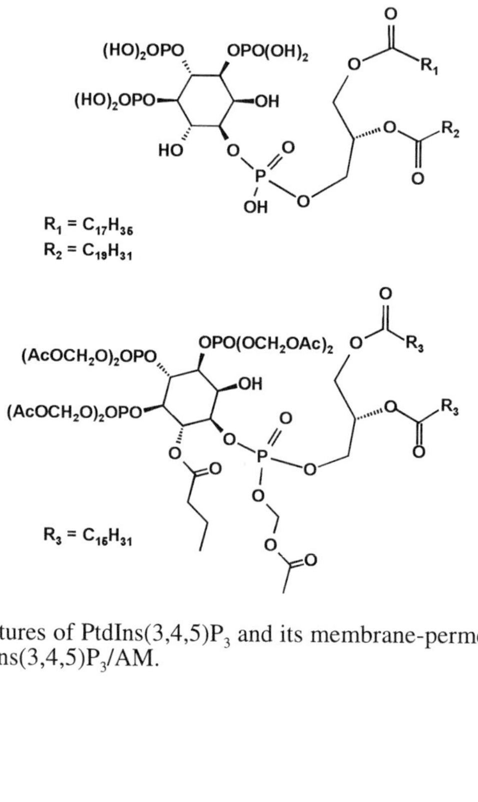

R$_1$ = C$_{17}$H$_{35}$
R$_2$ = C$_{19}$H$_{31}$

R$_3$ = C$_{16}$H$_{31}$

Figure 4. Structures of PtdIns(3,4,5)P$_3$ and its membrane-permeant derivative DiC$_{16}$-6-Bt-PtdIns(3,4,5)P$_3$/AM.

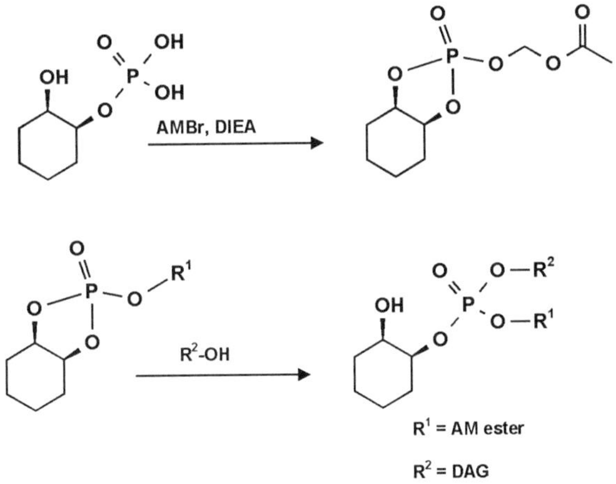

Figure 5. Cyclic intermediates generated by the introduction of AM-esters allow the subsequent lipid / inositol phosphate coupling.

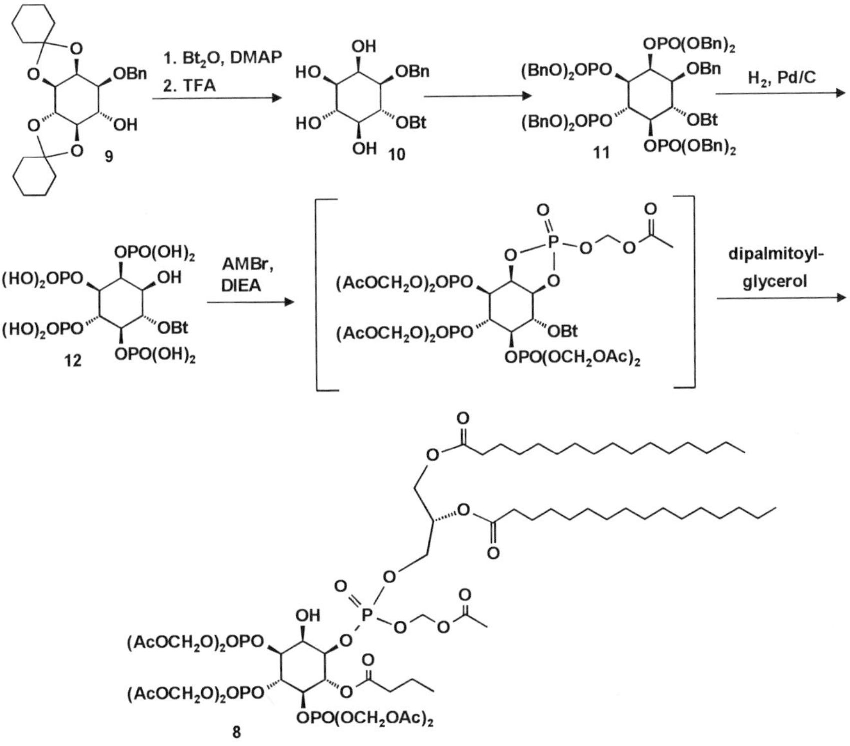

Figure 6. Synthetic pathway to DiC$_{16}$-6-Bt-PtdIns(3,4,5)P$_3$/AM.

afforded the free acid of the tetrakisphosphate (12). This compound has the desired free hydroxy group at the 1-position, ready to receive the cyclizing 2-O-phosphate. The alkylation with acetoxymethyl bromide in dry MeCN in the presence of an excess of diisopropylethylamine (DIEA) was performed in a NMR tube to follow the reaction by ^{31}P NMR. Formation of the expected pair of diastereomers could be monitored by appearance of resonances at 13.7 and 14.3 ppm. After 7 hours no more changes were observed and the mixture was transferred into a dry flask. The volatile compounds including DIEA were removed and the oily residue was dissolved in toluene. The excess of 1,2-dipalmitoyl-sn-glycerol (Sigma) was added and the mixture was stirred for 24 h at 4°C. The subsequent ^{31}P NMR analysis showed no more signal in the 14 ppm region. The mixture was brought to dryness and the residue was extracted with toluene and hexane to give the desired DiC$_{16}$-6-Bt-PtdIns(3,4,5)P$_3$/AM (8) in moderate yield. All compounds were examined by NMR and FAB-MS.

Membrane-permeant PtdIns(3,4,5)P$_3$ derivatives mimic EGF in its effect of inhibiting carbachol induced Cl⁻-secretion of T$_{84}$ cells. It has been shown before that epidermal growth factor (EGF) inhibits calcium-mediated chloride secretion (CaMCS) in T$_{84}$ cells, a human colonic epithelial cell line (28). This effect is most likely mediated by the activation of PtdIns 3-kinase, because the well-known PtdIns 3-kinase inhibitor wortmannin abolished the inhibitory response (29). We therefore chose the well-established Cl⁻-secretion assay to investigate whether our membrane-permeant PtdIns(3,4,5)P$_3$ derivative could mimic the effect of EGF mentioned above. T$_{84}$ cells, grown to confluency on snap-well inserts, were preincubated with DiC$_{16}$-6-Bt-PtdIns(3,4,5)P$_3$/AM (8) for 30 min. To avoid solubility problems, 8 was dissolved in DMSO containing 4% pluronic 127 (BASF, Germany) prior to use. The final DMSO concentration did not exceed 1%. After mounting the snap-wells into Ussing chambers chloride, secretion was monitored as short circuit current (ΔI_{SC}) as described before (30,31). Stimulation by the muscarinic agonist carbachol (100 µM) 10 min after mounting resulted in a transient elevation of Cl-secretion (Figure 7). Cells preincubated with 100 µM DiC$_{16}$-6-Bt-PtdIns(3,4,5)P$_3$/AM showed only an about 50% response to carbachol indicating that CaMCS is inhibited by DiC$_{16}$-PtdIns(3,4,5)P$_3$, or by one of its active metabolites generated through the intracellular hydrolysis of the bioactivatable protecting groups. Hence, we can conclude that the EGF signaling pathway employs PtdIns(3,4,5)P$_3$ (or one of its metabolites) as a second messenger in T$_{84}$ cells. In control experiments neither 1,2-dipalmitoyl-sn-glycerol nor PtdIns(3,4,5)P$_3$ itself exhibited any effect on CaMCS (data not shown). In a more detailed study we showed that compounds with shorter fatty acid chains such as lauroyl or octanoyl were equally potent, and that the effect of EGF, as well as that of DiC$_{16}$-6-Bt-PtdIns(3,4,5)P$_3$/AM, could be reversed by doses of a membrane-permeant derivative of Ins(1,4,5,6)P$_4$, thus restoring Cl⁻-secretion (32). Unfortunately, we cannot exclude the possibility that the signal generated by DiC$_{16}$-6-Bt-PtdIns(3,4,5)P$_3$/AM is mediated by a downstream metabolite of PtdIns(3,4,5)P$_3$. Therefore, membrane-permeant derivatives of other 3-phosphorylated phosphatidylinositides would be highly desirable for future experiments.

Acknowledgment. We are in indebted to the *Deutsche Forschungsgemeinschaft* and the NIH (to A.E.T.) for financial support.

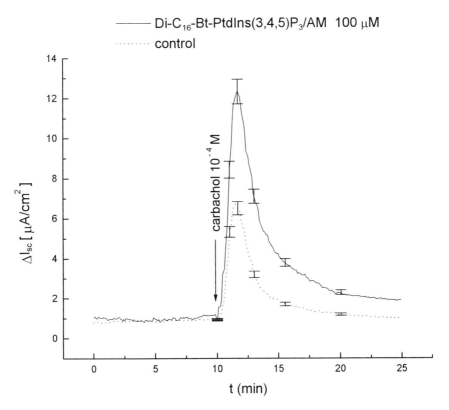

Figure 7. Extracellular doses of DiC$_{16}$-6-Bt-PtdIns(3,4,5)P$_3$/AM inhibit chloride secretion of T$_{84}$ cells, thus mimicking the effect of EGF.

Literature Cited

1. Thastrup, O.; Linnebjerg, H.; Bjerrum, P. J.; Knudson, J. B.; Christensen, S. B. *Biochem. Biophys. Acta* **1987**, *927*, 65-73.
2. Ui, M.; Okada, T.; Hazeki, K.; Hazeki, O. *Trends Biochem. Sci.* **1995**, 303-307.
3. Posternak, T.; Sutherland, E. W.; Denion, W. F. *Biochem. Biophys. Acta* **1962**, *65*, 558-564.
4. Jansen, A. B. A.; Russell, T. J. *J. Chem. Soc.* **1965**, 2127-2132.
5. Tsien, R. Y. *Nature* **1981**, *290*, 527-528.
6. Grynkiewicz, G.; Poenie, M.; Tsien, R. Y. *J. Biol. Chem.* **1985**, *260*, 3440-3450.
7. Sastry, J. K.; Nehete, P. N.; Khan, S.; Nowak, B. J.; Plunkett, W.; Arlinghaus, R. B.; Farquhar, D. *Mol. Pharmacol.* **1992**, *41*, 441-445.
8. Freed, J. J.; Farquhar, D.; Hompton, A. *Biochem. Pharmacol.* **1989**, *38*, 3193-3198.
9. Iyer, R. P.; Phillips, L. R.; Biddle, J. A.; Thakker, D. R.; Egan, W.; Aoki, S.; Mitsuga, H. *Tetrahedron Lett.* **1989**, *30*, 7141-7144.
10. Saperstein, R.; Vicario, P. P.; Strout, H. V.; Brady, E.; Slater, E. E.; Greenlee, W. J.; Ondeyka, D. L.; Patchett, A. A.; Hangauer, D. G. *Biochemistry* **1989**, *28*, 5694-5701.
11. Schultz, C.; Vajanaphanich, M.; Barrett, K. E.; Sammak, P. J.; Harootunian, A. T.; Tsien, R. Y. *J. Biol. Chem.* **1993**, *268*, 6316-6322.
12. Schultz, C.; Vajanaphanich, M.; Genieser, H.-G.; Jastorff, B.; Barrett, K. E.; Tsien, R. Y. *Mol. Pharmacol.* **1994**, *46*, 702-708.
13. Kruppa, J.; Keely, S. J.; Schwede, F.; Schultz, C.; Barrett, K. E.; Jastorff, B. *Bioorg. Med. Chem. Lett.* **1997**, *7*, 945-948.
14. Vajanaphanich, M.; Schultz, C.; Rudolf, M. T.; Wasserman, M.; Enyedi, P.; Craxton, A.; Shears, S. B.; Tsien, R. Y.; Barrett, K. E.; Traynor-Kaplan, A. E. *Nature* **1994**, *371*, 711-714.
15. Roemer, S.; Stadler, C.; Rudolf, M. T; Jastorff, B.; Schultz, C. *J. Chem. Soc. Perkin Trans. I*, 1683-1694.
16. Rudolf, M. T.; Li, W.; Wolfson, N.; Traynor-Kaplan A. E.; Schultz, C. *submitted.*
17. Rudolf, M. T.; Wolfson, N.; Traynor-Kaplan, A. E.; Schultz, C. *submitted.*
18. Berridge, M. J., *Nature* **1993**, *361*, 315-325.
19. Li. W.; Schultz, C.; Llopis, J.; Tsien, R. Y. *Tetrahedron* **1997**, *53*, 12017-12040.
20. Schultz, C.; Tsien, R. Y. *FASEB J.* **1992**, *6(5)*, Abstract A1924.
21. Mills, S. J.; Safrany, S. T.; Wilcox, R. A.; Nahorski, S. R.; Potter, B. V. L. *Bioorg. Med. Chem. Lett.* **1993**, *3*, 1505-1510.
22. Gillandt, H. H.; Berg, I.; Guse, A. H.; Mayr, G. W.; Schultz, C. *submitted.*
23. Benters, J.; Schäfer, T.; Beyersmann, D.; Hechtenberg, S. *Cell Calcium* **1996**, *20*, 441-446.
24. Berg, I.; Guse, A. H.; da Silva, C. P.; Gillandt, H. H.; Schultz, C.; Mayr, G. W. *in preparation.*
25. Traynor-Kaplan, A. E.; Harris, A. L.; Thompson, B. L.; Taylor, P.; Sklar, L. *Nature* **1988**, *334*, 353-356.
26. Toker, A.; Cantley, L. C. *Nature* **1997**, *387*, 673-676.
27. Schultz, C.; Metschies, T.; Gerlach B.; Stadler, C.; Jastorff, B. *Synlett* **1990**, *1*, 163-165.
28. Uribe, J. M.; Gelbmann J. M.; Traynor-Kaplan, A. E.; Barrett, K. E. *Am. J. Physiol.* **1996**, *271*, C914-C922.
29. Uribe, J. M.; Keely, S. J.; Traynor-Kaplan, A. E.; Barrett, K. E. *J. Biol. Chem.* **1996**, *271*, 26588-26595.
30. Dharmsathaphorn, K.; Pandol, S. J. *J. Clin. Invest.*, **1986**, *77*, 348-354.

31. Wasserman, S. I.; Barrett, K. E.; Huott, P. A.; Beuerlein, G.; Kagnoff, M.; Dharmsathaphorn, K. *Am. J . Physiol.* **1988**, *254,* C53-C62.
32. Eckmann, L.; Rudolf, M. T.; Ptasznik, A., Schultz, C.; Jiang, T.; Wolfson, N.; Tsien, R. Y., Fierer, J.; Shears, S. B.; Kagnoff, M. F.; Traynor-Kaplan, A. E. *Proc. Natl. Acad. Sci. USA* **1997**, *in press.*

Chapter 16

Synthesis of Inositol-Containing Glycophospholipids with Natural and Modified Structure

Alexander E. Stepanov and Vitaly I. Shvets

Department of Biotechnology, State Academy of Fine Chemical Technology, Moscow, 117571, Russia

Total syntheses of *myo*-inositol containing glycolipids related to the family of glycosylphosphatidylinositols anchors of membrane integral proteins have been developed. Directed protection-deprotection strategy of hydroxyl functions was used for the preparation of asymmetrically substituted *myo*-inositol intermediates. The enantiomerically pure *myo*-inositol derivatives were obtained by optical resolution of racemates via formation of diastereomers with monosaccharides or with chiral organic acids. The phosphorylated *myo*-inositol and sphingosine intermediates were prepared by employing phosphite chemistry. Formation of glycosidic bonds of target glycophospholipids was performed by using oxazoline and glycosyl fluoride methods of glycosylation. The described approaches resulted in the syntheses of phosphatidylinositol β-glucosaminides, phosphatidylinositol glucoside, ceramide inositol phosphate and the corresponding phosphorothionate, and the first total synthesis of inositol-containing glycophosphosphingolipid.

The structure, properties and biological function of *myo*-inositol-containing glycophospholipids have been a subject of intensive studies in recent years, owing to the recognition of the important role of these natural substances in functioning of a living cell. Thus, it was found that glycosylphosphatidylinositols (GPIs) are involved in fundamental biochemical processes such as "anchoring" of many proteins on the surface of the plasma membrane, modulation of physiological state and immunological status of cells (*1*). GPIs are also recognized as precursors of inositol glycans, which have been considered as second messengers in insulin signaling (*2*). Investigation of the biological function of GPIs is restricted by their minute content in natural sources. Therefore, development of synthetic methods leading to GPIs with natural structure, as well as their analogues, has become a vital part of broad interdisciplinary studies in our Laboratory.

All natural GPIs have a common structural fragment of 6-*O*-(2-amino-2-deoxy-β-D-glucopyranosyl)-D-*myo*-inositol-1-phosphate. This fragment has been reported to possess insulinomimetic activity (*1,2*). In some natural sources (yeast, fungi, bacteria, plants), numerous inositol-containing phospho- and glycophospholipids (phytoglycolipids) were found. These molecules feature long chain amino alcohols (sphingosine, phytosphingosine, dihydrophytosphingosine) as a structural fragment (*3,4*). In these lipids, the ceramide moiety is represented by the *N*-acylated 4-hydroxysphinganine; the inositol part of molecule is usually linked with branched oligosaccharide chains, and ceramide and inositol portions are joined together by means of asymmetric phosphoric acid diester. All phytoglycolipids have the common structural core of *myo*-inositol-phosphoryl-ceramide unit. As compared to intensive synthetic studies of glycerophosphate-based inositol lipids, the synthetic chemistry of phytoglycolipids is much less developed.   Meanwhile, in recent years it was discovered that sphingolipids are inhibitors of protein kinase C, and could be involved in control and regulation of the intracellular signal transduction pathways. This fact constituted the reason for our special interest in developing synthetic methods toward the series of inositol containing glycosphingophospholipids.

   This study included several main synthetic stages: (i) synthesis of asymmetrically substituted *myo*-inositol precursors and sphingosine derivatives in the racemic or chiral forms, with the protection pattern allowing the subsequent introduction of phosphate and carbohydrate moieties at predetermined positions; (ii) formation of phosphodiesters using trivalent phosphorus reagents; (iii) glycosylation of *myo*-inositol intermediates using glycosyl fluoride, thioglycoside and oxazoline methods.

Syntheses of Phosphatidylinositol β-D-Glucosaminides

Acylation of 2,3:5,6-di-*O*-isopropylidene-*myo*-inositol (**1**) [prepared according to a modified method of Gigg (*5*)] with levulinic acid in the presence of dicyclohexylcarbodiimide gave a mixture of mono- and diacyl derivatives **2** and **3** (Scheme 1), respectively, which were separated by column chromatography on silica gel. The monohydroxyl derivative **2** was further used as a suitable intermediate for the introduction of a carbohydrate residue at the C4-position of the cyclitol ring.

   In the natural GPIs the carbohydrate portion linked to *myo*-inositol is represented by the glucosamine with a free amino group. In the previous reports, formation of *myo*-inositol β-glucosaminides was achieved by using the Koenigs-Knorr (*6*), glycosyl fluoride (*7*) and trichloroacetimidate (*8*) methods, as well as *n*-pentenylglycosyl donors (*9*).   These methods gave high stereospecifity and satisfactory yields of aminoglycosyl-*myo*-inositols.   In order to prepare some β-glycosaminide analogues of natural GPIs we have developed new approach involving the oxazoline method of glycosylation. This method was originally used for the synthesis of oligosaccharides (*10*). Thus, 2-methyl-(3,4,6-tri-*O*-acetyl-1,2-dideoxy-α-D-glycopyranosyl)[2,1-d]-2-oxazoline (**4**) (*11*) was used as a glycosyl donor for coupling with the derivative **2**. The glycosylation was carried out using equimolar oxazoline-alcohol ratio in the presence of *p*-toluenesulfonic acid (ca. 1 hour in the refluxing mixture of nitromethane-toluene 1:1), and the resulting β-glucosaminide **5** was isolated by column chromatography on silica gel.   The glycosylation procedure was found to be stereospecific, forming exclusively the 1,2-*trans*-glycosidic product **5**. The structure and anomeric configuration of **5** were confirmed by ^{1}H- and ^{13}C-NMR spectroscopy. Selective removal of levulinoyl protecting group from **5** gave the monohydroxy derivative **6**, suitable for coupling with the phosphatidyl portion. For this purpose we chose the methods of trivalent phosphorus chemistry, which showed major advantages when applied to the synthesis of phospholipids (*12*).   The high reactivity of trivalent phosphorus derivatives makes it possible to form phosphoester linkages under mild reaction conditions with both primary and secondary hydroxyl groups. As a more convenient approach to *myo*-inositol phosphoesters, we have decided to use the H-phosphonate method. This methodology is attractive due to

246

experimental simplicity and easy availability of the reagents involved. Thus, the β-glycosaminide **6** reacted with triethylammonium salt of 1,2-dipalmitoyl-*rac*-glycero-3H-phosphonate (**7**) (*13*), and the resulting crude glycero-H-phosphonate was immediately oxidized into the corresponding protected phosphoinositide **8** (Scheme 1). The subsequent stepwise removal of the isopropylidene and acetyl groups gave the desired 1-*O*-(*rac*-1,2-dipalmitoylglycerophospho)-4-*O*-(2-amino-2-deoxy-β-D-gluco-pyranosyl)-*myo*-inositol (**9**) after purification by column chromatography on silica gel. The structure of glycophospholipid **9** was verified by standard spectroscopic methods, and was also confirmed by partial acidic hydrolysis. Refluxing **9** with 0.5 N HCl in methanol yielded a mixture of D-glucosamine and phosphatidylinositol that were chromatographically identified by comparing with authentic specimens.

General considerations suggest that *myo*-inositol phosphodiesters could be synthesized via two alternative reaction sequences: (i) condensation of diacylglycerol H-phosphonate with *myo*-inositol derivative, as has been demonstrated in Scheme 1; and (ii) coupling of *myo*-inositol H-phosphonate with the diacylglycerol component. In order to study the second way we have synthesized 1-*O*-(2-amino-2-deoxy-β-D-glucopyranosyl)-4)-(*rac*-1,2-dipalmitoylglycerophospho)-*myo*-inositol (**16**), which constitutes both an anomeric and structural isomer of natural GPIs because of the reversed location of the carbohydrate and phosphatidyl moieties at the 1- and 4-positions of the cyclitol ring.

The starting 1-*O*-β-benzoylpropionyl-2,3;5,6-di-*O*-isopropylidene-D-*myo*-inositol (**10**, Scheme 2) was prepared from the diketal **1** and β-benzoylpropionic acid in a similar way to the preparation of **2**. The pentasubstituted derivative **10** was phosphitylated at -10°C with a freshly prepared triimidazolylphosphite followed by aqueous work-up to give *myo*-inositol H-phosphonate **11** after chromatography, as an amorphous triethylammonium salt. The ^{31}P-NMR spectrum of **11** confirmed full regioselectivity of phosphitylation, and the absence of by-products. The subsequent coupling of phosphonate **11** with 1,2-dipalmitoyl-*rac*-glycerol (**12**) in the presence of 2,4,6-triisopropylbenzenesulfonyl chloride or pivaloyl chloride gave the corresponding H-phosphonate diester which was immediately oxidized with iodine in aqueous pyridine to afford the phosphate diester **13**. We found that pivaloyl chloride was the most convenient activating reagent due to very short reaction time and the absence of side processes. The next synthetic step required the selective removal of the β-benzoylpropionyl protecting group from **13**, with retention of the fatty acid residues, followed by introduction of an aminosugar portion at the 1-position of the cyclitol. The selective removal of the β-benzoylpropionyl group from **13** by means of hydrazine hydrate in ethanol gave the protected phosphatidylinositol **14**, having only one free hydroxyl group at the 1-position. By employing the oxazoline derivative **4**, and using condition analogous to those reported for synthesis of **5**, we obtained the product **15** which was converted into the final glycolipid **16** after the removal of the protecting groups. The above described syntheses of **9** and **16** demonstrate that combination of efficient phosphorylation using the trivalent phosphorus reagents and glycosylation using oxazoline methodology is a promising approach toward synthesis of complex *myo*-inositol containing glycophospholipids.

Synthesis of Phosphatidylinositol Glucoside. Phospholipid antigens isolated from pathogenic bacteria *Mycobacterium* were identified as phosphatidylinositol mannosides, and are used as components of diagnostic reagents for tuberculosis and leprosy (*14-16*). Continuing our program aimed at synthesis of novel biologically active *myo*-inositol derivatives, we took advantage of some recently developed methods in the chemistry of carbohydrates and lipids to accomplish the total synthesis of 1-*O*-(1,2-dipalmitoyl-*rac*-glycerophospho)-4-*O*-β-D-glycopyranosyl-*myo*-inositol (**22**) as a structural analog of the bacterial phosphatidylinositol mannoside and the

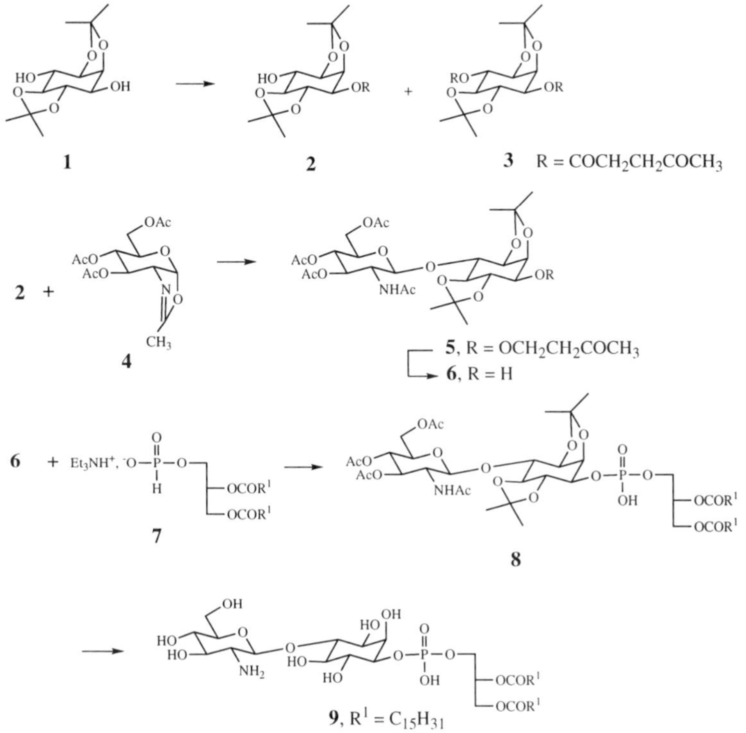

Scheme 1. Synthesis of 1-O-(rac-1,2-dipalmitoylglycerophospho)-4-O-(2-amino-2-deoxy-β-D-glucopyranosyl)-myo-inositol **9**.

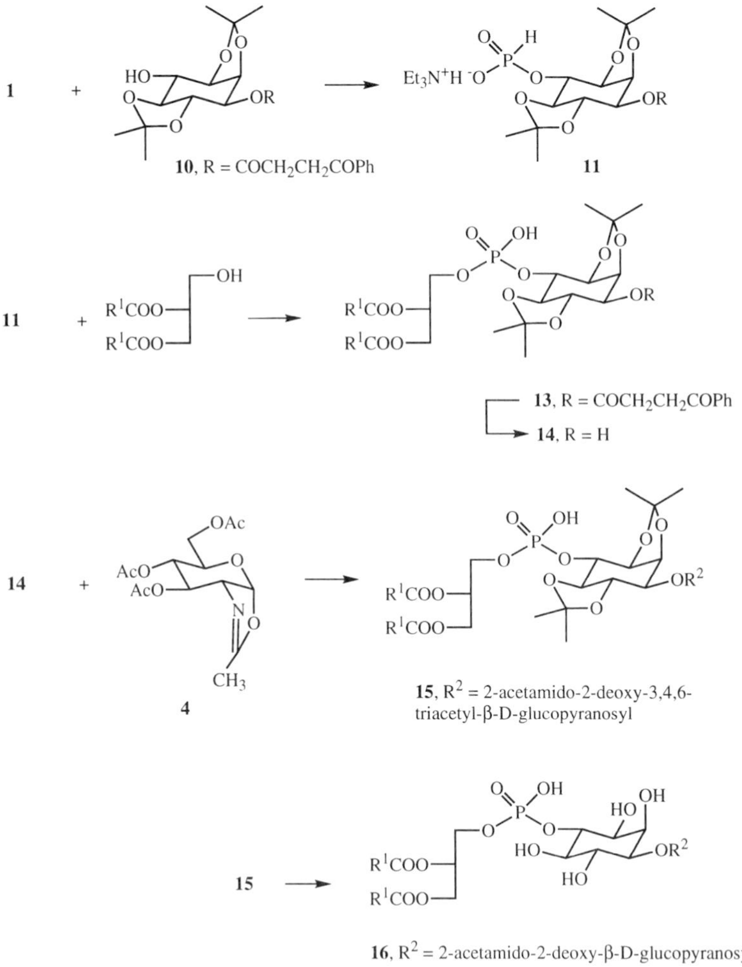

Scheme 2.   Synthesis of 1-*O*-(2-amino-2-deoxy-β-D-glucopyranosyl)-4-*O*-(*rac*-1,2-dipalmitoylglycerophospho)-*myo*-inositol **16**.

deaza-analog of GPIs. As a starting compound we used again the pentasubstituted *myo*-inositol **2** (Scheme 3), because the levulinoyl group is sufficiently stable under the conditions of glycosylation, and its removal can be achieved selectively without affecting the glycosidic bond and acetyl protecting groups of the carbohydrate moiety. The alcohol **2** was activated with a mixture of 1,1,1,3,3,3-hexamethyldisilazane and trimethylchlorosilane to give corresponding trimethylsilyl ester **17**, which was further glycosylated with 2,3,4,6-tetra-*O*-acetyl-β-D-glucopyranosyl fluoride (**18**) (*17*) in benzene with the catalytic amount of boron trifluoride etherate to give the protected glycoside **19**. Selective removal of the levulinoyl group from **19** gave the glycoside **20** having one free hydroxyl group. The 1,2-*trans*-configuration of the anomeric center in both **19** and **20** was established by physico-chemical methods. These results along with the syntheses of above-mentioned β-glycosaminides demonstrate the efficiency of the approach utilizing simultaneously two base-labile protecting groups (acetyl and levulinoyl) for directed synthesis of sophisticated glycolipids. The next step consisted of the formation of a phosphodiester with glycoside **20** and completion of the synthesis of the target glycophospholipid **22**. Thus, the coupling of the glycoside **20** with glycerol H-phosphonate **7** gave the protected phosphatidylinositol glucoside **21**, and the subsequent exhaustive removal of protecting groups yielded the phosphatidylinositol glucoside **22**. The results of this part of study demonstrate high efficiency of the glycosyl fluoride method for glycosylation of stereochemically hindered *myo*-inositol derivatives.

Synthesis of Ceramide Containing *myo*-Inositol Phosphate and Phosphorothionate. The first goal was to synthesize ceramide *myo*-inositol phosphate as a basic ubiquitous unit of more complex natural phytoglycolipids. The most accessible starting compound was *rac*-3-*O*-benzoylceramide (**23**, Scheme 4) (*18*). The reaction of **23** with bis(*N*,*N*-diisopropylamino)-2-cyanoethyl phosphoro-amidite in the presence of diisopropylammonium tetrazolide under argon atmosphere gave the phosphoramidite **24**, which was further coupled with 2,3,4,5,6-penta-*O*-acetyl-*myo*-inositol (*19*) in the presence of 1H-tetrazole to produce the phosphite **25**. The phosphite **25** was immediately oxidized and subjected to exhaustive deprotection to yield the diester **26**. The phosphorothionate analog **27** was synthesized analogously employing sulfurization of **25** with elemental sulfur. For compound **27**, the *Rp*- and *Sp*-diastereomers (chirality at phosphorus) were identified and separated by means of chromatography.

Synthesis of 1-*O*-(2-*N*-stearoyl-D-*erythro*-sphinganine-1-phosphoryl)-2-*O*-(α-D-mannopyranosyl)-D-*myo*-inositol. Synthesis of this compound consisted of three major elements: (i) synthesis of chiral protected *myo*-inositol mannoside, bearing one free hydroxyl group at the 1-position of the *myo*-inositol ring, (ii) phosphitylation of *myo*-inositol mannoside with phosphoramidite derivative of D-*erythro*-3-*O*-benzoylceramide (**40**), and (iii) oxidation to corresponding phosphate, deprotection and isolation of the target phytoglycolipid **44**.

For the preparation of the chiral protected *myo*-inositol mannoside, we have applied the earlier developed approach which include glycosylation of a racemic asymmetrically substituted *myo*-inositol with derivative of D-mannose, followed by isolation and separation of individual mannosides from the mixture of diastereomers (*20,21*). The racemic 1-*O*-crotyl-3,4,5,6-tetra-*O*-benzyl-*myo*-inositol (**28**) was glycosylated with 2-*O*-acetyl-3,4,6-tri-O-(α,β)-D-mannopyranosyl fluoride in the presence of boron trifluoride etherate to afford the diastereomeric mixture of the corresponding α-D-mannosides of **28**. The latter was treated with methanolic sodium methoxide to remove the acetyl protective group from monosaccharide moiety. The

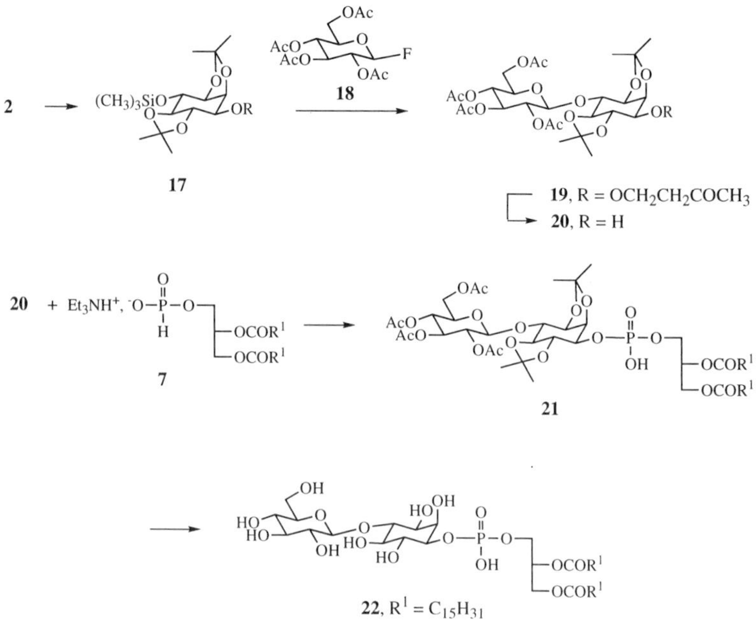

Scheme 3. Synthesis of 1-*O*-(*rac*-1,2-dipalmitoylglycerophospho)-4-*O*-(β-D-glucopyranosyl)-*myo*-inositol **22**.

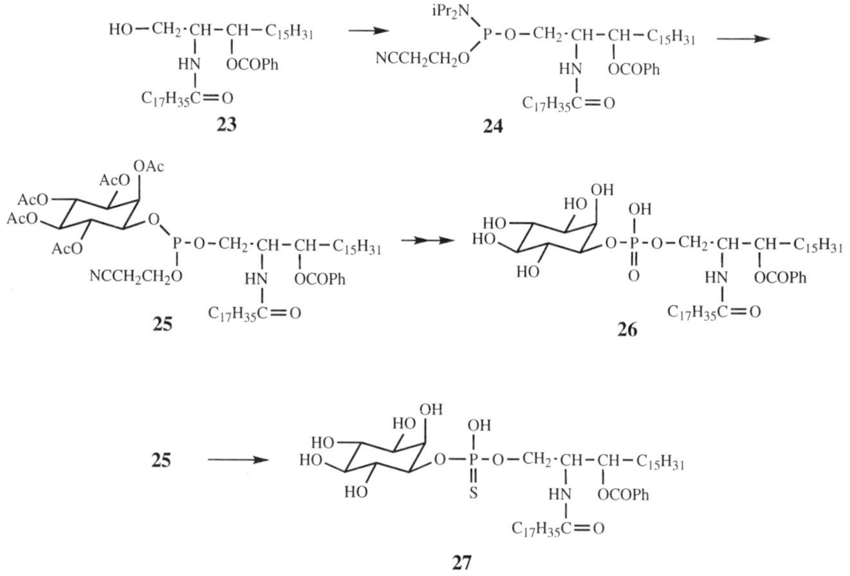

Scheme 4. Synthesis of ceramide-containing *myo*-inositol phosphate **26** and phosphorothioate analog **27**.

subsequent attempts to separate deacetylated diastereomers by column chromatography were unsuccessful due to negligible difference in polarity of the two mannosides. Likewise, we were unsuccessful in separation of the benzylated derivatives obtained after removal of the crotyl group.

Given the above unsuccessful attempt we have applied a different strategy using resolution of the inositol component prior to the mannosylation step. To realize this route, the racemic 3,4,5,6-tetra-O-benzyl-myo-inositol (**29**) was converted into the diastereomeric mixture of menthoxyacetates by the reaction with a (-)-l-menthoxyacetyl chloride, according the formerly described method (*22*). The diastereomer containing the D-inositol fragment was isolated from the mixture by crystallization. The subsequent saponification of this diastereomer with methanolic sodium methoxide afforded the desired 1D-3,4,5,6-tetra-O-benzyl-D-myo-inositol (**30**) with high enantiomeric purity as proven by comparison with the specimen obtained in the independent way (*23*). Further, the diol **30** was converted into 1-O-p-methoxybenzyl ether **31** (Scheme 5) by means of tin-mediated alkylation with 4-methoxybenzyl chloride. The compound **31** is a convenient chiral intermediate for mannosylation at the 2-position of the cyclitol ring, but because of the acid lability of the p-methoxybenzyl group, most of the standard glycosylation methods were not applicable in this case. Instead, mannosylation of **31** was carried out using thioglycoside donor, 2-O-benzoyl-3,4,6-tri-O-benzyl-1-ethylthio-α-D-manno-pyranoside (**32**) (*24*) (Scheme 5). The analysis of the reaction mixture by means of ^{13}C-NMR spectroscopy showed the presence of two anomeric products **33** and **34**. The removal of the benzoyl group at the C2 of the D-mannose fragment from **33** and **34** followed by chromatographic separation of the products allowed their structural verification by ^{13}C- and ^{1}H-NMR as 1,2-*trans*-mannoside **35** (61%) and 1,2-*cis*-isomer **36** (25%), respectively. The removal of the p-methoxybenzyl group from the **33** and **34**, and the subsequent chromatographic separation of the mixture gave individual α-mannoside **37** and β-mannoside **38** (3:1 α/β ratio).

Analogously as described above, the formation of phosphoester bonds in the final steps of the synthesis was carried out using phosphite triester methodology. Thus, treatment of D-*erythro*-3-O-benzoylceramide (**39**) with bis(*N,N*-diisopropylamino)-2-cyanoethoxyphosphine / diisopropylammonium tetrazolide gave the phosphoramidite **40**, which was found to be a mixture of *Rp*- and *Sp*-diastereomers differing in their configuration at phosphorus. The coupling of the diastereomeric mixture **40** with the protected mannoside **37** (Scheme 6) led to the corresponding phosphite triester intermediate **41**, which was oxidized at once to afford the phosphate triester **42**. This compound was selectively deblocked to yield the phosphodiester **43**, and finally subjected to the removal of benzyl and benzoyl protecting groups to give the target phytoglycolipid **44**.

In conclusion, we have described our recent synthetic studies which were aimed at providing reliable and convenient approaches to complex *myo*-inositol containing glycophospho- and glycosphingophospholipids with natural stereochemical configuration, as well as their analogues. The details of the syntheses can be found in our recent original publications (*25-30*).

Acknowledgments: We thank our coworkers who took part in carrying out the experimental work at various stages of the project: Dr. A. S. Bushnev, N. S. Shastina, L. I. Einisman and A. Y. Zamyatina. Stimulating collaboration with Prof. J. H. van Boom and his team at Gorlaeus Laboratories, Leiden, The Netherlands, is gratefully acknowledged. We extend special thanks to Dr. Svetlana V. Stepanova for assistance in the preparation of the manuscript.

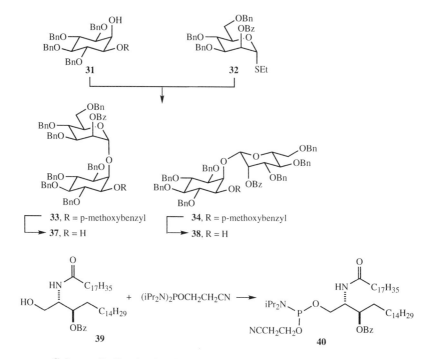

Scheme 5. Synthesis of protected *myo*-inositol-α-D-mannosides.

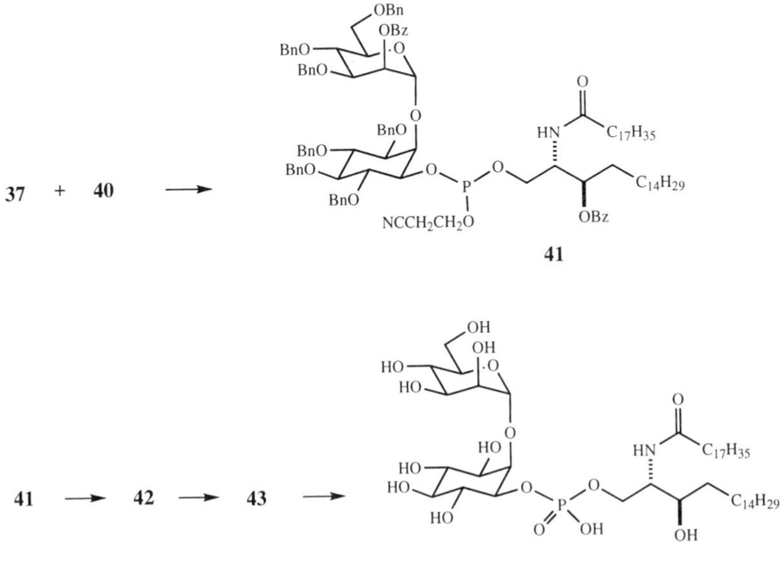

Scheme 6. Synthesis of 1-*O*-(2-*N*-stearoyl-D-*erythro*-sphinganine-1-phosphoryl)-2-*O*-(α-D-mannopyranosyl)-D-*myo*-inositol **44**.

254

Literature Cited

1. Ferguson, M. A. J. *Biochem. Soc. Trans.* **1992**, *20*, 243-256.
2. Low, M. G.; Saltiel, A. R. *Science* **1988**, *239*, 268-275.
3. Laine, R. A. *Chem. Phys. Lipids* **1986**, *42*, 129-135.
4. Puoti A.; Desponds, C. and Conzelmann, A. *J. Cell. Biol.***1991**, *113*, 515.
5. Gigg, J.; Gigg, R.; Payne, S. and Conant, R. *Carbohydr. Res.* **1985**, *142,*132-134.
6. Berlin, W. K.; Zhang, W.-S. and Shen, T.-Y. *Tetrahedron* **1991**, *47,* 1-20.
7. Murakata, C. and Ogawa T. *Carbohydr. Res.* **1992**, *235,* 95-114.
8. Zapata, A. and Martin-Lomas, M. *Carbohydr. Res.* **1992**, *234,* 93-106.
9. Udodong, U. E.; Madsen, R.; Roberts, C. and Fraser-Reid, B. *J. Am. Chem. Soc.* **1993**, *115*, 7886-7887.
10. Zurabyan, S. E.; Antonenko, T. S. and Khorlyn, A. Y. *Carbohydr. Res.* **1970**, *15,* 21-27.
11. Bovin, N. V.; Zurabyan, S. E. and Khorlin, A. Y. *Izv. Akad. Nauk SSSR, Ser. Khim.* **1981**, *12,* 2806-2808.
12. Lindh, I. and Stawinski, J. *J. Org. Chim* **1989**, *54,* 1338-1342.
13. Beaucage, S. L. and Iyer, R. P. *Tetrahedron* **1993**, *49,* 10441-
14. Pangborn, M. and McKinney, Y. A. *J. Lipid Res.* **1966**, *7,* 627-633.
15. Khuller, G. K. and Subrahmanyam, D. *Int. J. Leprosy* **1970**, *38,* 365-367.
16. Stepanov, A. E. and Shvets, V. I. *Usp. Biol. Khim.* **1979**, *20,* 152-168.
17. Voznyi, Y. V.; Kalicheva, I .S. and Galoyan, A. A. *Bioorg. Khim.* **1981**, *7,* 406-409.
18. Shapiro, D. *Chemistry of Sphingolipids;* Hermann, Paris, **1969**.
19. Angyal, S. J. and Tate, M. E. *J. Chem. Soc.* **1965**, *12,* 6949-6955.
20. Stepanov, A. E.; Shvets, V. I. and Evstigneeva, R. P. *Bioorg. Khim.* **1976**, *2,* 1618-1626.
21. Elie, C. J. J.; Dreef, C. E.; Verduyn, R.; van der Marel, G. A. and van Boom, J. H. *Tetrahedron* **1989**, *45,* 3477-3486.
22. Ozaki, S.; Watanabe. Y.; Ogasawara, T.; Kondo, Y.; Shiotani, N.; Nishii, H. and Matsuki, T. *Tetrahedron Lett.* **1986**, *27,* 3157-3160.
23. Shvets, V. I., Klyashchitskii, B. A.; Stepanov, A. E. and Evstigneeva, R. P. *Tetrahedron* **1973**, *29,*331-340.
24. Peters, T. *Liebigs Ann. Chem.* **1990**, 135-141.
25. Stepanov, A. E.; Krasnopolsky, Y. M.; and Shvets, V. I. *Physiologically Active Lipids* (in Russian). Nauka Publishing House, Moscow, **1991**, 1-136.
26. Zamyatina, A. Y.; Stepanov, A. E.; Bushnev, A. S.; Zvonkova, E. N. and Shvets V. I. *Bioorg. Khim.* **1993**, *19,* 347-353.
27. Shastina, N. S.; Einisman, L. I.; Kashiricheva, I. I.; Stepanov, A. E. and Shvets, V. I. *Bioorg. Khim.* **1995**, *21,* 641-650.
28. Zamyatina, A.Y. and Shvets, V. I. *Chem. Phys. Lipids* **1995**, *76,* 225-240.
29. Shastina, N. S.; Einisman, L. I.; Stepanov, A. E. and Shvets, V. I. *Bioorg. Khim.* **1996**, *22,* 446-450.
30. Stepanov, A. E. and Shvets, V. I. *Bioorg. Khim.* **1996**, *22,* 737-744.

Chapter 17

Inositol Phosphates: Inframolecular Physico-Chemical Studies: Correlation with Binding Properties

G. Schlewer[1], P. Guédat, S. Ballereau, L. Schmitt, and B. Spiess[1]

Laboratoire de Pharmacochimie Moléculaire UPR421 du CNRS, Faculté de Pharmacie, 74, route du Rhin, 67401 Illkirch, France

Potentiometric and ^{31}P NMR studies have been performed on Ins(1,4,5)P$_3$ and related synthetic analogues. The potentiometric analyses give a global view of the molecule and indicate competition between metal cations and acidic protons for phosphate binding sites. ^{31}P NMR allow submolecular (inframolecular) analyses. These inframolecular analyses reveal cooperative effects during protonation of neighboring phosphates, as well as participation of vicinal hydroxyls. For Ins(1,4,5)P$_3$, the 6-hydroxyl group seems to control the subtle arrangement of the proton equivalent in the molecule. Our results also suggest that, proton transfer from the Ins(1,4,5)P$_3$ to the active site, upon binding, may be involved in eliciting the biological action.

Structure-Activity Relationships (SAR) have been developed for different intracellular targets of inositol phosphates (IP). These SAR permit description of pharmacophore models defining structural determinants of IP activity, or inhibition by IP, for a given enzyme or a receptor site [1-4].

Classically, the pharmacophore models are built by establishing a direct correlation between pharmacological activity and the chemical structure of a given functional group. This correlation suggests that the modulation of activity is directly related to modification of the interactions of the functional group with its counterpart in the active / binding site. Such pharmacophore models often satisfactorily suggest the structure of new and more potent analogues [5-8].

For polyfunctional molecules such as IPs, modification of a given function can induce intramolecular changes, such as distribution of charges or electrostatic potential, stabilization processes and long distance effects. These intramolecular effects can be of importance and can help explain the observed SAR.

Moreover, since the biological data is obtained under standardized conditions which attempt to mimic the physiological ones as closely as possible, they must be considered as conditional data, largely dependent on reproducing such factors as pH, nature and ionic strength of the buffer, temperature, etc. Thus, the SAR usually offer a "pause on image" view of a mechanism, and do not take into account the dynamic phenomena that may occur at a cellular site when its ionic environment undergoes locally large changes in concentration. Due to their polyanionic nature, it can be expected that IP will strongly interact with all organic or inorganic cations present in the medium, leading to many complex species varying in structure. These species are in fast continuous interconversion in solution, but only one or a few of them act in a

---

[1]Corresponding authors.

specific biochemical process. Therefore, changes in concentration of the interfering cations will modify the concentration of the active IP species, and may lead to major changes in the biological response. This has been illustrated in many studies where the dependence of binding of IP to a receptor, or calcium release activity of IP, versus pH, concentration of alkali-earth cations or polyamines was investigated [9-13]. It has also been reported that metal complexation to IP can be a source of experimental artifacts, especially at non-physiological cation concentrations [14,15].

The additional interactions of IP with cations may be to induce coordination of a specific geometry, or stabilization of a particular conformation that fits a binding site. For instance, in the case of inositol monophosphatase, it has been clearly shown that Ins(1)P-Mg^{2+}-enzyme complex must be considered to explain the mechanism of hydrolysis [2,3,16-18].

As a complementary approach to SAR, this chapter reports part of a comprehensive study aimed at determining the nature, stability and biological significance of complexes formed between IP and various cations. Here, we mainly focus on interactions of the proton with *myo*-inositol 1,4,5-trisphosphate (1), the well known second messenger, and some of its analogues, in order to probe environmental (presence of various cations), electronic and structural factors involved in molecular recognition of 1 by its endoplasmic reticulum receptor.

Synthesis of 1 was the subject of numerous publications, either as optically active compound or as racemate [7,19-29]. In general, our studies do not require optically active compounds, and therefore, our syntheses have been conducted to give racemic products [30-32].

Macroscopic Protonation Constants

Macroscopic acid-base properties are customarily expressed as overall protonation or dissociation constants, which characterise a molecule as a whole. These constants, easily obtained from the interpretation of potentiometric titrations, allow the calculation of the distribution of protonated species versus pH. Figures 1a and 1b show such curves obtained for IP 1 in 0.1 M Et$_4$NClO$_4$ at 25°C and 0.2 M KCl at 37°C, respectively. The former medium enables the study of the intrinsic acid-base properties of the ligand, whereas the latter describes more precisely its ionisation state under physiological conditions. From these curves it can be seen that at a given pH the protonation state largely depends on the medium. Specifically, in the presence of potassium cations, the curves are shifted to more acidic pH, which is a sign of the competition between acidic protons and potassium cations of the medium [33]. At the physiological pH, in the absence of interfering cations (Figure 1a), mainly mono- and diprotonated species are present, while, in a biomimetic medium (Figure 1b), the monoprotonated trisphosphate coexists with the fully deprotonated one. It is interesting to note, that many previous works showed a drastic increase in the binding profile and calcium mobilisation by trisphosphate 1 upon increasing pH of the assay [9,10]. This could be related to the increase in the concentration of the fully deprotonated trisphosphate when pH rises. Such alterations can be of biological and functional significance since intracellular pH varies in subcellular compartments following, for instance, phospholipase C activation or action of trophic factors.

Although very often used as electronic descriptors in QSAR studies, the protonation constants of polyfunctional molecules do not provide information on the ionization state of each individual group. Therefore, for the IP, an inframolecular study, i.e. a study at a level of individual phosphate, was undertaken employing combination of potentiometric and ^{31}P NMR titrations. Such an investigation seemed especially worthwhile for trisphosphate 1, in order to localize the position of the first proton equivalent in the ligand.

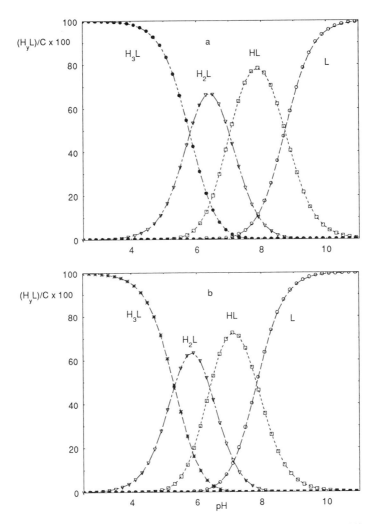

Figure 1. Distribution of *myo*-Inositol 1,4,5-trisphosphate species at different pH. (a) 25°C, 0.1 M tetraethylammonium perchlorate; (b) 37°C, 0.2 M KCl. Reproduced with permission from reference [*33*].

Inframolecular NMR Analysis

To gain insight into the site-by-site protonation process of a trifunctional molecule such as **1**, the microprotonation equilibria shown in Scheme 1 have to be discerned. Such a distinction becomes possible by performing ^{31}P NMR titrations since the observed chemical shift for any resonance δ^{obs}_i is the weighted average of shifts for all possible protonated and deprotonated forms. Accordingly $f_{i,p}$, i.e. the fraction of protonation of the phosphate the position i can easily be calculated from equation 1.

$$ f_{i,p} = \frac{\delta^{obs}_i - \delta_{i,d}}{\delta_{i,p} - \delta_{i,d}} \tag{1} $$

where $\delta_{i,p}$ and $\delta_{i,d}$ are the chemical shifts of the protonated and deprotonated phosphate, respectively. It should be noted, that only the first protonation step for each phosphate is considered, since the second protonation occurs below pH 2.5. Scheme 1 has been entirely resolved for Ins(1,2,6)P$_3$, a regioisomer of **1** [34]. For the latter, an unexpected shielding of phosphorus at the 1-position 1 (1-P) upon protonation at higher pH prevented such a resolution. Below, we discuss several examples of ^{31}P NMR titration curves of less phosphorylated IP, or analogues of **1**.

As a first simple example, Figure 2 shows the ^{31}P-NMR titration curve obtained for *myo*-inositol 1-monophosphate (**2**) over the pH range 2 to 12. The ^{31}P NMR-derived curve is monophasic, similarly to the potentiometric one, and resembles those of various other monophosphate esters. Such a curve can be explained a model in which the addition of one proton equivalent affects only a single phosphate group. In the case of a vicinal bisphosphate such as the *myo*-inositol 4,5-bisphosphate **3**, the curves for each phosphate (Figure 3) appear clearly biphasic, indicating that the first proton is almost equally shared by 4- and 5-phosphate groups [31].

The titration curves become more complicated for the trisphosphate **1** (Figure 4). The curves are different from those obtained by superposition of the curves for compounds **2** and **3**, showing an important cooperative effect among the three phosphates upon protonation. The initial downfield shift observed for 1-P on going from pH 11 to pH 8 cannot be explained by phosphate deprotonation. Therefore, the observed variations in chemical shifts in this pH region most likely result from effects other than the electronic effects accompanying protonation of the phosphate. It is known that conformational changes of the inositol ring, or variation of the O-P-O bond angle, would affect the chemical shifts. For **1**, it was demonstrated that no major change in the phosphate valence angle and the six-membered ring conformation occur [35]. On the other hand, the $^3J_{HCOP}$ coupling constant for 1-P varies from less than 2.0 Hz to 9 Hz on changing pH from 11.7 to 2.4, which indicates large changes in the P-O-C-H torsion angle. Thus, the unexpected low initial chemical shifts may be attributed to the formation of a strong hydrogen bond between 1-P and 2-OH in a *cis* configuration. At high pH, in the totally deprotonated form of **1**, 4-phosphate and 5-phosphate repel each other, so that 1-P is pushed towards 2-OH. Subsequently, as the first added proton is bound between the two vicinal phosphates, the hydrogen bond becomes weaker, 1-P moves more freely and can interact with 6-hydroxyl group. This shows that pH variations may modulate the biological activity of **1** by altering the population of conformers differing in their phosphate orientations. In addition, these conformer populations will be affected by the cationic content of the medium, as a result of the change in the apparent basicity of phosphates. Indeed, under the biomimetic conditions and at the physiological pH, the phosphates at positions 1-, 4- and 5- are protonated at about 5%, 24% and 51%, respectively. In the absence of interfering cations, the degree of protonation of the same groups reaches 48%, 34%

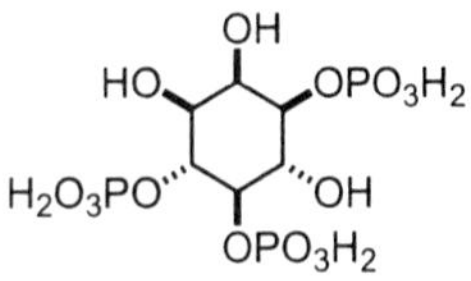

1

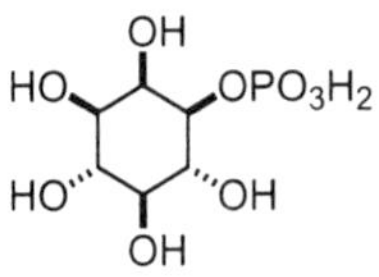

2

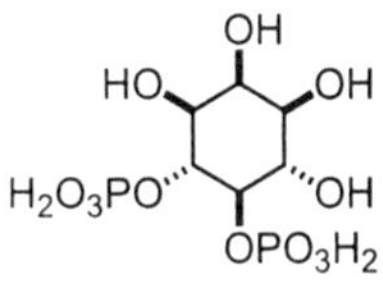

3

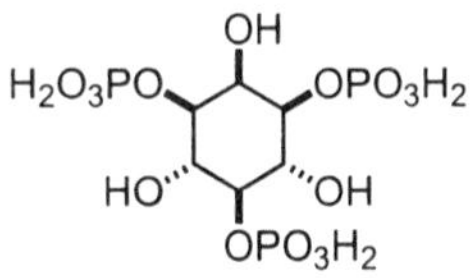

4

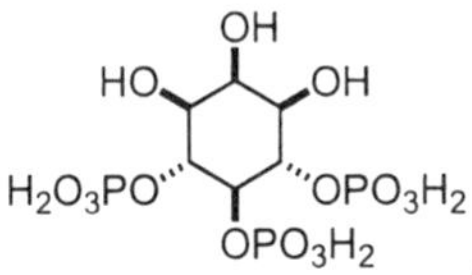

5

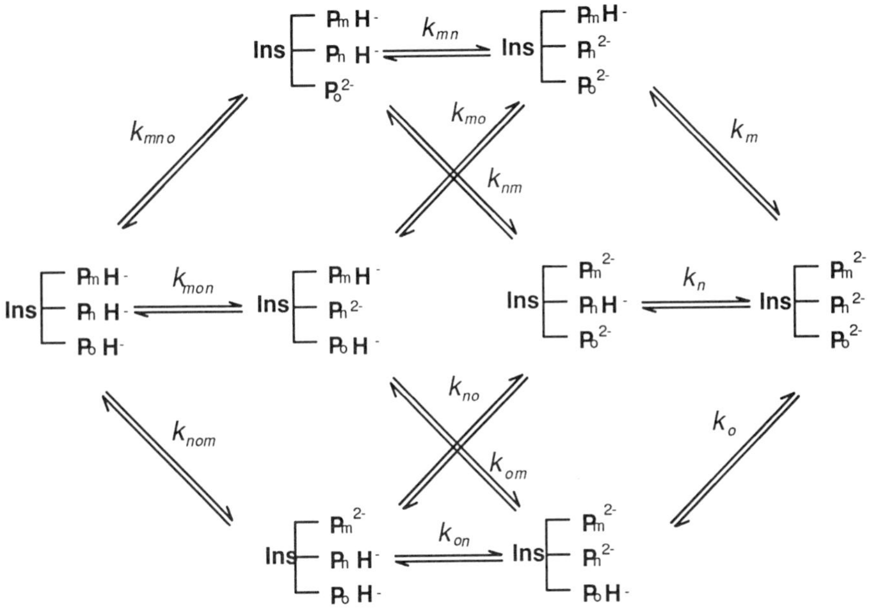

Scheme 1. Microscopic protonation scheme for inositol trisphophate. Positions of the phosphates on the inositol ring are designated as m,n and o. Reproduced with permission from reference [*34*].

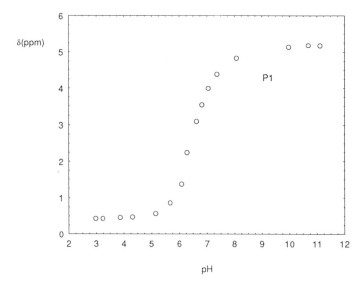

Figure 2. pH dependence of ^{31}P NMR chemical shifts for *myo*-inositol 1-phosphate in 0.2 M KCl at 37°C. Reproduced with permission from reference [*31*].

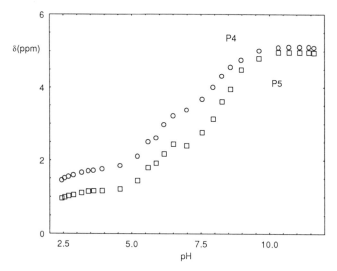

Figure 3. pH dependence of ^{31}P NMR chemical shifts for *myo*-inositol 4,5-bisphosphate in 0.2M KCl at 37°C. Reproduced with permission from reference [*31*].

and 70%, respectively. Moreover, changes in the cationic environment influence each phosphate to a different extent, and may alter the electrostatic potential distribution.

It is interesting to note, that under biomimetic conditions, the variations of the ionization state of 5-P follow the variations of the binding properties, which indicates a crucial role of 5-P in binding of 1 to its receptor [31].

Observation of the cooperative effect in protonation of IP 1 prompted us to examine the behavior of other trisphosphates, such as Ins(1,3,5)P$_3$ 4 and Ins(4,5,6)P$_3$ 5 [36]. The corresponding titration curves for compounds 4 and 5 are shown in Figures 5 and 6. Only two resonances are observed in both cases due to their *meso* structure. Compound 4 features three alternating phosphates, all in an equatorial orientation. For this compound, both curves are monophasic and very similar, showing an independent titration of the phosphates. For compound 5, the three phosphates are vicinal and their protonation process appears very peculiar. In a 0.2 M KCl medium at 37°C, 5-P remains partly protonated even at pH 12, indicating high basicity of this phosphate placed in an area with a high density of the negative charge. Upon titration, 5-P becomes fully protonated at pH 9.5, while 4-P and 6-P remain unprotonated. Between pH 9.5 to 7.5 no protonation occurs for the latter moieties. 4-P and 6-P begin to protonate at pH 7.5, surprisingly partly at the expense of 5-P. Indeed, between pH 7.5 and 6.5, 5-P looses about 20% of its protons in favor of 4-P and 6-P, which will be ultimately recovered only at pH 4. It should be noted, that the significance of the shielding of 5-P in 5 is different from that of 1-P in 1. In the former case it originates from a proton exchange between the phosphates, whereas in the second case it seems to result from the formation of a strong hydrogen bond.

Since the phosphate configuration in 1 is neither totally alternating nor totally vicinal, the cooperative effects in this compound are likely to involve the hydroxyl groups of the inositol ring. The analysis of the behavior of 2,3,6-trideoxy-analog 6 supports this notion, as discussed below. Synthesis of this compound is shown in Figure 7. In brief, Birch reduction of anisole 7 yielded the enol ether 8. Treatment of 8 with phenylene-1,2-dimethanol transformed the enol ether into the acetal 9. The double bond of 9 was oxidized with *m*CPBA to yield the epoxide 10, which was opened to give the monoacetate 11, and further protected as diacetate 12. The acetal group was removed by hydrogenolysis to regenerate the ketone 13. Stereoselective reduction of the carbonyl followed by hydrolysis of the diacetate yielded the trihydroxy derivative 14. Phosphorylation by the phosphite method [37] afforded the protected trisphosphate 15. Final deprotection by hydrogenolysis provided the expected 2,3,6-trideoxy-*myo*-inositol 1,4,5-trisphosphate [1(RS),2(RS),4(RS)-cyclohexanetriol trisphosphate, 6] [38].

As compared with 1, this derivative was 2 orders of magnitude less potent agonist of the ER receptor. Such a dramatic decrease in activity would classically be interpreted as resulting from the loss of interactions with hydroxyl, however, the inframolecular approach using ^{31}P-NMR titrations performed on 6 can support a different explanation of this pharmacological result. The titration curves of 6 (Figure 8) exhibit very different shapes as compared to these of the parent compound 1. At pH 7.5 the phosphates equivalent to 1-P, 4-P and 5-P of 1 are protonated at 51%, 53% and 58%, respectively. These percentages are significantly different from those of 1 in the same medium, and presumably result from different distribution of electrostatic potential in both molecules. In addition, titration curves for the deoxy derivative 6 can be closely reproduced by a superposition of individual curves of an isolated 1-monophosphate and vicinal bisphosphate. Therefore, the loss of the cooperative effects between the two phosphate groups may be ascribed to the absence of the hydroxyls of the inositol ring.

In order to see if all the three hydroxyls are required to induce the cooperative effect, and to study how the hydroxyl groups generate such an effect, the behavior of

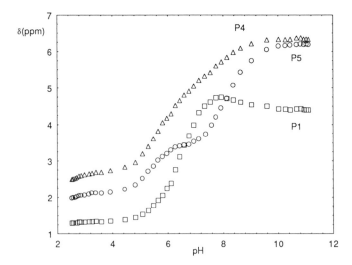

Figure 4. pH dependence of ^{31}P NMR chemical shifts for *myo*-inositol 1,4,5-tris(phosphate) in 0.2 M KCl at 37°C. Reproduced with permission from reference [*31*].

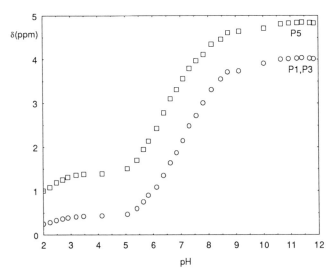

Figure 5. pH dependence of ^{31}P NMR chemical shifts for *myo*-inositol 1,3,5-tris(phosphate) in 0.1 M tetraethylammonium perchlorate at 37°C. Reproduced with permission from reference [*36*].

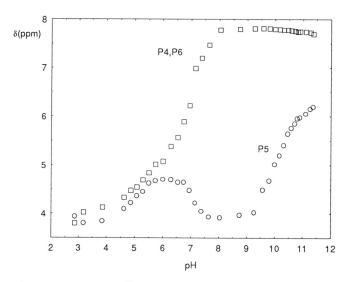

Figure 6. pH dependence of ^{31}P NMR chemical shifts for *myo*-inositol 4,5,6-tris(phosphate) in 0.2 M KCl at 37°C. Reproduced with permission from reference [*36*].

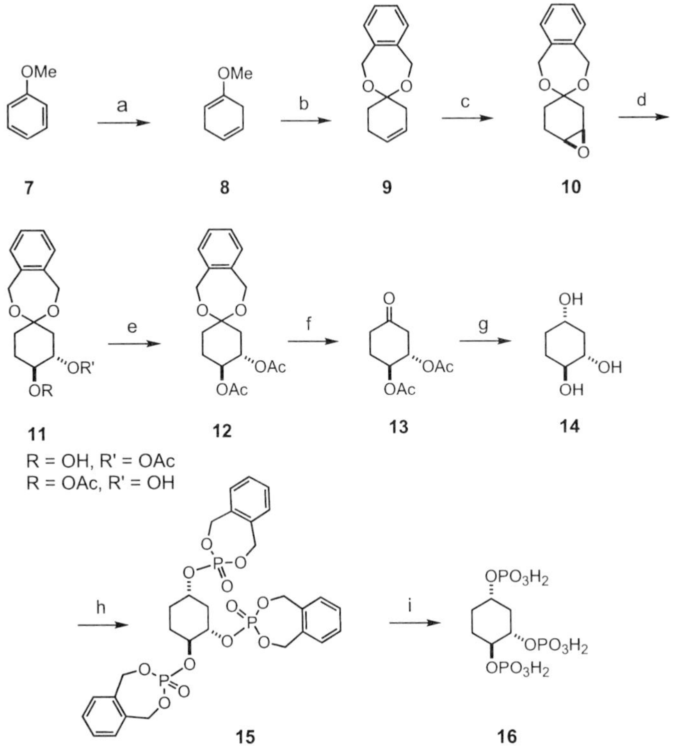

Figure 7. Synthesis of 2,3,6-trideoxy-*myo*-inositol trisphosphate: (a) Na/liq. NH$_3$; (b) *O*-xylenediol,Amberlyst H$^+$; (c) *m*CPBA, CH$_2$Cl$_2$, 0°C; (d) AcONa, AcOH, 70°C; (e) AcCl, pyridine, CH$_2$Cl$_2$, room temperature; (f) H$_2$, Pd/C, EtOH; (g) NaBH$_4$,EtOH, 0°C; (h) *1H*-tetrazole, *N,N*-diethyl-*O*-xylylenephosphoramidite,THF; *m*CPBA, CH$_2$Cl$_2$,-40°C; (i) H$_2$,Pd/C, EtOH, room temperature. Reproduced with permission from reference [*38*].

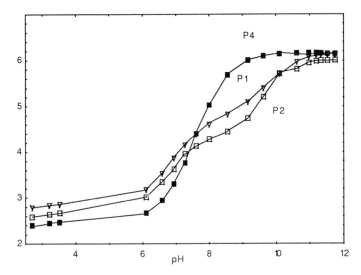

Figure 8. pH dependence of ^{31}P NMR chemical shifts for 2,3,6-trideoxy-*myo*-inositol 1,4,5-trisphosphate in 0.2 M KCl at 37°C. Reproduced with permission from reference [*31*].

6-deoxy-6-fluoro-*myo*-inositol 1,4,5-trisphosphate (**16**) was examined. The replacement of the hydroxyl by a fluorine atom is a classical isosteric change which also eliminates the hydrogen bond at the modified position.

The synthesis of this compound was previously reported by Ley, who used enzymatic reactions to obtain the enantiomerically pure derivative [*27,39*]. As the optical activity was not necessary for our studies, we started from 1,2-*O*-cyclohexylidene-*myo*-inositol **17** (Figure 9) prepared according to Massy [*40*]. The tetraol **17** was treated with benzyl bromide in the presence of dibutyltin oxide[*41*], to give a mixture of two tri-*O*-benzyl derivatives **18** and **19**, which were separated by chromatography. Fluorination of the alcohol **18** with DAST gave the fluoro-derivative **20**. This is an unexpected result , since usually fluorination with DAST proceeds with inversion of configuration of the reacting carbon [*42,43*]. In this particular case retention was observed, probably due to involvement of the neighboring benzyl group [*44,45*]. The benzyl protective groups were removed by hydrogenolysis, and the triol product was phosphorylated by the phosphite method giving the protected trisphosphate **21**. The trisphosphate was subjected to hydrogenolysis on Pd/C to give the expected 6-deoxy-6-fluoro-*myo*-inositol 1,4,5-trisphosphate **16** [*46,47*].

This compound has been found some 80 times less potent than the compound **1** [*47*]. Such a significant loss of activity can be interpreted as resulting from the inability of the 6-position in analog **16** to form a hydrogen bond with the receptor. Here also, the ^{31}P-NMR titration curves (Figure 10) provide an additional information at an inframolecular level. As can be seen, the slight modification of the 6-position deeply affects the general shape of the curves which, similarly as for **6**, appear to be the result of the superimposition of two independent monophosphorylated (1-P) and vicinal bis-phosphorylated (4-P and 5-P) systems. Thus, the absence of the cooperative effect seems to be due to the absence of an intramolecular hydrogen bond with the position 6. Also, the negative charge distribution in compound **16** at the physiological pH is different from that from **1**. Thus, in a 0.2 M KCl medium, 1-P, 4-P and 5-P are protonated at 18%, 47% and 65%, respectively. It is noteworthy, that 5-P in **16**, the ionization of which seems crucial for binding to the receptor, is ca. 15% less ionized than in **1**.

In summary, physico-chemical studies reported here can account for significant differences in biological activities of inositol phosphates.

Supramolecular Interactions

Polyamines such as spermine play a physiological role in the cell, and are often found in the millimolar concentrations. Since structural analysis of the Ins(1,4,5)P$_3$ receptor sites revealed areas rich in basic amino-acids [*48,49*], polyamines can also be considered as a rough approximation of the receptor site. Therefore, we have studied the spermine-Ins(1,4,5)P$_3$ system [*50*]. The comparison of the titration curves obtained for spermine/Ins(1,4,5)P$_3$ complex (concentration ratio of 1.42) and trisphosphate **1** alone is shown in Figure 11. Surprisingly, in the presence of spermine all the curves become monophasic, with the main alteration being experienced by 5-P, suggesting that 5-P is a primary site involved in binding to the sperminium cation. The shift towards the lower pH indicates the decreased basicity of the phosphates. The cooperative effect between the phosphates present in **1** has disappeared in the complex. At physiological pH there is about one proton bound on **1**, mainly located at 5-P. Upon complex formation with spermine, this proton moves from the Ins(1,4,5)P$_3$ moiety to the spermine, causing strengthening of the complex. It can also be expected that the removal of the proton from the Ins(1,4,5)P$_3$ by spermine should favor the conformer having 1-P close to 2-hydroxyl. The same kind of mechanism may also take place within the binding site of the ER IP$_3$ receptor.

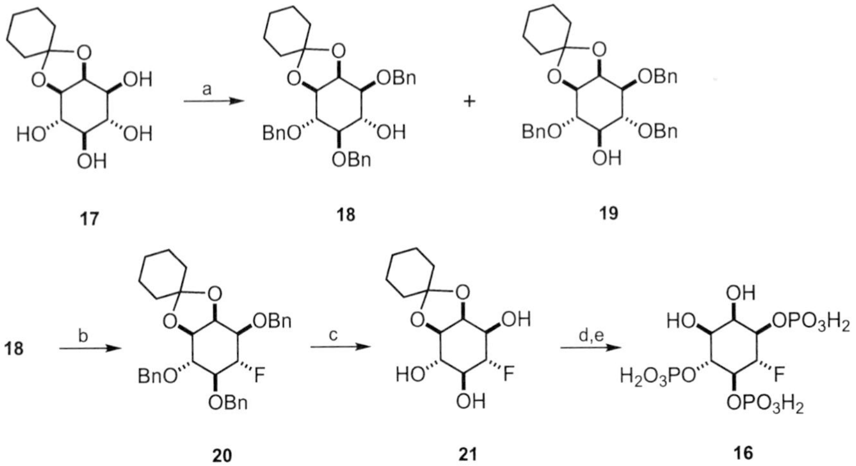

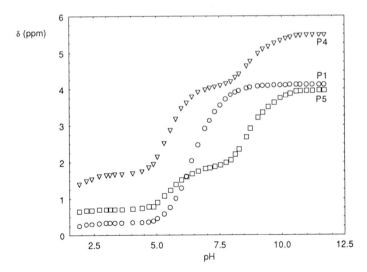

Figure 9. Synthesis of 6-deoxy-6-fluoro-*myo*-ionsitol 1,4,5-trisphosphate: (a) BnBr, *n*Bu₂SnO; (b) DAST; (c) H₂,Pd/C; (d) *1H*-tetrazole, *N,N*-diethyl-*O*-xylylenephosphoramidite; *m*CPBA; H₂Pd/C, MeOH, H₂O. Reproduced with permission from reference [47].

Figure 10. pH dependence of ^{31}P NMR chemical shifts for 6-deoxy-6-fluoro-*myo*-inositol 1,4,5-trisphosphate in 0.2 M KCl at 37°C. Reproduced with permission from reference [47].

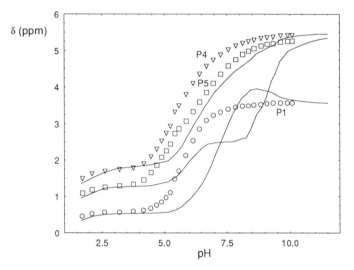

Figure 11. pH Dependence of ^{31}P NMR chemical shifts for *myo*-inositol 1,4,5-trisphosphate alone (solid line) and in the presence of spermine (discrete points). Reproduced with permission from reference [*50*].

Conclusion

The ^{31}P NMR titrations performed on Ins(1,4,5)P$_3$ and several of its analogues revealed the effect of the hydroxyl groups on the basicity of the phosphate groups. For Ins(1,4,5)P$_3$, the 6-hydroxyl group seems to control the subtle arrangement of the first proton equivalent in the molecule, which in turn affects the rotamer distribution of the phosphate groups. Intracellular pH changes, along with modification of the ionic environment in the vicinity of the receptor may modulate the relative concentrations of the active species. Moreover, the intermolecular interaction between, Ins(1,4,5)P$_3$ and spermine receptor model, suggest that upon binding to its receptor, the transfer of a proton from the Ins(1,4,5)P$_3$ to the binding site of this receptor may be involved in triggering the biological response.

Literature Cited

1. Jiang, C.; Moyer, J. D.; Baker, D. C. *J. Carbohydr. Chem.* **1987,** *6,* 319.
2. Bone, R.; Frank, L.; Springer, J. P.; Atack, J. R. *Biochemistry* **1994,** *33,* 9468.
3. Cole, A. G.; Gani, D. *J. Chem. Soc. Perkin Trans 1* **1995,** 2685.
4. Kozikowski, A. P.; Ognyanov, V. I.; Fauq, A. H.; Nahorski, S. R.; Wilcox, R. A. *J. Am. Chem. Soc.* **1993,** *115,* 4429.
5. Ter Laak, A. M.; Venhorst, J.; Donné-Op der Kelder, G. M.; Timmerman, H. *J. Med. Chem.* **1995,** *38,* 3351.
6. Zhang, W.; Koehler, K. F.; Harris, B.; Skolniak, P.; Cook, J. M. *J. Med. Chem.* **1994,** *37,* 745.
7. Reitz, A. B. *Inositol phosphates and derivatives. Synthesis, biochemistry and therapeutic potential.* A. B. Reitz; Washington, 1991.
8. N'Goka, V.; Schlewer, G.; Linget, J. M.; Chambon, J. P.; Wermuth, C. G. *J. Med. Chem.* **1991,** *34,* 2547.
9. Worley, P. F.; Baraban, J. M.; Supattapone, S.; Wilson, V. S.; Snyder, S. H. *J. Biol. Chem.* **1987,** *262,* 12132.
10. Parent, A.; Quirion, R. *Eur. J. Neurosci.* **1994,** *6,* 67.
11. White, A. M.; Varney, M. A.; Watson, S. P.; Rigby, S.; Changsheng, L.; Ward, J. G.; Reese, C. B.; Graham, H. C.; Williams, R. J. P. *Biochem. J.* **1991,** *278,* 759.
12. Sayers, L. G.; Michelangeli, F. *Biochem. Soc. Trans* **1994,** *22,* S152.
13. Guillemette, G.; Segui, J. A. *Mol. Endocrin.* **1988,** *2,* 1249.
14. Luttrell, B. M. *Cell Signal.* **1994,** *6,* 355.
15. Luttrell, B.M. *J. Biol. Chem.* **1993,** *268,* 1521.
16. Cole, A. G.; Gani, D. *J. Chem. Soc. Chem. Commun.* **1994,** 1139.
17. Wilkie, J.; Cole, A. G.; Gani, D. *J. Chem. Soc. Perkin Trans 1* **1995,** 2709.
18. Piettre, S. R.; Ganzhorn, A.; Hoflack, J.; Islam, K.; Hornsperger, J. *J. Am. Chem. Soc.* **1997,** *119,* 3201.
19. Dreef, C. E.; Van der Marel, G. A.; Van Boom, J. H. *Recl. Trav. Chim. Pays-Bas* **1987,** *106,* 512.
20. Vacca, J. P.; deSolms, S. J.; Huff, J. R. *J. Am. Chem. Soc.* **1987,** *109,* 3478.
21. Cooke, A. M.; Potter, B. V. L.; Gigg, R. *Tetrahedron Lett.* **1987,** *28,* 4179.
22. Reese, C. B.; Ward, J. D. *Tetrahedron Lett.* **1987,** *28,* 2309.
23. Stepanov, A. E.; Runova, O. B.; Schlewer, G.; Spiess, B.; Shvets, V. I. *Tetrahedron Lett.* **1989,** *30,* 5125.
24. Dreef, C. E.; Tuinman, R. J.; Elie, C. J. J.; Van der Marel, G. A.; Van Boom, J. H. *Recl. Trav. Chim. Pays-Bas* **1988,** *107,* 395.
25. Falck, J. R.; Yadagiri, P. *J. Org. Chem.* **1989,** *54,* 5851.

26. Watanabe, Y.; Fujimoto, T.; Shinohara, T.; Ozaki, S. *J. Chem. Soc. Chem. Commun.* **1991**, 428.
27. Ley, S. V.; Parra, M.; Redgrave, J. L.; Sternfeld, F. *Tetrahedron* **1990**, *46*, 4995.
28. Billington, D. C. *The inositol phosphates. Chemical synthesis and biological significance.* VCH Publishers; VCH; Weinhein, New York, Basel, Cambridge, 1993.
29. Ozaki, S.; Watanabe, Y.; Ogasawara, T.; Kondo, Y.; Shiotani, N.; Nishii, H.; Matsuki, T. *Tetrahedron Lett.* **1986**, *27*, 3157.
30. Shvets V. I.; Stepanov A. E.; Schmitt L.; Spiess B.; Schlewer G. *Inositolphosphates and derivatives. ACS Symp. Ser. 463*, Washington, D.C., 1991, p 156.
31. Schmitt, L.; Bortmann, P.; Schlewer, G.; Spiess, B. *J. Chem. Soc. Perkin Trans 2* **1993**, 2257.
32. Schmitt, L.; Bortmann, P.; Spiess, B.; Schlewer, G. *Phosphorus Sulfur and Silicon* **1993**, *76*, 147.
33. Schmitt, L.; Schlewer, G.; Spiess, B. *J. Inorg. Biochem.* **1992**, *45*, 13.
34. Mernissi-Arifi, K.; Schmitt, L.; Schlewer, G.; Spiess, B. *Anal. Chem.* **1995**, *67*, 2567.
35. Barrientos, L. G.; Murthy, P. P. N. *Carbohydr. Res.* **1996**, *296*, 39.
36. Schmitt, L.; Spiess, B.; Schlewer, G. *Bioorg. Med. Chem. Lett.* **1995**, *5*, 1225.
37. Perich, J. W.; Johns, R. B. *Tetrahedron Lett.* **1987**, *28*, 101.
38. Schmitt, L.; Spiess, B.; Schlewer, G. *Tetrahedron Lett.* **1992**, *33*, 2013.
39. Ley, S. V. *Pure Appl. Chem.* **1990**, *62*, 2031.
40. Massy, D. J.; Wyss, P. *Helv. Chim. Acta* **1990**, *73*, 1037.
41. Gigg, J.; Gigg, R.; Martin-Zamora, E. *Tetrahedron Lett.* **1993**, *34*, 2827.
42. Offer, J. L.; Voorheis, H. P.; Metclafe, J. C.; Smith, G. A. *J. Chem. Soc. Perkin Trans 2* **1992**, 953.
43. Anderson, G. *J. Sci. Food. Agric.* **1963**, *14*, 352.
44. Yang, S. S.; Chiang, Y. C. P.; Beattie, T. R. *Carbohydr. Res.* **1993**, *249*, 259.
45. Yang, S. S.; Beattie, T. R.; Shen, T. Y. *Synth. Commun.* **1986**, *16*, 131.
46. Ballereau, S.; Guédat, P.; Spiess, B.; Rehnberg, N.; Schlewer, G. *Tetrahedron Lett.* **1995**, *36*, 7449.
47. Guédat, P.; Poitras, M.; Spiess, B.; Guillemette, G.; Schlewer, G. *Bioorg. Med. Chem. Lett.* **1996**, *6*, 1175.
48. Mignery, G. A.; Sudhof, T. *EMBO J.* **1990**, *9*, 3893.
49. Mignery, G. A.; Newton, C. L.; Archer III, B. T.; Sudhof, T. C. *J. Biol. Chem.* **1990**, *21*, 12679.
50. Mernissi-Arifi K.; Imbs, I.; Schlewer, G.; Spiess, B. *Biochim. Biophys. Acta* **1996**, *1289*, 404.

Chapter 18

5-Deoxy-5-fluoro-1-O-phosphatidylinositol: Evidence for the Enzymatic Incorporation of a Fraudulent Cyclitol into the Phosphatidylinositol Cycle

A Mass Spectrometric Study

David C. Baker[1], Stephen C. Johnson[1,3], James D. Moyer[2,4], Nancy Malinowski[2], Frank Tagliaferri[1], and Albert Tuinman[1]

[1]Department of Chemistry, The University of Tennessee, Knoxville, TN 37996
[2]Laboratory of Biological Chemistry, National Cancer Institute, National Institutes of Health, Bethesda, MD 20892

Synthetic analogues of *myo*-inositol have been employed as biochemical probes to aid in the elucidation of discrete steps of the phosphatidylinositol (PtdIns) pathway, and have been designed as potential chemotherapeutic agents, especially for the chemotherapy of cancer. For these compounds to be effective, they must become incorporated into the phosphatidylinositol pathway. The first step in that process, i.e. the conversion of the cyclitol to its 1-O-phosphatidyl derivative, is carried out by PtdIns synthase that normally transfers a phosphatidyl moiety from CDP-diacylglycerol to the 1D position of *myo*-inositol. Evidence of the incorporation of a fraudulent inositol has traditionally involved the quantitative measurement of CMP release, or comparisons of the TLC mobilities of the products to those of natural phospholipids. In the present study, we provide the first definitive spectrometric evidence for the incorporation of a fraudulent cyclitol analogue, 5-deoxy-5-fluoro-*myo*-inositol (5dFIns), into the PtdIns cycle by way of its 1-O-phosphatidyl derivative. The structure of the product, a 5dFPtdIns conjugate, was determined via negative-ion FAB and electrospray-ionization mass spectrometric analyses of a crude lipid extract. The MS evidence, together with known enzymatic preferences for 1D 1-O-phosphatidylation, leads us to conclude that we indeed observe the title conjugate.

Synthetic analogues (*1*) of *myo*-inositol have been designed as biochemical probes to aid in the elucidation of discrete steps of the phosphatidylinositol (PtdIns) pathway (*2*), or to serve as potential chemotherapeutic agents, especially in the treatment of cancer. Over the past decade significant effort has been expended to demonstrate the ability of synthetic analogues to act as fraudulent substrates or inhibitors of the various enzymes of the PtdIns pathway, either by incorporation of fraudulent cyclitols into the pathway enzyme of the PtdIns pathway, (*3-5*), or via presynthesized

[3]Current address: Ibbex, 2800 Milan Court, Birmingham, AL 35294.
[4]Current address: Pfizer Central Research, Groton, CT 06340.

phosphatidylinositol derivatives (6). The first enzyme of the PtdIns pathway, PtdIns synthase [CDPdiacylglycerol:*myo*-inositol 3-phosphatidyltransferase, EC 2.7.8.11], normally transfers a phosphatidyl moiety from CDP-diacylglycerol to the 1D-position of *myo*-inositol (7). Evidence for the enzymatic incorporation of a synthetic analogue (5-deoxy-5-fluoro-*myo*-inositol, 5dFIns, below, for example) into the pathway has traditionally been the quantitative measurement of cytidine monophosphate (CMP) release (3, 4, 8). In addition, the enzymatic incorporation of ^{3}H-labeled synthetic monodeoxyfluoro-*myo*-inositol analogues into phospholipids has been demonstrated by comparisons of the TLC mobilities of the products to that of their natural phospholipid counterparts (3, 4, 9). What is desired, however, is a more definitive method for the structural characterization of these novel PtdIns products. Since several earlier studies involving fast-atom-bombardment mass spectrometry (FABMS) of PtdIns analogues (10–12) had shown that an abundance of structural data could be obtained from a relatively small amount of sample, we decided to apply this technique to the analysis the components of a crude lipid extract from incorporation studies of [^{3}H]-5dFIns using a preparation of rat-brain microsomal PtdIns synthase.

5dFIns

Historically, work in this area has principally involved the synthesis (4, 5, 13–16) of a number of modified cyclitols and their evaluation (3–5) as substrates of PtdIns synthase. The earliest work (13) was devoted to the probing of structural modifications that might give rise to incorporation of fraudulent cyclitols into the PtdIns pathway via PtdIns synthase. Among the analogues evaluated, 5dFIns was determined to have about 26% of the substrate activity of *myo*-inositol, at equal concentrations, based on CMP release data obtained from experiments using rat-brain microsomal PtdIns synthase (3, 4). This level of incorporation, which has been reconfirmed in independent studies (5), has not been realized with other modified cyclitols, including the 3-deoxy-, 3-aminodeoxy-, and 3-deoxyhalogeno compounds that were specifically designed to limit activity of PtdIns 3-kinase (5). Hence, Kozikowski, Powis, and their co-workers have focused their approach, especially with the 3-modified cyclitols, on synthetic phosphatidylinositol analogues that do not require PtdIns synthase for activation (6). These phosphatidylinositol analogues have limitations as pharmacologic agents because they are relatively unstable as well as difficult to synthesize and purify. Simple cyclitols that produce inhibitory phospholipids *in situ* by metabolic incorporation into lipids may, therefore, have significant advantages.

In a previous study [³H]-5dFIns was used as a substrate for PtdIns synthase, and treatment of the derived phospholipid with phospholipase D yielded a radiolabeled product that was chromatographically identified as 5dFIns, in agreement with the formation of a 5dFPtdIns conjugate as the product (4). Further investigations have shown the ³H-labeled analogue to be taken up by intact L1210 murine leukemia cells, where it was incorporated into a phospholipid having lipophilic properties consistent with that of a modified PtdIns(4)P, but not of the more highly phosphorylated species, which would be a modified PtdIns(4,5)P₂ (3, 4). Therefore, because of the lack of a hydroxyl function at C-5 in 5dFIns, phosphorylation in vivo by PtdIns(4)P 5-kinase [EC 2.7.1.68] did not occur. These results were later supported by Offer et al. (9) in experiments conducted with 5dFIns and PtdIns in electropermeabilized thymocytes.

In the present report, we provide compelling evidence for the incorporation of a fraudulent cyclitol (5dFIns) into the PtdIns cycle. The structure of the product, a 5dFPtdIns conjugate, was determined via negative-ion FAB- and electrospray-ionization tandem mass spectrometric analyses of a crude lipid extract. The MS evidence, together with known enzymatic preferences for 1D phosphatidylation, leads us to conclude that we indeed observe the title conjugate, 5dFPtdIns (below).

PtdIns   X = OH
5dFPtdIns   X = F

Materials and Methods

5-Deoxy-5-fluoro-*myo*-inositol (5dFIns) was prepared from *myo*-inositol as described in a previous report from these laboratories (13, 14). The ³H-labeled 5dFIns was prepared by Moravek Biochemicals (Brea, CA) using a modification of the procedure for the Raney nickel-mediated deuteration of inositols in general (*17*) and 5dFIns in particular (*14*).

PtdIns synthase was prepared from rat-brain microsomes and solubilized by 0.5% Triton X as described by Rao and Strickland (*18*). The enzyme used in this study had a specific activity of 25–41 nmol/min/mg protein when assayed with 5 mM *myo*-inositol and protein as described in a previous report from these laboratories (*3*). The conjugate product (1.2 mg containing ca. 0.29 mg of 5dFPtdIns based on the radiolabel and specific activity used) was purified by preparative TLC (silica gel, 50:25:8:2.5 $CHCl_3$–CH_3OH– HOAc–0.9% NaCl) and had a total radioactivity of about 0.5 μCi.

Negative-ion FABMS was carried out on a VG ZAB-EQ instrument (VG Analytical, Manchester, UK) using 8 kV Xe atoms as the bombarding species. For these experiments a matrix/analyte stock solution was prepared by adding a small amount of crude lipid extract in 1:1 $CHCl_3$–CH_3OH to triethanolamine. High-energy collision-induced dissociation (HE–CID) experiments were conducted by selecting a precursor ion with the magnetic sector of the ZAB and activating it via 8 keV collisions with He atoms (He pressure to give 50% attenuation of the precursor signal) prior to analysis in the electrostatic sector.

Negative-ion electrospray experiments were carried out on a Quattro-II triple-quadrupole instrument (Micromass, Inc., Manchester, UK) by infusing a solution of the extract in 1:1 CH_3OH–H_2O containing 0.7% Et_3N at 5 μL/min. Nitrogen was used as the nebulizing and drying gas at 20 and 300 L/min, respectively. Precursor ions for the low-energy CID experiments were generated with 2.5 kV capillary voltage and 35 V cone voltage. CID was performed at 40 eV collision energy with argon at 3.7 x 10^{-4} mBar.

Results and Discussion

Negative-ion FAB mass spectrometry was conducted on the crude lipid extract. A full-scan spectrum (Figure 1) shows high-mass peaks at 811.5 and 953 Da. Scans to higher masses (1500 Da; not illustrated) did not provide significant additional peaks. The m/z 811 ion is consistent with that expected for 5dFPtdIns having the *sn*-1 and *sn*-2 positions substituted with palmitoyl groups ($C_{15}H_{31}CO_2$). (For a definition of *sn* nomenclature, see ref 21.)

Verification of the 5dFPtdIns conjugate as a component of the crude lipid extract was achieved predominantly by high-energy collision-induced dissociation (HE–CID) as described in the Materials and Methods section. Jensen et al. (*10*) and Murphy et al. (*12*) have employed this technique in identifying characteristic CID fragmentation patterns for structurally analogous phospholipids, including PtdIns. Van Breemen et al. (*19*) have used the related technique of linked scanning of the magnetic- and electrostatic sectors of a double focusing mass spectrometer to achieve the same type of analysis with better resolution.

HE–CID of the anion at m/z 811 produced the spectrum depicted in Figure 2a. The data and the assigned fragment structures presented in Table I represent HE–CID processes known to occur for PtdIns analogues as reported by Jensen et al. (*10*) and Sherman et al. (*20*). The technique determines mass analysis based on the kinetic energy of the fragments and usually provides peaks that are broadened (due to energy release that translates into kinetic energy of the fragments on bond fission) and shifted to lower apparent mass (resulting from kinetic energy loss from the primary ions on collision). Hence, the determination of the correct mass of product ions can be problematic. In this experiment, the observed m/z values of the fragmentation are 0.5–3.0 Da lower than those expected for the transitions assigned to them in Table I.

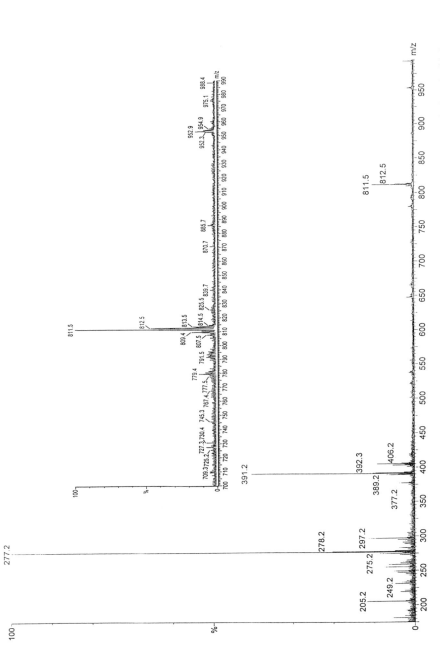

Figure 1. A full-scan, negative-ion fast-atom bombardment mass spectrum of the crude lipid extract with insert depicting the higher mass region.

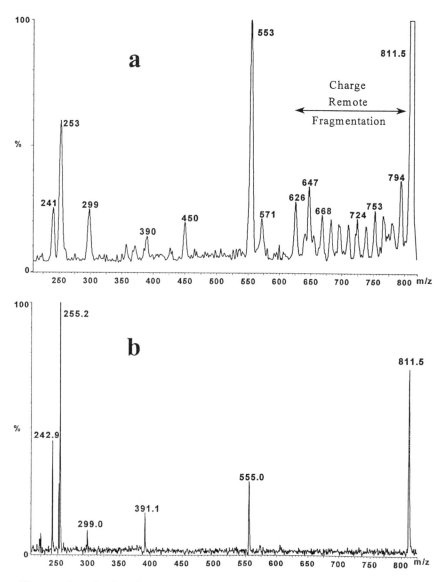

Figure 2. Product ions produced by collision-induced dissociation of the precursor at *m/z* 811.5 by (a) high-energy CID (HE–CID) and (b) low-energy CID (LE–CID).

Table I. Observed Masses and Structural Assignments from Collision Induced Dissociations of Precursor m/z 811.5

Assignments[a,b]	Expected Mass [Da]	Observed HE–CID Mass [Da]	Observed LE–CID Mass [Da]
Hf - H_2O	243.0	241.1	242.9
c	255.2	253.3	255.2
P - (2 x Hc)	299.0	298.5	299.0
Hg - Hc	391.2	390.5	391.1
Hd	555.3	553.4	555.0
Hb	573.3	571.4	------
P - $C_{13}H_{28}$	627.4	625.7	------
P - $C_{12}H_{26}$	641.4	640.8	------
Hg [c]	647.5	647.3	------
P - $C_{11}H_{24}$	655.4	654.6	------
P - $C_{10}H_{22}$	669.4	668.2	------
P - C_9H_{20}	683.4	682.5	------
P - C_8H_{18}	697.4	695.1	------
P - C_7H_{16}	711.5	710.5	------
P - C_6H_{14}	725.5	724.4	------
P - C_5H_{12}	739.5	738.2	------
P - C_4H_{10}	753.5	752.8	------
P - C_3H_8	767.5	765.9	------
P - C_2H_6	781.5	780.1	------
P - CH_4	795.5	793.6	------

[a] Unless otherwise stated, all CID fragmentation types have been previously identified for Ptdlns by Jensen et al. (*10*). The expected mass assignments presented have been adapted to the proposed structure of 5dFPtdlns. [b] "**P**" represents the 5dFPtdlns precursor anion at m/z 811.5. Other boldfaced letters refer to fragments of 5dFPtdlns as depicted in Figure 3. [c] This HE–CID fragment type was not identified by Jensen et al. (*10*), although the corresponding mass peak appears to be present in the published spectra.

HE–CID of m/z 811 (Figure 2a, Table I) results in abundant product ions at m/z 255 (observed 253), representing the carboxylate anion derived from the sn-1- and sn-2-substituted acyl chains, and m/z 555 (observed at 553) derived from the loss of a 16:0 acid (palmitic acid) from the (M - H) precursor anion. Additional fragments still retaining the fluoroinositol moiety include m/z 243 (fluoroinositol phosphate - H_2O; observed at 241), m/z 299 (loss of both palmitic acids from the precursor; observed at 298), and m/z 573 (loss of palmitic aldehyde; observed at 571). CID product ions that no longer encompass the fluoroinositol are m/z 647 (the phospholipid) and m/z 391 (phospholipid - palmitic acid). No logical assignment could be found for m/z 450 on the basis of the proposed structure. It may be that the precursor ion at m/z 811 encompasses not only the quasimolecular ion of 5dFPtdIns, but also an ion of different structure that derives from a further constituent of the crude lipid extract and produces a fragment at m/z 450.

In addition to the fragments described above, HE–CID of m/z 811 produces a series of high-mass product ions formed by the loss of C_nH_{2n+2} neutrals from the fatty acid substituents via a specific 1,4-elimination of H_2 as previously described by Jensen et al. (22). This charge remote fragmentation (CRF) is useful in the structural characterization of variously substituted alkyl chains including palmitic acid (22). The relatively simple fragmentation pattern resulting from CRF, in addition to the presence of only one peak attributable to a $C_nH_{2n+1}CO_2$ anion (m/z 255, Figure 2a), suggest that the sn-1 and sn-2 carboxylate chains of the 5dFPtdIns conjugate are identical or at least isomeric. A plausible fragmentation pattern for the molecule as the anion is shown in Figure 3.

To overcome the uncertainty of mass assignments associated with HE–CID, low-energy CID (LE–CID) experiments were also conducted, comprising the isolation of the precursor ion m/z 811 in the first quadrupole of the Quattro-II electrospray instrument, with subsequent drift through a gas cell where low-energy (40 eV) collisions with argon atoms occur (see Materials and Methods). A further quadrupole analyzer determines the resulting product-ion masses to within 0.3 Da. In contrast to the HE–CID described above, the product ion peaks are narrower, and the observed masses are more accurate. However, the degree of observed fragmentation is less, because collisions are less energetic. In particular, CRF is not usually observed in LE–CID experiments. The results of the LE–CID experiment (Figure 2b and Table I) verify the assignments made for the HE–CID peaks, but only for those fragments common to both techniques.

To further confirm the structure of the $C_{15}H_{31}CO_2-$ chain, the relatively weak m/z 255 ion in the FAB spectrum (see Figure 1) was selected and subjected to HE–CID. The relatively smooth intensity progression for ions representing C_nH_{2n+2} losses induced by CRF suggests the saturated and unbranched nature of the carboxylate acid chain (Table II). Comparison of the data with those obtained by Jensen et al. for authentic palmitic acid (22) provides conclusive evidence that the sn-1- and sn-2-substituted carboxylate chains are indeed palmitic acid.

Conclusions

HE– and LE–CID mass spectrometric studies on a crude lipid extract obtained from the treatment of a synthetic myo-inositol analogue (5dFIns) with rat-brain microsomal PtdIns synthase under the conditions known to provide a PtdIns product, have been used to establish the structure of a 5dFPtdIns conjugate as a component of the crude

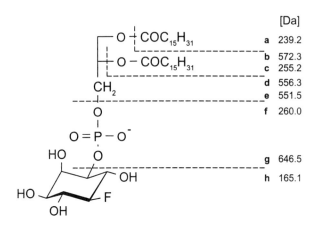

	[Da]
a	239.2
b	572.3
c	255.2
d	556.3
e	551.5
f	260.0
g	646.5
h	165.1

Figure 3. Masses and fragment ions that would arise from the formal cleavage of selected bonds in 5dFPtdIns.

Table II. Observed Masses and Structural Assignments from High-Energy Collision Induced Dissociation of Precursor m/z 255.

Assignments (F = $C_{16}H_{31}O_2$)	Expected Mass [Da]	Observed Mass [Da]	Relative Intensity [%]
F - CH_4	239.2	238.4	100
F - C_2H_6	225.2	224.6	79
F - C_3H_8	211.2	210.6	82
F - C_4H_{10}	197.2	196.3	95
F - C_5H_{12}	183.1	182.8	82
F - C_6H_{14}	169.1	168.6	66
F - C_7H_{16}	155.1	154.5	72
F - C_8H_{18}	141.1	140.1	72
F - C_9H_{20}	127.1	126.3	46
F - $C_{10}H_{22}$	113.1	112.8	40

lipid extract. This analysis provides definitive evidence for the incorporation of a fraudulent cyclitol analogue into the PtdIns pathway. It should be noted, however, that the mass spectrometric study cannot conclusively establish the position of the attachment of the phospholipid specifically at 1D of 5dFPtdIns. However, previous work with PtdIns synthase by Agranoff and Fisher (7), as well as abundant work conducted by Gross and co-workers (10, 22) on the FAB mass spectrometric identification of natural PtdIns analogues, provides ample support for the argument that the attachment of the phospholipid is indeed at the 1D position of 5dFPtdIns.

Acknowledgments

This work was supported in part by Contract No. N01-CM-27571 and Grant No. R01-CA45795 provided by the National Cancer Institute of the National Institutes of Health. The UT-K Chemistry Mass Spectrometry Center is funded by The Science Alliance, a State of Tennessee Center of Excellence. The National Science Foundation Chemical Instrumentation Program contributed to the acquisition of the Quattro-II (grant no. BIR-94-08252).

Literature Cited

1. For a preliminary report, see: Johnson, S. C.; Tuinman, A.; Tagliaferri, F.; Moyer, J. D.; Baker, D. C. *Abstracts of Papers*, 212th National Meeting of the American Chemical Society, Orlando, FL; Aug 25–29, 1996; American Chemical Society: Washington, DC, 1996; CARB-011.
2. Vance, D. E. In *Biochemistry,* 2nd ed.; Zubay, G., Ed.; Macmillan: New York, 1988, pp 698–702.
3. Moyer, J. D.; Reizes, O.; Ahir, S.; Jiang, C.; Malinowski, N.; Baker, D. C. *Mol. Pharmacol.* **1988**, *3*, 683–689.
4. Moyer, J. D.; Reizes, O.; Malinowski, N.; Jiang, C.; Baker, D. C. In *Fluorinated Carbohydrates: Chemical and Biochemical Aspects*; Taylor, N. F., Ed.; ACS Symposium Series No. 374; American Chemical Society: Washington, DC, 1988, pp 43–58.
5. Johnson, S. C.; Dahl, J.; Shih, T.-L.; Schedler, D. J. A.; Anderson, L.; Benjamin, T. L.; Baker, D. C. *J. Med. Chem.* **1993**, *36*, 3628–3635.
6. Kozikowski, A. P.; Qiao, L.; Tuckmantel, W.; Powis, G. *Tetrahedron* **1997**, *53*, 14903–14914, and references cited therein.
7. Agranoff, B. W.; Fisher, S. K. In *Inositol Phosphates and Derivatives*; Reitz, A. B., Ed.; ACS Symposium Series No. 463; American Chemical Society: Washington, DC, 1991, pp 20–32.
8. Benjamins, J. A.; Agranoff, B. W. *J. Neurochem.* **1969**, *16*, 513–527.
9. Offer, J.; Metcalfe, J. C.; Smith, G. A. *Biochem. J.* **1993**, *293*, 553–560.
10. Jensen, N. J.; Tomer, K. B.; Gross, M. L. *Lipids* **1987**, *22*, 480–489.
11. Roberts, W. L.; Santikarn, S.; Reinhold, V. N.; Rosenberry, T. L. *J. Biol. Chem.* **1988**, *263*, 18776–18784.
12. Murphy, R. C.; Harrison, K. A. *Mass Spectrom. Rev.* **1994**, *13*, 57–75.
13. Jiang, C.; Moyer, J. D.; Baker, D. C. *J. Carbohydr. Chem.* **1987**, *6*, 319–355.
14. Jiang, C.; Schedler, D. J. A.; Morris, P. E., Jr.; Zayed, A.-H. A.; Baker, D. C. *Carbohydr. Res.* **1990**, *207*, 277–285.
15. Johnson, S. C.; Tagliaferri, F.; Baker, D. C. *Carbohydr. Res.* **1993**, *250*, 315–321.

16. Tagliaferri, F.; Johnson, S. C.; Seiple, T. F.; Baker, D. C. *Carbohydr. Res.* **1995**, *266*, 301–307.
17. Angyal, S. J.; Odier, L. *Carbohydr. Res.* **1983**, *123*, 13–22.
18. Rao, R. J.; Strickland, K. P. *Biochim. Biophys. Acta* **1974**, *348*, 306–314.
19. Van Breemen, R. B.; Wheeler, J. J.; Boss, W. F. *Lipids* **1997**, *25*, 328-334.
20. Sherman, W. R.; Ackerman, K. E.; Bateman, R. H.; Green, B. N.; Lewis, I. *Biomed. Mass Spectrom.* **1985**, *12*, 409–413.
21. IUPAC-IUB Commission on Biochemical Nomenclature, *Biochem. J.* **1978**, *171*, 21–35.
22. Jensen, N. J.; Tomer, K. B.; Gross, M. L. *J. Am. Chem. Soc.* **1985**, *107*, 1863–1868.

INDEXES

Author Index

Subject Index

Bestsellers from ACS Books

The ACS Style Guide: A Manual for Authors and Editors (2nd Edition)
Edited by Janet S. Dodd
470 pp; clothbound ISBN 0–8412–3461–2; paperback ISBN 0–8412–3462–0

Writing the Laboratory Notebook
By Howard M. Kanare
145 pp; clothbound ISBN 0–8412–0906–5; paperback ISBN 0–8412–0933–2

Career Transitions for Chemists
By Dorothy P. Rodmann, Donald D. Bly, Frederick H. Owens, and Anne-Claire Anderson
240 pp; clothbound ISBN 0–8412–3052–8; paperback ISBN 0–8412–3038–2

Chemical Activities (student and teacher editions)
By Christie L. Borgford and Lee R. Summerlin
330 pp; spiralbound ISBN 0–8412–1417–4; teacher edition, ISBN 0–8412–1416–6

Chemical Demonstrations: A Sourcebook for Teachers, Volumes 1 and 2, Second Edition
Volume 1 by Lee R. Summerlin and James L. Ealy, Jr.
198 pp; spiralbound ISBN 0–8412–1481–6
Volume 2 by Lee R. Summerlin, Christie L. Borgford, and Julie B. Ealy
234 pp; spiralbound ISBN 0–8412–1535–9

The Internet: A Guide for Chemists
Edited by Steven M. Bachrach
360 pp; clothbound ISBN 0–8412–3223–7; paperback ISBN 0–8412–3224–5

Laboratory Waste Management: A Guidebook
ACS Task Force on Laboratory Waste Management
250 pp; clothbound ISBN 0–8412–2735–7; paperback ISBN 0–8412–2849–3

Reagent Chemicals, Eighth Edition
700 pp; clothbound ISBN 0–8412–2502–8

Good Laboratory Practice Standards: Applications for Field and Laboratory Studies
Edited by Willa Y. Garner, Maureen S. Barge, and James P. Ussary
571 pp; clothbound ISBN 0–8412–2192–8

For further information contact:
Order Department
Oxford University Press
2001 Evans Road
Cary, NC 27513
Phone: 1-800-445-9714 or 919-677-0977

Highlights from ACS Books